AF848874

ADVANCES IN FOOD SAFETY AND FOOD MICROBIOLOGY

PROGRESS ON QUANTITATIVE APPROACHES OF THERMAL FOOD PROCESSING

ADVANCES IN FOOD SAFETY AND FOOD MICROBIOLOGY

Dr. Anderson de Souza Sant'Ana
and Dr. Bernadette D. G. M. Franco
(Series Editors)
Department of Food and Experimental Nutrition,
Faculty of Pharmaceutical Sciences, University of São Paulo,
São Paulo, SP, Brazil

Food safety is a specific area of food science focused in the study of microbiological, chemical and physical hazards in foods and beverages.Several tools, systems and approaches have been developed last years to systematize the management of food production chain aiming at protecting public health.Food Microbiology is an interesting and exciting field of microbiology concerned with the study of foodborne microorganisms, their occurrence, interactions and responses to the environmental found in foods and beverages. Depending mainly on the intrinsic or extrinsic properties of foods and beverages, one or more of microorganisms such as protozoan, viruses, yeasts, moulds and bacteria or toxins will be of significance.

Food safety is a specific area of food science focused in the study of microbiological, chemical and physical hazards in foods and beverages. Several tools, systems and approaches have been developed in recent years to systematize the management of food production chain aiming at protecting public health.

Food Microbiology is an interesting and exciting field of microbiology concerned with the study of foodborne microorganisms, their occurrence, interactions and responses to the environmental found in foods and beverages. Depending mainly on the intrinsic or extrinsic properties of foods and beverages, one or more of microorganisms such as protozoan, viruses, yeasts, moulds and bacteria or toxins will be of significance.

In fact, Food Microbiology and Food Safety can be considered a "boiling and changing" field of microbiology and food science, respectively. These fields have risen to that status mainly in the last 30 years. Data from epidemiological studies of foodborne disease outbreaks or from investigations of spoilage episodes have highlighted relevant changes in the pillars supporting the microbial ecology of foods: consumer habits, practices and consumers health status, food preservation techniques, responses of microorganisms to preservation methods, microbial expression of virulence factors and food production chain systems.

In addition, a strong trend in food microbiology and safety is the application of computational, statistical, molecular biology and chemical approaches, resulting in a deeper approach to study and understand the microbial interactions with foods and among them. Trends indicate that the insertion of these four sciences into food safety and microbiology will make an important difference in the evolving of studies of microbial ecology and food safety. Based on this scenario, it is the main aim of the Series "Advances in Food Safety and Food

Microbiology" to cover "hot topics" in the field of Food Microbiology and Food Safety. Coverage will be given to applied food microbiology and safety with a collection of texts focused on the study of spoilage, pathogens and industrial applications of foodborne microorganisms. Focus will be given to responses of microorganisms to preservation methods, consumer food safety practices, advances in microbiological methods and on the applications of statistical, molecular biology, computational and the "omics" in food microbiology and safety. Risk analysis, integration between epidemiology and food microbiology, chemical and physical contaminants of foods are also given consideration.

We hope that the books in this Series will be widely used by food microbiologists and those concerned with food safety in their studies or as references for new approaches to be considered in an effort to continuously evolve food microbiology and food safety from farm to fork, from tradition to technology.

Probiotic and Prebiotic Foods: Technology, Stability and Benefits to Human Health
Nagendra P. Shah, Adriano Gomes da Cruz, and Jose de Assis Fonseca Faria (Editors)
2011. ISBN: 978-1-61668-842-4 (Hardcover)
2010. ISBN: 978-1-61728-825-8 (e-book)

Advances in Post-Harvest Treatments and Fruit Quality and Safety
Manuel Vázquez and José A. Ramírez de Leon
2011. ISBN: 978-1-61122-973-8 (Hardcover)
2011. ISBN: 978-1-61470-700-4 (e-book)

Bacteriophages in Dairy Processing
Andrea del Luján Quiberoni and Jorge Alberto Reinheimer (Editors)
2012. ISBN: 978-1-61324-517-0 (Hardcover)
2012. ISBN: 978-1-62257-090-4 Hardcover)

Stress Response of Foodborne Microorganisms
Hin-chung Wong (Editor)
2012. ISBN: 978-1-61324-410-4 (Hardcover)
2012. ISBN: 978-1-61122-810-6 (e-book)

New Trends in Marine and Freshwater Toxins: Food and Safety Concerns
Ana G. Cabado and Juan Manuel Vieites (Editors)
2012. ISBN: 978-1-61470-324-2 (Hardcover)
2011: ISBN: 978-1-61470-398-3 (e-book)

Clostridium Botulinum: A Spore Forming Organism and a Challenge to Food Safety
Christine Rasetti-Escargueil, and Susanne Surman-Lee (Editors)
2012. ISBN: 978-1-61470-575-8 (Hardcover)
2011. ISBN: 978-1-61470-653-3 (e-book)

Enterococcus and Safety
Teresa Semedo-Lemsaddek,
Maria Teresa Barreto-Crespo, and Rogério Tenreiro (Editors)
2012. ISBN: 978-1-61470-569-7 (Hardcover)
2011. ISBN: 978-1-61470-664-9 (e-book)

Essential Oils as Natural Food Additives: Composition, Applications, Antioxidant and Antimicrobial Properties
Luca Valgimigli (Editor)
2012. ISBN: 978-1-62100-241-3 (Hardcover)
2012. ISBN: 978-1-62100-282-6 (e-book)

On-Farm Strategies to Control Foodborne Pathogens
Todd R. Callaway and Tom S. Edrington (Editors)
2012. ISBN: 978-1-62100-411-0 (Hardcover)
2012. ISBN: 978-1-62100-480-6 (e-book)

Molecular Typing Methods for Tracking Foodborne Microorganisms
Steven L. Foley, Rajesh Nayak, and Timothy J. Johnson (Editors)
2012. ISBN: 978-1-62100-643-5 (Hardcover)
2012. ISBN: 978-1-62100-728-9 (e-book)

Pathogenic Vibrios and Food Safety
Yi-Cheng Su (Editor)
2012. ISBN: 978-1-62100-866-8 (Hardcover)
2012. ISBN: 978-1-62100-903-0 (e-book)

Foodborne Protozoan Parasites
Lucy J. Robertson and Huw V. Smith (Editors)
2012. ISBN: 978-1-61470-008-1 (Hardcover)
2012. ISBN: 978-1-61942-569-9 (e-book)

Predictive Mycology
Philippe Dantigny and Efstathios Z. Panagou (Editors)
2012. ISBN: 978-1-61942-675-7 (Hardcover)
2012. ISBN: 978-1-61942-684-9 (e-book)

Progress on Quantitative Approaches of Thermal Food Processing
Vasilis P. Valdramidis and Jan F. M. Van Impe (Editors)
2012. ISBN: 978-1-62100-842-2 (Hardcover)
2012. ISBN: 978-1-62100-899-6 (e-book)

ADVANCES IN FOOD SAFETY AND FOOD MICROBIOLOGY

PROGRESS ON QUANTITATIVE APPROACHES OF THERMAL FOOD PROCESSING

VASILIS P. VALDRAMIDIS
AND
JAN F. M. VAN IMPE
EDITORS

Nova Science Publishers, Inc.
New York

Copyright © 2012 by Nova Science Publishers, Inc.

All rights reserved. No part of this book may be reproduced, stored in a retrieval system or transmitted in any form or by any means: electronic, electrostatic, magnetic, tape, mechanical photocopying, recording or otherwise without the written permission of the Publisher.

For permission to use material from this book please contact us:
Telephone 631-231-7269; Fax 631-231-8175
Web Site: http://www.novapublishers.com

NOTICE TO THE READER

The Publisher has taken reasonable care in the preparation of this book, but makes no expressed or implied warranty of any kind and assumes no responsibility for any errors or omissions. No liability is assumed for incidental or consequential damages in connection with or arising out of information contained in this book. The Publisher shall not be liable for any special, consequential, or exemplary damages resulting, in whole or in part, from the readers' use of, or reliance upon, this material. Any parts of this book based on government reports are so indicated and copyright is claimed for those parts to the extent applicable to compilations of such works.

Independent verification should be sought for any data, advice or recommendations contained in this book. In addition, no responsibility is assumed by the publisher for any injury and/or damage to persons or property arising from any methods, products, instructions, ideas or otherwise contained in this publication.

This publication is designed to provide accurate and authoritative information with regard to the subject matter covered herein. It is sold with the clear understanding that the Publisher is not engaged in rendering legal or any other professional services. If legal or any other expert assistance is required, the services of a competent person should be sought. FROM A DECLARATION OF PARTICIPANTS JOINTLY ADOPTED BY A COMMITTEE OF THE AMERICAN BAR ASSOCIATION AND A COMMITTEE OF PUBLISHERS.

Additional color graphics may be available in the e-book version of this book.

Library of Congress Cataloging-in-Publication Data

Progress on quantitative approaches of thermal food processing / editors, Vasilis P. Valdramidis, Jan F.M. Van Impe.
p. ; cm.
Includes bibliographical references and index.
ISBN 978-1-62100-842-2 (hardcover)
1. Food--Effect of heat on. 2. Food--Microbiology. 3. Food industry and trade. I. Valdramidis, Vasilis P. II. Impe, Jan F. van.
[DNLM: 1. Food Handling--methods. 2. Food Microbiology--methods. 3. Food Safety--methods. 4. Hot Temperature. 5. Microbial Viability. 6. Models, Theoretical. TP 371.2]
TP371.P76 2011
363.19'2--dc23

2011039231

Published by Nova Science Publishers, Inc. † New York

This book is dedicated to our parents

CONTENTS

PREFACE

Heat treatment is one of the most common practices used to produce safe and shelf stable foods or otherwise stated, to reduce the probability of survival and/or growth of the microorganisms in a particular food to a tolerable level. This book covers the advances in thermobacteriology, including technological and engineering aspects of thermal processes targeting on the production of food safe products. Target microorganisms of relevance in thermally processed foods are reviewed while mathematical relationships describing the kinetics of thermal destruction of microorganisms within the frame of predictive microbiology as well as heat transfer phenomena of solid and liquid foods are outlined. Additionally, model structure selection criteria and kinetic parameter estimation approaches are under consideration. Specific focus is given in all relevant statistical aspects during parameter identification and model validation. Induced microbial adaptation phenomena and their impact on the microbial heat resistance are discussed while rigorous modelling approaches to describe this kinetics are presented. Finally, issues related to quantitative modelling risk assessment in thermal processes are outlined.

Overall the objective of this book is to provide a comprehensive overview of innovations in assessing thermal processes while considering integrated information from the field of microbiology of thermal processes and engineering of these processes. The book has a strong focus on statistical and mathematical methods in order to be a useful reference for food microbiologists, food technologists and engineers.

It could be used as a professional reference for those working in food processing, food technology, food microbiology, bioengineering research areas in academia as well as to the food safety authorities and food industry.

Vasilis P. Valdramidis and Jan F.M. Van Impe

Msida (Malta) and Leuven (Belgium), December 2011

In: Progress on Quantitative Approaches in Thermal … ISBN: 978-1-62100-842-2
Editors: V. P. Valdramidis and J. F. M. Van Impe © 2012 Nova Science Publishers, Inc.

Chapter 1

MICROORGANISMS OF RELEVANCE IN THERMALLY PROCESSED FOODS

J. T. Rosnes[1], P. S. Fernández[2], P. M. Periago[2] and T. Skåra[1,3]
[1]Nofima AS, Richard Johnsens gate 4, Stavanger, Norway
[2] Universidad Politecnica de Cartagena,
Departamento de Ingeniería de Alimentos y Equipamiento Agrícola, Spain
[3]BioTeC - Chemical and Biochemical Process Technology and Control, Department of Chemical Engineering, Katholieke Universiteit Leuven, Leuven (Heverlee), Belgium

Abstract

Thermal processing is one of the most common methods for achieving safe convenience food products with an extended shelf life. Designing a thermal process for such products is challenging, with a wide range of heat processes, from a low pasteurization process (70 °C for 2 min) to a full commercial sterilization process (higher than 3 min at 121 °C). This chapter aims at identifying the relevant microorganisms for the main groups of heat processed foods of, meat, fish and vegetables. Special emphasis is made on selected target pathogens like *Clostridium botulinum*, *Clostridium perfringens*, *Bacillus cereus* and *Listeria monocytogenes*, although some spoilage microorganisms are also considered, due to their very strong economic implications for food stability. The heat load required for inactivating target microorganisms may cause undesirable quality changes and loss of nutrients, and D- and *z*-values must therefore be considered for each specific food commodity. Concern about the safety of some food products exists, particularly when considering the potential abuse caused by storage temperature.

New methods that focus on minimal heating or rapid heating of food products are of vital importance. The main aim for new developments is to reduce the overall thermal load by reducing the temperature gradients in the product or by targeting specific potentially infected areas. Dielectric heating, like microwaves and radiofrequency heating as well as ohmic heating, are technologies that meet these aims, and are available for industrial applications. Another relevant approach is minimal processing, which often relies on the use of multiple sub-lethal stresses or processes to achieve a similar level of microbial control such as that traditionally achieved by using a single lethal stress to increase product quality. Most minimally processed products require refrigerated storage and distribution to maintain food safety.

1. Introduction

Heat resistant microorganisms are one of the main threats for the stability and safety of heat treated foods. Therefore, to know and understand the target microorganisms for each group of foods is of the utmost importance in order to establish adequate preservation treatments to guarantee both their safety and stability. There is a wide range of heat processes, from low pasteurization (in the range of 70 °C for 2 min) typically combined with refrigeration storage, to a full commercial sterilization, more than 3 min at 121 °C for shelf-stable products. Changing lifestyles have caused a deep impact on food consumption habits in most developed countries. An increasing demand for quality, variety and service characteristics has triggered the development of many new products designed to meet the need created by the fact that more people are eating their meals outside their homes (Kennedy and Wall 2007), as well as new and more efficient production systems being introduced in hospital food service. In addition, there is a huge increase in retail ready meals with a shelf life of one to four weeks. In such packaged meals, both vegetables and muscle foods, e.g. fish, meat or poultry, may be combined although individual heat treatments have to be used for each raw material, since the impact of the substrate is of major importance for the apparent heat resistance of microorganisms. An appropriate heat treatment designed to achieve a specified lethality of microorganisms is influenced by many factors, some of which can be attributed to the inherent resistance of microorganisms, while others are due to environmental influences. It has been shown that factors such as pH or a_w can significantly reduce the *D*-values, therefore it is necessary to perform studies in the specific matrix of the product to ensure that it is adequate to control pathogenic microorganisms and to ensure its stability during its shelf life against spoilage agents.

Thermal processing is the primary method for adding value and ensuring microbial safety of food products. Although numerous technologies, e.g. irradiation, ultra high pressure, pulsed electric fields, use of bacteriocins are new promising technologies for the food industry, the application of heat will certainly continue as the dominant means to impact desirable characteristics, add economic value and ensure product safety. Dielectric heating, like microwaves and radiofrequency heating and ohmic heating are technologies that are available for industrial applications. They can provide a more efficient application of heat through different mechanisms, which imply reducing the overall intensity of heat treatment in the product, still maintaining the same safety levels. Minimal processing often relies on the use of multiple sublethal stresses or processes to achieve a similar level of microbial control such as that traditionally achieved by using a single lethal stress. Additionally, major shifts in consumer demand and regulatory burden have increased the importance of thermal processing of these areas. During the development of any new heat treated product it is essential to assess the combined effects of the total system such as heat process, preservatives, packaging and storage conditions in order to ensure that the product is of a good microbiological standard and does not present any food safety hazard.

This chapter aims at identifying the relevant microorganisms for some main food commodities like meat and poultry, fish and vegetables. Special emphasis is made on pathogenic bacteria, although some spoilage microorganisms are also considered, due to their very strong economic implications in food stability.

2. Heat Processing Related to the Food Substrate

The heat resistance of bacteria is described by two parameters: *D*- and *z*-values. The *D*-value is the time at a particular temperature necessary to destroy 90% of the viable cells or spores of a specific organism, and it is a measure of the death rate of the organism. The *z*-value is the change in heating temperature needed to change the *D*-value by 90% or 1 log cycle, and gives information on the relative resistance of an organism at different destructive temperatures.

The *D*- and *z*-values of foodborne pathogens are influenced by the food in which it is examined. Foods are complex ecosystems composed of intrinsic factors inherent to the food, e.g. pH, water activity and nutrients, and extrinsic factors external to it, e.g. temperature, gaseous environments and the presence of other bacteria. Both intrinsic and extrinsic factors can be manipulated to preserve food by designing conditions in the food that limit bacterial growth. For example microorganisms are most resistant to heat at their optimum pH of growth, which is generally about 7.0, and a pH of < 4.6 has traditionally been considered as safe with respect to growth of the target organism *Clostridium botulinum.* Thus pH is a factor that both affect the heat inactivation kinetics and ability for the microorganisms to grow after heating, and thereby defining a safe heat process for the specific food product.

2.1. Shelf Stable, Sterile Food Products

These foods receive a severe heat treatment that provides a long shelf life at room temperature, typically at least one year. In terms of the heat treatment received, they can be classified as acidic products (those with a pH value lower than 4.6) or low acid products (those with a pH value equal or higher than 4.6).

2.1.1. Acidic Foods

Acidic foods are processed at target temperatures in the range of 80-100 °C in order to guarantee a pasteurization process. The main microorganisms of concern are spoilage ones. For pH values up to 4.0 these include yeasts, moulds and lactic acid bacteria, which can be inactivated at temperatures in the range of 90 °C (Barrett et al., 2004; Nelson and Tressler, 1980; Odlaug and Pflug, 1978; Reichart and Mohácsi-Farkas, 1994), and sporeforming bacteria, such as *Clostridium pasteurianum* and *Bacillus coagulans*, that require more severe heat treatments and can pose risk of spoilage in foods within the pH range of 4.0-4.6, such as tomato products. These microorganisms present typical *D*-values of 0.01-0.07 min at 121 °C and 0.1-0.5 min at 100 °C respectively (Larousse and Brown, 1997; Lynch and Potter, 1989; Palop et al., 1996; Stumbo, 1973).

A sporeforming microorganism, *Alycyclobacillus acidoterrestris*, has been described as able to grow at lower pH values (3.0 to 4.0) with a high heat resistance ($D_{100\,°C}$ in the range of 2.6-3.2 min in orange juice with a *z* value of 9.5 °C, (Conesa et al., 2009)), which can challenge the heat treatments applied to commercial fruit juices. It causes off flavors in the final product. Although its presence is relatively low in the raw products, several cases of spoiled fruit juices have been described and if its incidence would spread, a considerable increase in the heat treatment of acidic products would be needed. Different control measures of this organism in acidic fruit products have been developed (Silva and Gibbs, 2004) and combination of heat and supercritical carbon dioxide have been suggested to minimize quality reduction (Bae et al., 2009).

2.1.2. Low Acid Foods

The main concern for this range of products is the presence of very heat resistant sporeforming bacteria belonging to *Bacillus* spp., *Geobacillus* spp. and *Clostridium* spp. This group includes pathogenic bacteria, as well as spoilage microorganisms.

2.2. Not Shelf Stable, Refrigerated Foods

2.2.1. Acidic Foods

For acidic foods, target microorganisms to be considered would be yeasts, moulds and lactic acid bacteria. In this case, psychrotrophic strains, able to grow at refrigeration temperatures, are the major cause of microbial spoilage. Apart from that, the same considerations made for this group in the heat stable, acidic foods apply. There are no sporeforming bacteria that can develop and spoil these products, since the ones affecting shelf stable foods are mesophilic and therefore will not grow at low temperatures. This implies that the heat treatments can be lower, which has beneficial implications for the organoleptic and nutritional quality of the final product.

2.2.2. Low Acid Foods

The target microorganisms to be considered are very different from those in shelf stable foods since only psychrotrophic strains or spp. will grow at low temperatures (below 8 °C). The most relevant microorganisms will include both sporeforming and vegetative bacteria.

2.3. Death Rates in Different Substrates

The heat resistance of spores of *C. botulinum* has been studied more thoroughly than that of most other pathogens, in which also the differentiation in acidic and low acid foods also has its origin. Since pre-fabricated meals that rely on refrigerated storage for food safety have seen a steady growth over the last decades in the European marketplace, with an increasing focus on quality and minimal processing, heat resistant vegetative pathogens have attained increased focus. *L. monocytogenes* is used as an example, information about other microorganisms can be found later on in the chapter.

2.3.1. Heat Inactivation of *Listeria monocytogenes* in Different Substrates

Comparing the heat resistance of some pathogens, e.g. *L. monocytogenes*, *Salmonella* spp., and *E. coli* O157:H7, Juneja (2002) found that *L. monocytogenes* is relatively more heat resistant. Therefore, in this chapter, among all the heat resistant microorganisms, *L. monocytogenes* is used to show the influence different food commodities (meat and poultry, fish, and vegetables) have on the inactivation kinetics.

L. monocytogenes is found in soil, water and decaying plant material. Although it has been a known pathogen since the 1920ies, its status as a food borne organism is relatively recent (Buchrieser and Rocourt, 2007). The organism's omnipresence, its ability to establish itself in production environments and to multiply at refrigeration temperatures, combined with a high thermal resistance among the non-spore formers, makes *L. monocytogenes* an important food pathogen. A number of factors influence the heat resistance of *L. monocytogenes*. With respect to the strains, there is considerable natural variation. Lunden et al. (2008) subjected 40 *L. monocytogenes* strains in tryptic soy broth (TSB) to heat (55 °C for 40 min) and found large variation in heat tolerance (3 log units). Interestingly also, the strains showed even larger variation (6 log units) in acid tolerance (pH 2.4, adjusted with HCl, for 2 h). Lianou et al. (2006) used a similar approach subjecting 25 *L. monocytogenes* strains in TSB supplemented with 0.6% yeast extract, to heat (55 °C for 240 min) and lactic acid (pH 3, for 2 h). The largest variation was found in acid tolerance, and it was observed that serotype 4b isolates (9) showed a lower heat resistance than did all the other strains (16) tested. Furthermore the age of the organisms as well the growth and recovery conditions are of great importance. Mcmahon et al. (2000) found that stationary phase cells of *L. monocytogenes* serotype 4b (NCTC 11994) were significantly more thermotolerant than exponential cultures under identical processing conditions.

The thermotolerance of *L. monocytogenes* can be substantially increased (one to eight-fold) when subjected to heat shock (Sergelidis and Abrahim, 2009). But the trend is not completely uniform. According to Lin and Chou (2004) the survival of *L. monocytogenes* at 55 °C for 1 h varied with strains and the conditions of the heat shock. These authors saw effects of 45 °C for 1 h whereas 48 °C for 10 min did not exhibit any significant difference from control cells. Using a 10-strain composite of *L. monocytogenes* Skandamis et al. (2008) found that the highest thermotolerance was achieved after acid (pH 5 w/ HCl) and heat shock (T= 45 °C), followed by exposure to osmotic shock (10% NaCl), as well as by a combination of osmotic and heat shock.

A study on the effect of water activity between 0.99-0.90 (in sucrose solutions) on thermal resistance of *L. monocytogenes* showed increasing thermal resistance with decreasing a_w from 0.99 to 0.93, and reduced linearity and increased tailing of survival curves, with decreasing water activity (Fernández et al., 2007). Osmotic stress is achieved in salted fish products. Lorentzen et al. (2010) found that the salt-curing and rehydration of cod inoculated with *L. monocytogenes* led to a 1 $\log_{10}$ CFU reduction, but a 10-100 fold increase could be seen after 5-10 days of storage of the rehydrated product at 4 °C.

Sörqvist (2003) compiled heat resistance data for *L. monocytogenes* by plotting $\log_{10}D$ values, and found a good linear relationship for values obtained between 50-75 °C. Based on this line a *z*-value of 5.7 could be derived for a number of unstressed *L. monocytogenes* strains in a variety of media (Figure 1).

Numerous studies indicate that *L. monocytogenes* is more heat resistant in foods than when suspended in laboratory media (Boyle et al., 1990; Casadei et al., 1998; Foegeding and Stanley, 1991; Jorgensen et al., 1999). This is further described in the following sections.

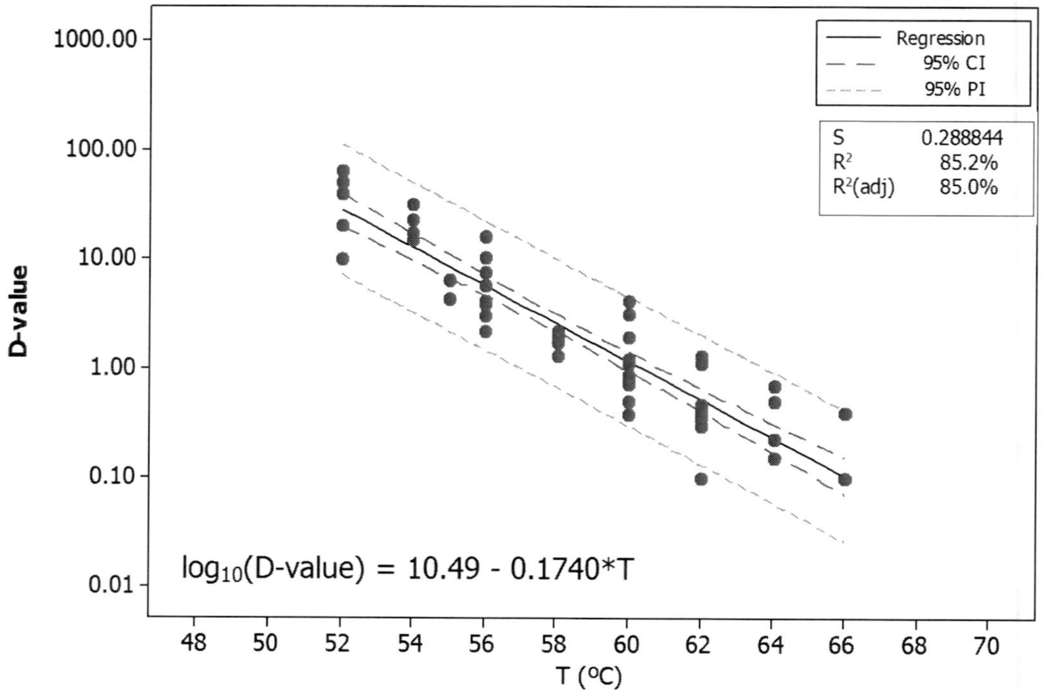

Figure 1. *D*-values of *L. monocytogenes* strains obtained in growth media. Adapted from Doyle et al.(2001). CI = confidence interval and PI = prediction interval at a confidence level of 95%. *z*=5.74.

Data taken from:

Golden et al. (1988) Test medium; tryptose phosphate broth (TPB). Survivors counted on tryptose phosphate agar (TPA) after growth for 3 days at 30 °C.

Foegeding and Stanley (1991) Test medium: phosphate buffer (PB). Survivors counted on TPA after growth for 5 days at 35 °C.

Sörqvist (1993) Test medium: physiological saline (PS). Survivors counted on blood agar after 5 days at 48 °C followed by 7 days at 37 °C.

Sörqvist (1994) Test medium: PS. Survivors counted on blood agar after 5 days at 48 °C followed by 7 days at 37 °C.

Patchett et al. (1996) Test medium: PB. Cells grown in batch or chemostat cultures. Survivors counted on coryneform agar after growth for 2 days at 30 °C.

Casadei et al. (1998) Test medium: trypticase soy broth (TSB). Survivors counted on TSA after growth for 6 days at 30 °C.

Quintavalla and Campanini (1991) Survivors were counted on TSA with pyruvate after growth for 3 days at 37 °C.

3. Target Organisms of Specific Food Commodities

3.1. Heat Resistant Microorganisms in Meat and Poultry Products

Meat is a rich nutrient matrix that provides a suitable environment for proliferation of meat spoilage microorganisms and common foodborne pathogens. Major pathogens that need to be controlled in fresh meat include *Salmonella*, *Campylobacter*, and enterohaemorrhagic *E.*

coli O157:H7. A great effort is being done for controlling these organisms, but some of these pathogens will continue to be of concern still for a long time (Sofos and Geornaras, 2010). *L. monocytogenes* will also continue to be the number one target for control in RTE meat and poultry products, considering its ubiquitous presence, potential to contaminate products after processing, and the ability to multiply even at cold temperatures (Mataragas et al., 2008; Thevenot et al., 2006; Tompkin, 2002). Additional pathogens may emerge and become of concern in meat products in the future (Sofos, 2008) and include non-O157 shigatoxin producing *E. coli* serotypes, *Mycobacterium avium* subsp. paratuberculosis, *Escherichia albertii*, *Clostridium difficile*, etc. Consumers are increasingly demanding fresher foods with high nutritive and sensory qualities, and thus, minimal heat processing and limited use of preservatives are desired.

The presence of pathogenic bacteria in food raw materials poses great challenges on the food industry. New strategies, like surface pasteurization of raw material have been used in order to decrease a possible contamination and survival during processing. Steam pasteurization process consists of exposing meat carcasses and meat products to water vapour at 82–97 °C inside a chamber at atmospheric pressure during 6–12 s. Normally, the treatment includes water removal, steam-pasteurization and a rapid chilling. In some equipment the steam can be combined with pressure to enhance the efficiency of the treatment. The industrial process was developed in the US and the steam-pasteurization treatment of carcasses was approved by the FDA in 1995 for whole and parts of carcasses that are to be further processed. The efficiency of the treatment was demonstrated in a series of articles; Nutsch et al., (1998) reported a reduction of 1.0 log CFU/cm^2 of the total viable count after a treatment of 6.5 s at 82.2 °C on beef carcasses. Retzlaff et al. (2004) had a significant reduction (>1.0 log CFU/cm^2) of *E. coli* O157:H7, *Salmonella* Typhimurium, and *L. innocua* in beef carcass with a 93.3 °C steam treatment during more than 6 s. Recently, Trivedi et al. (2008) showed a reduction of 4.38 log CFU/cm^2 on pork skin inoculated by *L. monocytogenes*, using commercial steam cleaner during 30 s.

3.1.1. *Clostridium botulinum*

The most relevant microorganism for shelf stable, low acid foods is *C. botulinum*. It is an anaerobic rod in which the spore develops in a swollen sporangium. It produces botulinum toxin, which can kill consumers even when it is present at very low concentrations. Relevant strains in foods can be classified according to their physiological characteristics as shown in Table 1.

Spores of proteolytic strains of *C. botulinum* are the more heat resistant ones among Clostridia and they can be considered therefore as the target pathogenic microorganism for this food group. If the environmental conditions of the product support its growth, it is necessary to inactivate any spores present by an effective, correctly estimated heat treatment. Foodborne botulism has often been associated with home-preserved foods such as cured meat and canned vegetables, where group I (proteolytic) *C. botulinum* prevails.

Due to its relevance for heat processing of ambient stable foods, the heat resistance of proteolytic *C. botulinum* spores has been studied extensively. The first studies were conducted by Esty and Meyer (1922), who showed that the heat resistance depended on the strain and the substrate in which the spores were suspended, after studying over 100 isolates in different food matrixes. They found that the most heat resistant spores had a $D_{121.1\ °C}$ value of 0.21 min. These results were used by Stumbo et al. (1975) to establish the basis of heat processing in the canning industry that considered that the heat treatment in ambient stable foods supporting *C. botulinum* growth would need to achieve 12 logarithmic reductions in the population, which was known since then as a 12D process. Therefore, this requirement would be achieved by heating for a minimum time of 2.52 min at a reference temperature of 121.1 °C in the coolest point of the container or batch of product. In the industry, to give an extra safety margin, this time has been rounded to 3 min at 121.1°C, which is known as F_0 value. To estimate the equivalent lethality of a process at temperatures other than the reference one (121.1 °C), the *z*-value of proteolytic *C. botulinum* spores is considered to be 10 °C (Lund and Peck, 2000). The main inactivation parameters are summarized in Table 1. Other Clostridia have been described to produce foodborne botulism, these are *C. baratii* and *C. butyricum*, although their spores would not challenge the established heat treatments for proteolytic *C. botulinum* (Stringer and Peck, 2008).

This heat treatment can be too high for some food products. Therefore, it is necessary to consider a different approach, by using a combination of preservative factors that will allow reducing the severity of the heat treatment. In this way, each inactivation parameter or growth limiting factor can reduce the probability of growth by a certain amount, ensuring that the same standard of safety (12D) is achieved. Examples of this are the acidification of canned vegetables or the use of nitrite and salt in canned meat products (Roberts et al., 1981a; Roberts et al., 1981b).

Table 1. Inactivation parameters and limiting growth factors of *C. botulinum* associated with foodborne botulism. Compiled from: (Stumbo et al., 1975; ACMSF, 1992; Anniballi et al., 2002; Peck and Stringer, 2005)

	Group I (proteolytic strains)	**Group II (non proteolytic strains)**
Spores heat resistance (*D* value)	$D_{121\,°C}$= 0.2 min	$D_{100\,°C}$<0.1 min
Recommended heat treatment	2.52 min at 121.1 °C	10 min at 90 °C
Minimum growth temperature (°C)	10-12 °C	3.0 °C
Minimum growth pH	4.6	5.0
Typical foods involved in outbreaks	Commercial or home-made low acid canned foods	Fermented marine products

3.1.2. Non-proteolytic *Clostridium botulinum*

This organism is the major concern for anaerobic, mildly heat treated, refrigerated foods. Spores can survive short heat treatments in the range of 90-100°C and cells can produce botulinum toxin at temperatures as low as 3°C but it has not been demonstrated at 2.5 °C (Graham et al., 1997). They does not produce toxin or grow at pH lower than 5.0 or at a NaCl concentration at about 5% (Graham et al., 1996; Hauschild, 1989; Lund and Peck, 2000). The main inactivation parameters are listed in Table 1.

Over the last years, with an accelerating trend in ready-to-eat food consumption, the number of botulism cases caused by these strains has increased. Concern of botulism as a result of consumption of REPFED products first emerged in the 1960s, and cases due to group II *C. botulinum* have been reported in the last 30 years all over the world. A comprehensive overview of human botulism due to group II *C. botulinum* in commercial food products (1980 – 2004), prevalence in raw foods, and heat resistance of spores in various media is presented by Lindstrom et al. (2006).

The *D*-values of *C. botulinum* group II spores are greatly affected by the food matrix, and the *z*-values, which are extrapolated from the thermal death-time curves formed by *D*-values measured at different temperatures, are therefore equivalent influenced. Reported *z*-values vary from 4.2 °C in oyster homogenate to 16.5 °C in pork and pea broth.

There is a wide variability in their heat resistance although *D*-values have been generally reported within a range of 0.5 to 2.4 min at 82.2 ºC (Stringer and Peck, 2008). However, the apparent heat resistance of spores varies depending on the food substrate and its composition. For example, an estimated *D*-value of 18 min at 82.2ºC was reported in beef (Fernandez and Peck, 1997) in the absence of lysozyme, but this value increased to 139 min in the same conditions, in the presence of lysozyme in the substrate (Fernandez and Peck, 1999). Lysozyme has been shown to induce germination of heat damaged spores by hydrolysing peptidoglycan in the spore cortex (Gould, 1989) and in the case of non proteolytic *C. botulinum*, can increase significantly its apparent heat resistance. The heat resistance of lysozyme is strongly dependent on the heating medium and is not well determined in foods. Alderton et al. (1974) have shown that lysozyme was quite stable to heat when it was tested at pH 3.0 but very heat labile in alkaline reactions. In studies of lysozyme stability to heat at pH 7.3, Garibaldi et al., (1968) showed that more than 15% of the activity was lost after 3.5 minutes at 60 °C. Even when low concentrations of lysozyme (25 mg/ml) were used in meat medium and heated to temperatures as high as 90 °C for 20 min, sufficient activity remained to have an effect on the heat resistance of nonproteolytic spores of *C. botulinum* (Gorris and Peck, 1999; Graham et al., 1997).

The recommended heat treatments for processing of group II *C. botulinum* spores vary from 36 min (ACMSF, 1992) to 52 min (ECFF, 1996; 2006) at 85°C based on reference *z*-values of 7°C (ECFF, 1996) and 9°C (ACMSF, 1992), respectively. The Advisory Committee on the Microbial Safety of Food (ACMSF, 1992) in the UK proposed a heat treatment of 90 ºC for 10 min would provide a 6D reduction, which was considered an appropriate reduction for this microorganism in refrigerated foods. However, with these recommendations a deviation from the estimated heat processes in the food industry might occur. Before making mathematical predictions for safe processing of foods, it is therefore of the utmost importance that the *D*- and *z*-values used in the calculations are for the food type in question instead of averages from the literature.

3.1.3. *Clostridium perfringens*

Gastroenteritis associated with *C. perfringens* in beef or poultry following inappropriate handling temperatures is one of the most common foodborne diseases in the U.S. (Mead et al., 1999). Rather than killing the spores, the heat stimulates dormant spores to germinate. When the food is consumed, a large number of the cells survive the low pH of the stomach and begin sporulating in the small intestine. It is during the sporulation process that the *C. perfringens* enterotoxin is produced. A reduction of risk of foodborne illness due to *C. perfringens* can be obtained by avoiding exposure of the food products to temperatures between 10°C and 47°C. Much of the minimally processed, refrigerated foods include cook-chill foods, e.g. sous vide, that are mildly pre-cooked, rapidly chilled, portioned, and distributed for refrigeration prior to reheating and consumption by consumers. Unlike frozen foods, cook-chill foods are more susceptible to abuse temperatures above 10°C for extended periods of time, allowing pathogens to grow to sufficiently high numbers (Taormina et al., 2003). *C. perfringens* is able to grow in a temperature range of 15 to 50°C with generation times of 6.3 min at 43 to 47 °C in meat or thioglycollate broth (Labbe and Huang, 1995). The result may be fatal as was demonstrated by an outbreak attributed to pea soup in a home for the elderly in Sweden in 2002, resulting in 64 cases of food poisoning and two deaths (de Jong et al., 2004).

Pasteurization temperatures (72 °C) for 5–10 min and routine well-cooking procedures readily inactivate vegetative cells of this organism (Heredia and Labbé, 2001). There is a greater concern of the heat resistance of their spores. This resistance varies depending on the strain, for example, in one study, the *D*-values at 95 °C for the heat resistant spores were between 17.6 and 63 min; however, in some reports, it has been as high as 200 min, compared with decimal reduction values between 1.3 and 2.8 min for the heat sensitive spores (Ando et al., 1985). Juneja et al. (2003) demonstrated an increased thermotolerance of *C. perfringens* spores following sublethal heat shock. *D*-values at 100°C for *C. perfringens* spores ranged from 15.5 to 21.4 min. Heat shocked spores with 9 out of 10 strains had significantly higher ($p < 0:05$) *D*-values at 100°C than unstressed spores. Acquired thermotolerance is of substantial practical importance to food processors and should provide useful information for designing thermal treatments to eliminate *C. perfringens* spores in ready-to-eat foods.

The levels of *C. perfringens* spores in raw meats will therefore be more important than total populations, which include heat sensitive vegetative cells. Microbiological data from the FSIS, Food safety and Inspection Service (2001), indicate a "worst case" of approximately 10^4 per gram density of *C. perfringens* on the raw product. For raw beef carcasses, there were 5 out of 4191 samples analyzed (0.12%) with results that were greater than 10^4 but less than 10^5 CFU/cm^2. There is still a lack of epidemiological evidence linking *C. perfringens* food poisoning to specific industry practices. In spite of this the U.S. Department of Agriculture Food Safety and Inspection Service (FSIS) has promulgated a regulation that a production process must result in ''no more than a 1 log10 multiplication of *C. perfringens* within the product'' (Food safety and Inspection Service, 2001). In appendix B of that document, the FSIS stated that the maximum internal temperature of certain cured cooked meats should be reduced from 54.4 to 26.7 °C in 5 h and from 26.7 to 7.2 °C in 10 h (15-h total cooling time) to meet the performance standard.

Akhtar et al. (2009) suggested an alternative strategy to inactivate *C. perfringens* spores in meat products using novel high pressure processing technology in combination with heat. The current study aimed to develop an inactivation strategy for *C. perfringens* spores in meat through a combination of spore activation at low pressure (100–200 MPa, 7 min) and elevated temperature (80 °C, 10 min), spore germination at high temperatures (55, 60 or 65 °C) and inactivation of germinated spores with elevated temperatures (80 and 90 °C, 10 and 20 min) and high pressure (586 MPa, at 23 and 73 °C, 10 min). Collectively, such studies demonstrate the feasibility of alternative and novel strategies to inactivate *C. perfringens* spores in meat products.

3.1.4. *Listeria monocytogenes*

Dairy, meat, egg and poultry constitute the product categories most often used to obtain data for *L. monocytogenes*. Plotting of reported $\log_{10}D$-values against temperatures obtained for model systems of some of these product categories (Figure 2 and 3) brings a somewhat different perspective from that of inactivation in media (Figure 1).

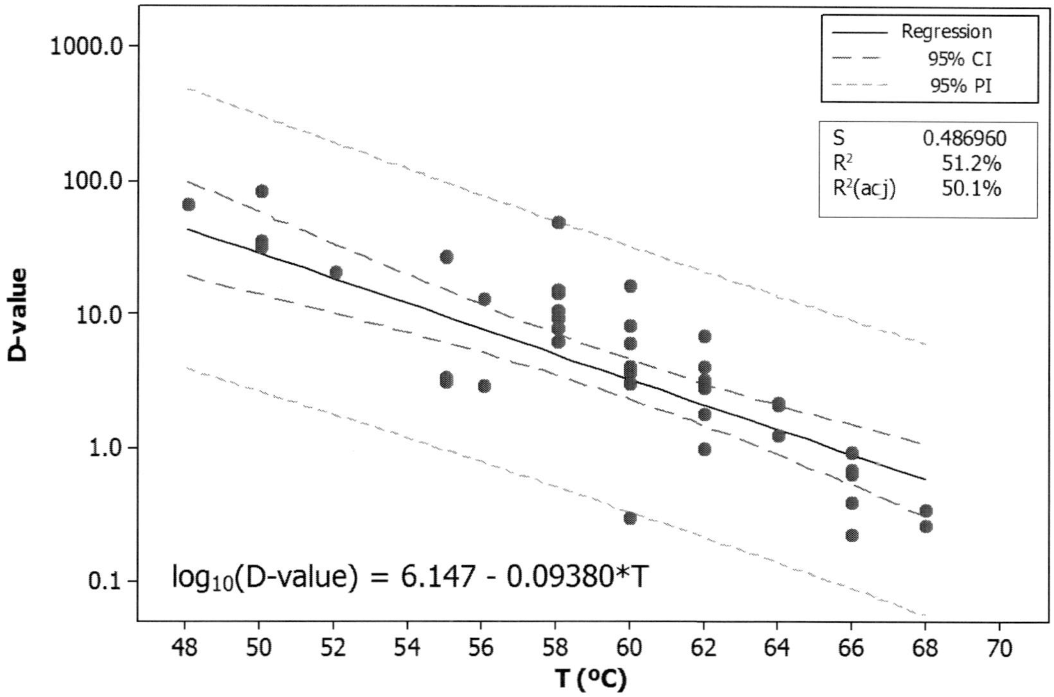

Figure 2. *D*-values of *L. monocytogenes* strains obtained in meat products. Adapted from (O'Bryan et al., 2006). z =10.7. CI = confidence interval and PI = prediction interval at a confidence level of 95%.

Data taken from:

Farber(1989): Inoculated ground meat placed in pouches in wire racks in water bath, phase of *L. monocytogenes* (LM) growth not stated.

Bolton et al.(2000): Uninoculated minced beef in vacutainers, placed in a water bath, allowed to reach temp then inoculated with stationary phase cells, Strips of beef inoculated with stationary phase cells minced, vacuum-packed, placed in water bath, allowed to reach temperature., Strips of beef inoculated with stationary phase cells, minced, vacuum-packed, and processed in commercial retort. Mazzotta and Gombas (2001): Batter prepared, inoculated with stationary phase cells, placed in pouches, sealed, rolled flat, placed in water bath, Batter prepared, inoculated with starved cells, placed in pouches, sealed, rolled flat, placed in water bath. Juneja (2003): Ground beef inoculated, mixed, placed in pouches, rolled flat, heat sealed, placed in water bath, phase of growth of *L. monocytogenes* not stated. Gaze et al. (1989): Beef macerated, autoclaved, placed in universal bottles, allowed to reach temperature, inoculated with stationary phase cells, mixed and held. Mackey et al. (1990): Ground beef inoculated with stationary phase cells, mixed, placed in pouches, rolled flat, heat sealed. placed in water bath.

The *D*-values reported for meat are from a wide temperature range (48-68 °C), but most abundant around 58-62 °C. The fit is rather poor (R^2 =0.52) and the variation is considerable, especially at 60 °C where a range between 0.15 – 16.7 can be seen. At this temperature the experimental conditions also vary to a great extent, and this may well give rise to the observed variation. Furthermore it appears that the experiments carried out in the range between 58-66 °C show inactivation kinetics that would contribute to a lower *z*-value.

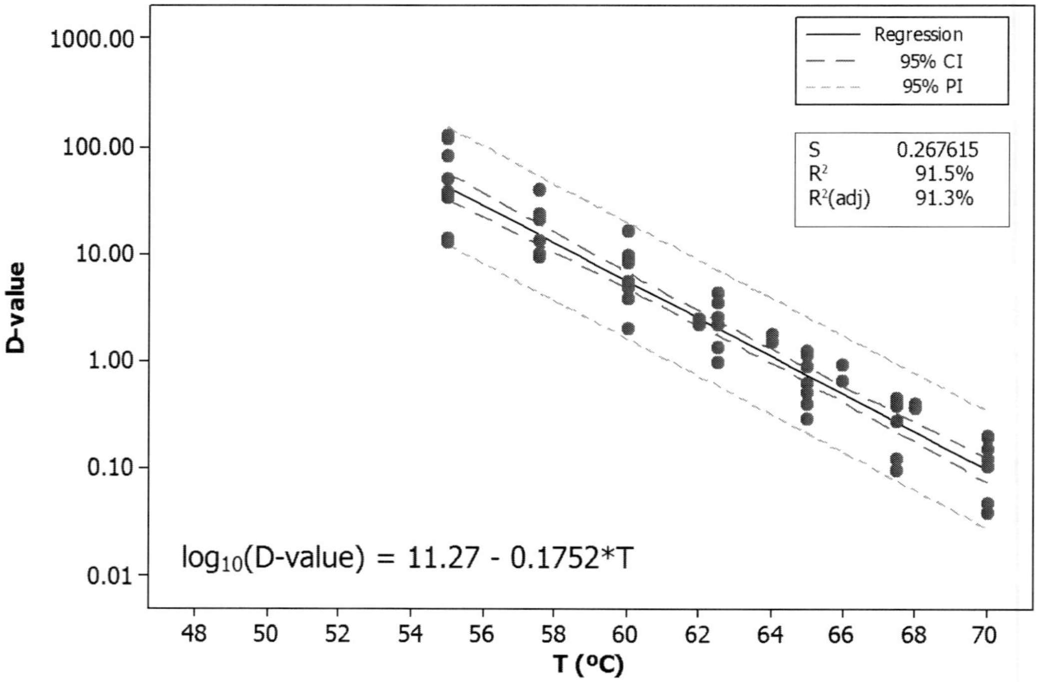

Figure 3. *D*-values of *L. monocytogenes* strains obtained in poultry products. Adapted from O'Bryan et al. (2006). CI = confidence interval and PI = prediction interval at a confidence level of 95%. *z*=5.7.

Data taken from:

Murphy et al. (2004a): Samples inoculated with stationary phase cells sealed in bag, rolled flat, placed in water bath at temperature for various times, assumed immediate come up. Murphy et al. (2004b): Samples inoculated with stationary phase cells sealed in bag, rolled flat, placed in water bath at temperature for various times, assumed immediate come up. Gaze et al. (1989): Samples placed in universal bottles, equilibrated to temperature in water bath, inoculated with stationary phase cells and held at temperature for various times. Murphy et al.(2003b): Samples inoculated with stationary phase cells, sealed in bag, rolled flat, placed in water bath at temperature for various times, assumed immediate come up. Murphy et al. (2003a): Cooked leg quarters inoculated with an "overnight culture" by injection, vacuum-packed, steamed in oven. Mackey et al. (1990): Chicken macerated, autoclaved, placed in universal bottles, allowed to reach temperature, inoculated with stationary phase cells, mixed, and held.

The heat resistance of *L. monocytogenes* in culture media and foods has been reviewed by Doyle et al. (2001). Important factors in the food product itself are water activity, acidity and the presence of other inhibitors. Ben Embarek and Huss (1993) also found that the *z*-value was higher in salmon (fatty species) than in cod (lean species), indicating an influence of fat content of the food matrix on the thermal resistance of *L. monocytogenes*. Selby et al. (2006) concluded that fat did not affect *L. monocytogenes* heat resistance in bologna with fat levels of 26 and 3.2%. Whereas Casadei et al. (1998) found that the D_{60}-value of *L. monocytogenes* Scott A grown in butter was 7.14 times higher than in broth.

Figure 3 shows a higher fit (R^2 = 91.5%) and linearity of the poultry data than that of the meat data. This may be attributed to more comparable model systems (4 datasets by the same group), and the more limited temperature range investigated (55 – 70 ºC). Also, the *z*-value obtained from poultry is quite similar to that obtained in liquid media.

3.2. Heat Resistant Microorganisms in Fish Products

Several pathogens are indigenous in the environments for fish and seafood and may be a health hazard for low acid pasteurized foods. The toxin producing bacteria are represented by psychrothrophic non-proteolytic *C. botulinum* type B, E and F, and psychrotolerant histamine producing bacteria (photobacteria). The infective microorganisms are *L. monocytogenes*, *Vibrio cholerae*, *Vibrio parahaemolyticus*, *Vibrio vulnificus*, *Aeromonas hydrophila*, *Plesiomonas shigelloides*. The role of toxin producing *B. cereus* are not well described for raw fish materials, but it may be a risk factor from ingredients in mixed or minced fish products. Bacteria from the environment, like psychrotropic *C. botulinum* type E and *L. monocytogenes* may easily contaminate fish and shellfish. *L. monocytogenes* has been considered as a leading cause of death amongst the foodborne bacterial pathogens (Paoli et al., 2005). The prevalence of *L. monocytogenes* in ready-to-eat foods varies with the products type and the stage in the production-to-consumption chain at which it is monitored. Cooked ready-to-eat products which may be consumed without further heating is obviously of concern, but as the levels of *L. monocytogenes* associated with contamination of these products are typically low, the risks are minimal if multiplication does not or cannot occur during storage, distribution and preparation. Epidemiologic data indicate that foods involved in listeriosis outbreaks are those in which the organism has multiplied and in general have contained levels significantly higher than 100 CFU/g (Buchanan et al., 1997; ICMSF, 2002). Based on risk assessment for *L. monocytogenes* in RTE foods, the Codex Alimentarius recommended that the maximum contamination level for *L. monocytogens* in food at consumption should be less than 100 CFU/g (Codex Alimentarius, 2002).

Analysis of foodborne outbreak data is one approach to estimate the proportions of human cases of specific enteric diseases attributable to a specific food item. In a microorganism-specific food source profile for foodborne outbreaks reported internationally between 1988 and 2007, it was *C. botulinum, Vibrio spp.* and *L. monocytogenes* that constituted the highest proportion (%) in seafoods (Greig and Ravel, 2009) and a thermal process should target a safe destruction of these pathogens.

3.2.1. *Listeria monocytogenes*

Inactivation of *L. monocytogenes* has been suggested as a criterion for minimum heat treatment by the European Chilled Food Federation (ECFF, 1996; 2006). *Listeria* is regarded as the most heat resistant of the non sporulating pathogens, and *Vibrio* species will be more effectively heat inactivated if present (Johnston and Brown, 2002; Wong et al., 2002). Thermal processes designed for shelf lives of more than 10 days at chilled conditions, requires at least a 6 log inactivation (6D) of non-proteolytic *C. botulinum* corresponding to 90 °C for 10 min (European Commission, 1997), Table 2.

A large numbers of heat resistance values have been published and different recommendations have been made on a satisfying listericidal treatment (ECFF, 1996; NACMCF (U.S.National Advisory Committee on Microbiological Criteria for Foods), 1992; Schellekens and Martens, 1993). Reviewing the published data, it seems that the organism cannot survive a proper pasteurization process but different results have been found on the effect of particular environmental conditions (fat and salt content, heating media, heating profile, temperature history) on the heat resistance of *L. monocytogenes*. Lou and Yousef (1999) reviewed in detail the large expanse of literature on the thermal inactivation of *L. monocytogenes*. ICMSF (1996) have provided extensive lists of thermal inactivation times under different conditions and food types and heat tolerance of *L. monocytogenes* can be maximized by prior sub-lethal shocks, stress or applying to cells having reached stationary phase. These effects, and effects on subsequent lag and growth, have been studied and modelled (Augustin et al., 1998; Breand et al., 1997; Breand et al., 1999; Stephens et al., 1994).

Table 2. Application of heat and hurdle principles of mild heat treated fish products

Target organism	**Heat resistance (min)** [a]	**Recommended heat treatment - *6D* of target organism**	**Surviving pathogens**	**Storage conditions to inhibit surviving organisms**
Psychrotropic *C.botulinum* type E	$D_{90} = 1.5$	90 °C for 10 min	Pathogens producing heat resistant spores survive	< 10 °C [b]
L. monocytogenes and other non-spore-forming pathogens	$D_{70} = 0.3$	70 °C for 2 min	Sporeforming pathogens survive	< 3 °C [c]
V. parahaemolyticus	$D_{70} = 0.001$	70 °C for 2 min	Sporeforming pathogens survive	< 3 °C
B. cereus	$D_{100} = 1\text{-}36$	100 °C for 48 min	Heat resistant spore- formers survive	< 4 °C

[a] Most heat resistant species in the target group (ECFF 1996; ECFF 2006)

[b] 10 °C – lowest growth temperature for proteolytic *C. botulinum*. If food is likely to support *B. cereus* growth, the limit should be lowered to 4 °C, which is the lowest growth threshold for psychrotrophic *B. cereus*

[c] 3.0 °C lowest growth threshold for non-proteolytic *C. botulinum* (Graham et al., 1997)

In cases where *L. monocytogenes* is selected as target organism, a 6D process equivalent to 70 °C for 2 min, is generally advised (Table 2). FDA's risk assessment on *L. monocytogenes* indicates that approximately 7% of raw fish are contaminated with from 1 to 10^3 CFU/g, and that approximately 92% are contaminated at less than 1 CFU/g. Less than 1% of raw fish are contaminated at levels greater than 10^3 CFU/g, and < 0.1 % at levels greater than 10^6 CFU/g.

The seafood *D*-values are also obtained in a wide temperature range, with quite diverse sample systems. Despite this the fit is quite good (R^2 = 89.7%) and the *z*-value obtained from the pooled seafood data is a little higher that than obtained from the media and the poultry data, but it can be seen that the amount of available data for seafoods is much less than for the other animal protein categories.

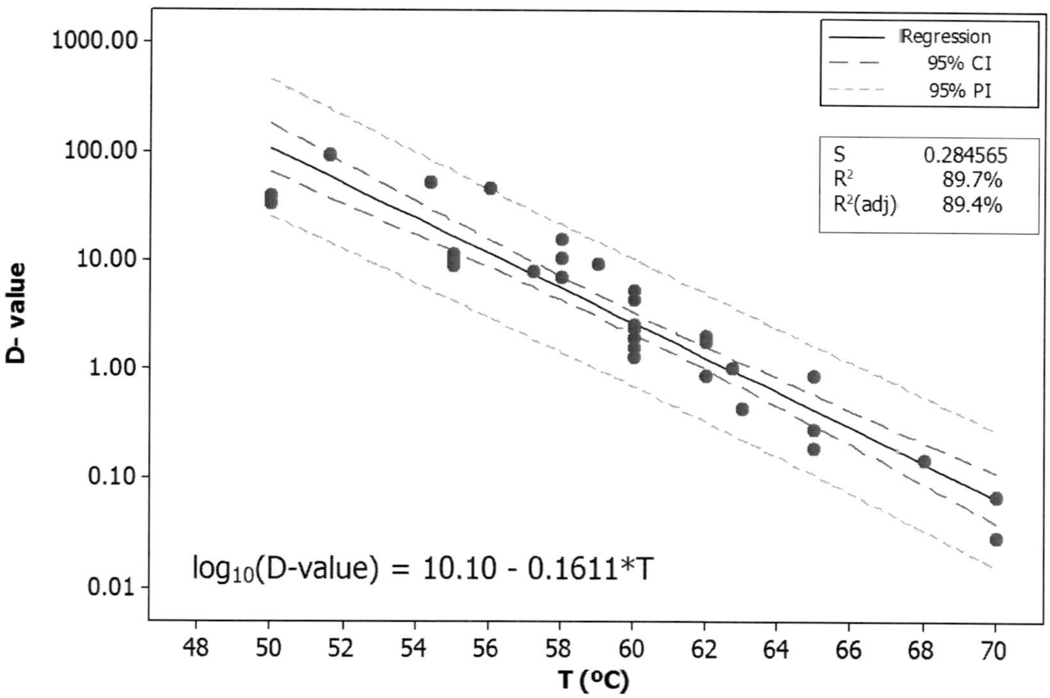

Figure 4. *D*-values of *L. monocytogenes* strains obtained in seafood products. Adapted from Doyle et al. (2001). CI = confidence interval and PI = prediction interval at a confidence level of 95%. *z* = 6.2.

Data taken from:

Harrison and Huang (1990): Canned crabmeat was blended, inoculated, packaged in sausage casing and heated in a waterbath. Budu-Amoako et al. (1992): Cooked lobster was blended, inoculated with five-strain mixture and heated in a waterbath. Ben Embarek and Huss (1993): Fish fillets were inoculated and vacuum packed in plastic bags. After heating in a waterbath, samples were stored at 28 °C for 3 weeks simulating storage for sous vide cooked food and allowing anaerobic resuscitation of injured cells. Dorsa et al. (1993): Cooked crawfish tail meat was blended and inoculated with three-strain mixture and heated in a waterbath. Bremer and Osborne (1995): Mussels were heated, brined, inoculated, blended and heated a waterbath. Miettinen et al. (2005): Inoculated rainbow trout roe kept at 10 °C for 5 days, vacuum packaged in glass jars and heated in autoclave.

3.2.3. *Clostridium botulinum*

Several authors have shown the prevalence of *C. botulinum* in seafood (Cann et al., 1966; Fach et al., 2002; Hielm et al., 1998; Huss, 1980; Hyytia et al., 1998; Korkeala et al., 1998). When found in seafood products from cold-water areas such as Scandinavia, Canada, Alaska, Russia and some parts of Japan, *C. botulinum* type E is reported to be the most prevalent type (Huss, 1997) and is considered to be a true aquatic organism. The amount of toxin necessary to cause human illness is very small (nanograms) and if an environment favourable to germination and growth of spores from this organism exists in fish, before or during processing, there is a potential risk of illness.

Many traditionally preserved fish products are dried, smoked or salted. The sodium chloride concentration is an important factor in controlling the outgrowth and toxin production of *C. botulinum*. It is generally accepted that the inhibitory concentration for non-proteolytic *C. botulinum* is 5.0% water-phase salt (WPS), equivalent to a water activity of 0.97. A 10% WPS concentration is necessary to inhibit the proteolytic strains. The hurdle principle will apply in such a situation, and it should be stressed that these NaCl limits apply under optimal conditions for bacterial growth, and that other factors as the presence of oxygen, low temperature and high or low pH will reduce the salt tolerance of *C. botulinum* in a practical situation. Huss (1994) concluded that in fish products stored at temperatures below 10 °C, a water phase salt concentration of 3 % is sufficient to inhibit the growth of *C. botulinum* for at least 30 days. Lynt et al. (1977) studied the heat resistance of five strains of *C. botulinum* type E heated in crab meat at 82.2 °C and reported *D*-values ranging from 0.49 to 0.74 min with *z*-values of 7.2 to 8.3 °C. The heat resistance at 90 °C can be calculated to be approximately 0.005 to 0.07 min using the higher *z* value of 8.3 °C. These data were lower than the one found by Gaze and Brown (1990) of type E in cod homogenate; the *D*-value at 90 °C was 0.79.

Studies by Eklund et al. (1988) and Peterson et al. (1997) showed that the type *C. botulinum* type B spores were more heat resistant in pasteurized seafoods than in buffer. Growth and toxin production by nonproteolytic type B strains occurred in salmon after a heat treatment of 88.9 °C for 46 min (Eklund et al., 1988). This may be related in part to the 5 to 8% lipid content of the salmon used. The microbial risks in regard to mild hot smoking of fish has been reviewed by Sikorski and Kolodziejska (2002) and thermal inactivation of nonproteolytic *C. botulinum* type E spores was investigated in rainbow trout and whitefish media at 75 to 93 °C by Lindstrom et al. (2003). In the latter study it was shown that a time-temperature combination of 42 min at 85 °C (fish surface temperature) with > 70% relative humidity prevented growth from 10^6 spores in vacuum-packaged hot smoked rainbow fillets and whole whitefish stored at 5 weeks at 8 °C. When Dungeness crab meat (lipid content 0.8 to 3.0%) was used as the heating medium, the *D*-value ranged from 12.9 min at 88.9 °C to 2.9 min at 94.4 °C (Peterson et al., 1997). These data on the survival and growth of type B are similar to those reported when lysozyme was included in the meat medium (Graham et al., 1997). Lysozyme at levels up to 200 mg/g has been reported in fish and seafood (Lie et al., 1989; Lund and Notermans, 1992) and may have contributed to the greater heat resistance in salmon and crab meat.

3.2.4. *Bacillus cereus*

B. cereus has been associated with food service operations because of the heat resistance of their spores. Fish as a commodity has rarely been examined for the presence of *B. cereus* in particular (Abeyta, 1983). *Bacillus* species has been isolated from *sous vide* cod fillets (Ben Embarek, 1994) and in many other *sous vide* products (Nissen et al., 2002). In an examination of 347 fresh and processed retail seafood samples Rahmati and Labbe (2008) found 62 *B. cereus* isolates from separate samples at levels ranging from 3.6 to > 1,100 CFU/g. Thirty (48%) of 62 isolates produced both the hemolysin BL (HBL) and nonhemolytic (NHE) enterotoxins, and 58 (94%) and 31 (50%) produced NHE or HBL toxins, respectively. Studies on the microbiology of spices have demonstrated that *B. cereus* and *C. perfringens* are the sporeformers commonly found in spices (Banerjee and Sarkar, 2004). A study of dried spices and herbs from retail and production premises to determine the microbiological status of such products was undertaken in the UK during 2004. Overall, 3.0% of the herbs and spices contained high counts of *B. cereus* (1%, $\geq 10^5$ CFU g^{-1}), *C. perfringens* (0.4%, $\geq 10^3$ CFU g^{-1}) and/or *E. coli* (2.1%, $\geq 10^2$ CFU g^{-1}) (Sagoo et al., 2009).

Complex fish products added various ingredients, e.g. vegetables, spices, milk and flour, may contain different heat resistant microorganisms. The mild heat treatment of many ready meals may be sufficient to inactivate spores of non-proteolytic *C. botulinum*, but not for the more heat resistant spores of *B. cereus*.

3.3. Heat Resistant Microorganisms in Vegetable Substrates

In general, vegetable products are relatively low in protective compounds and they present a pH value in the range of 4.5 to values just below 7.0. Therefore, in many cases a reduction in the apparent heat resistance of sporeforming bacteria has been described. Cameron et al. (1980) described a reduction in the *D*-values of *C. sporogenes* suspended in pea puree compared to phosphate buffer, which increased when the puree was acidified, similar to the finding in artichokes by Rodrigo and Martinez (1988). Brown and Martinez (1992) also found lower *D*-values in *C. botulinum* suspended in mushroom extract than in phosphate buffer. Fernández et al. (1995) reported a decrease of *D*-values in *B. stearothermophilus* suspended in natural mushroom extract (with a pH of 6.7) compared to bidistilled water.

Acidification of the recovery medium to the same pH value of the substrate decreases the apparent heat resistance of spores of *B. stearothermophilus* (Fernández et al., 1995), due to the inhibition of heat damaged spores. In *B. sporothermodurans* the influence of the vegetable substrate has also been evidenced reducing heat resistance (van Zuijlen et al., 2010). Other factors present in the recovery medium will in addition influence the recovery of spores. The impact of CO_2 concentration, pH and NaCl on nonproteolytic *C. botulinum* has been studied (Fernandez et al., 2001b). It has been shown that *Alycyclobacillus acidoterrestris*, can grow at typical pH values of fruit juices (3.0 to 4.0) surviving the typical heat processes applied. It causes off flavors in the final product Although its presence is relatively low in the raw products, several cases of spoiled fruit juices have been described and if its incidence would spread, a considerable increase in the heat treatment of acidic products would be needed. Different control measures of this organism in acidic fruit products have been developed (Silva and Gibbs, 2004) and combination of heat and supercritical carbon dioxide have been suggested to minimize quality reduction (Bae et al., 2009).

The presence of oil in the suspension medium has a protective effect, increasing spore heat resistance. This behavior has been reported by Ababouch and Busta (1987) in *C. sporogenes*. The influence of free fatty acids in heating medium or in recovery medium on the heat resistance of spores of *B. cereus* and *C. sporogenes* has also been studied. The presence of free fatty acids in the recovery medium highly decreases bacterial spore apparent heat resistance, particularly with unsaturated fatty acids.

However, the impact of the substrate and pH levels on *z*-values is not consistent through the literature. Some studies have reported no changes in *z* value due to substrate or moderate acidification, such as Xezones and Hutching (1965) in *C. botulinum* or Silla Santos et al. (1992) in *C. sporogenes* PA3679. Others have reported a decrease in the *z*-values, for example Cameron et al. (1980) in *C. sporogenes*. And there are also reports indicating an increase in *z*-values, such as Brown and Martinez (1992) in *C. botulinum*, Fernández et al. (1995) in *B. stearothermophilus* or Condon and Sala (1992) in *B. subtilis*. Therefore, the impact of the food substrate on the *z*-value seems to be substrate dependent and it also determines the effect of moderate acidification on it. It is necessary to use information as close as possible to the real substrate in order to establish accurately the heat processes required (Figure 5).

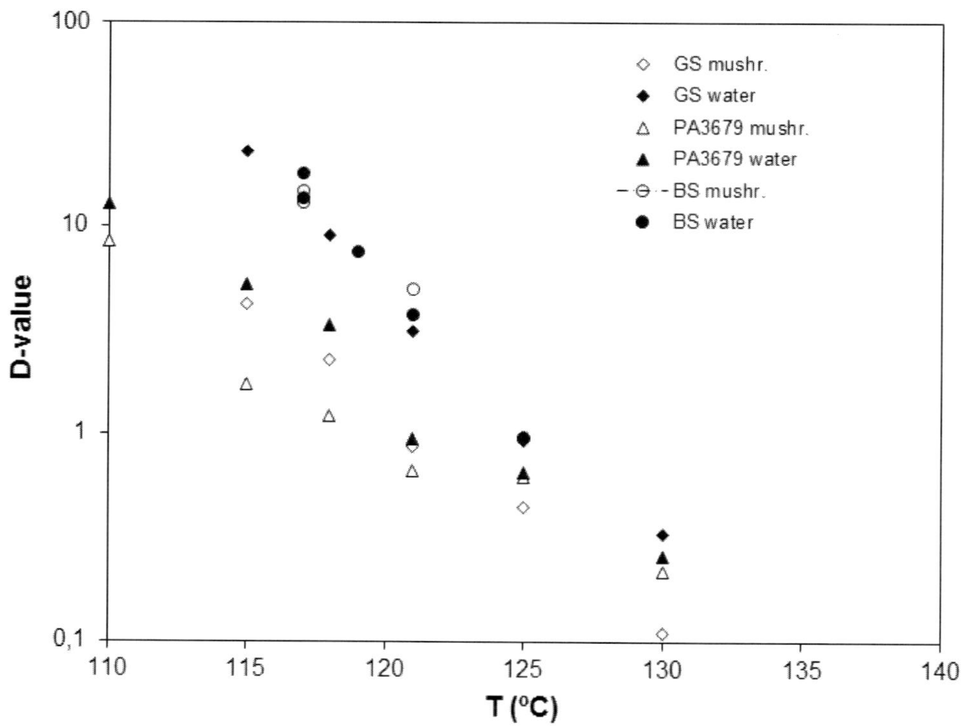

Figure 5. Representation of *z* values of sporeforming bacteria in water or buffer and in mushroom substrate . GS, *Geobacillus stearothermophilus* (---) (Fernández et al., 1995; Rodrigo et al., 1997), *C. sporogenes* PA3679 (····) (Ocio et al., 1994; Sanchez et al., 1995) and *B. sporothermodurans* (—)(van Zuijlen et al., 2010).

3.3.1. Psychrotrophic Strains of *Bacillus cereus*

B. cereus spores and cells are common in soil, in water, the gastro-intestinal tract of eukaryotic organisms and also in ingredients commonly used in food production (Granum, 2007; Hong et al., 2009; Jensen et al., 2003; Swiecicka, 2008). This wide and ubiquitous distribution in the environment, particularly on ingredients of plant origin, together with the formation of endospores resistant to pasteurization temperatures, chemical and other environmental stresses, favor *B. cereus* contamination in ready meals, refrigerated processed foods of extended durability (REPFED) and pasteurized products such as milk or fruit juices. Its spores present relatively low heat resistance, since they can be readily inactivated at 100 °C (Fernandez et al., 2001a), therefore, for shelf stable foods they do not mean a risk, although they can pose a hazard in not shelf-stable, refrigerated foods.

Heating and the resultant killing of the majority of the food microflora allow the storage of products at refrigeration temperatures for weeks and, in some cases, months (Lund and Peck, 1994). Psychrotrophic strains of *B. cereus* can pose a serious threat to this range of foods since they can grow and produce toxins at relatively low temperatures and they produce spores that can survive moderate heat treatments. *B. cereus* spores heat resistance in various media show $D_{95\,°C}$ in a range from 1.5 to 36.2 min and *z*-values from 6.7 to 10.1 °C (ICMSF, 1996). The diversity of *B. cereus* has been acknowledged as a source of difficulty for risk assessment in foods (EFSA, 2005), and a quantitative exposure assessment in cooked, pasteurized foods has been performed (Afchain et al., 2008). *B. cereus* have been classified into seven phylogenetic groups (Guinebretiere et al., 2008) and their main characteristics of growth temperature, heat resistance and association to cases of foodborne disease have been described. Cardinal growth parameters of the *Bacillus cereus sensu lato* phylogenetic groups have been established (Carlin et al., 2006) and there are three groups described (II, V and VI) that can grow at refrigeration temperatures (in the range of 2.2-5.5°C) (Christiansson et al., 1989; Van Netten et al., 1990). The more psychrothropic groups seem to be less heat resistant (Guinebretiere et al., 2008) and Carlin et al. (2006) also observed different percentages in spore germination at chilled temperatures. The virulence of each strain is not clearly documented, but some genetic groups potentially more virulent, while some of the psychrothropic strains seem to have low ability to cause gastro-enteritis (Carlin et al., 2006; Choma et al., 2000; Guinebretiere et al., 2008). *B. cereus* is a common contaminant of milk (Christiansson et al., 1989). However, psychrotrophic *B. cereus* from pasteurized milk appeared to have a low enterotoxin production potential, and they were not producers of emetic toxin or cytotoxin K, and therefore may be less likely to cause illness than mesophilic strains (Svensson et al., 2007).

3.3.2. Highly Heat Resistant Bacillus Spp. and Geobacillus Spp.

This group includes the widest range of spoilage microorganisms, particularly frequent in soil, vegetables and spices. According to the heat process applied to the food, the most relevant ones are *Geobacillus stearothermophilus* and *Bacillus sporothermodurans*. *Geobacillus stearothermophilus* is a thermophilic, high heat resistant microorganism that causes flat-sour spoilage in a wide range of foods, including vegetables and dairy products (Stumbo, 1973). Heat resistance values described can be a $D_{121\,°C}$ of 3-4 min with a *z*-value of 7 °C (Fernandez et al., 1995). It is particularly relevant for shelf stable products produced, distributed or exported in tropical or subtropical areas, where temperatures during distribution and storage can be above 40 °C. The impact of food substrate (pH, NaCl and other components) on *G. stearothermophilus* has been studied extensively, and predictive models describe the expected *D*-value according to several environmental factors (Periago et al., 1998). When performing tests with this microorganism, sporulation conditions have to be chosen carefully in order to obtain high heat resistant spores. This microorganism has been extensively used for validation studies in heat processed foods.

Bacillus sporothermodurans is a mesophilic sporeforming bacterium with a very high heat resistance. $D_{121°C}$ values in the range of 3 to 5 min have been described (van Zuijlen et al., 2010). Its apparent heat resistance is affected by the substrate, although it can still survive and grow at pH values of 4.7.

4. Target Organisms of Novel Thermal Inactivation Processes

The ideal processing technology for the food producers would be the one that meets the following demands (taken from (Raso et al., 2005); i) improvement of the shelf life and safety by inactivating enzymes, spoilage and pathogenic microorganisms, ii) no changes in organoleptic and nutritional attributes, iii) no residues left in the food, iv) convenient to apply, v) cheap, vi) no objections from consumers and legislators.

During the last years, new thermal methods are tested in order to overcome or at least minimise undesirable quality changes. Examples of such methods are HTST (high temperature short time), ohmic heating, infra-red heating and high frequency (radio frequency heating and microwave heating). Microwave and radio frequency heating refers to the use of electromagnetic waves of certain frequencies to generate heat. Microbial cells in food may be differently heated compared with the food matrix, which could result in specific inactivation.

4.1. Target Organisms for Novel Thermal Processes

The main inactivation mechanism for dielectric heating (microwave and radiofrequency heating) and ohmic heating seems to be thermal in nature. There are studies pointing in the direction of specific non-thermal effects, however, the influence on *D*- and *z*-values still lack complementary data. The target organisms for novel heating processes will for the time being, therefore be similar as for conventional heating technologies, i.e. related to a combination of substrate, type of packaging and shelf life.

4.2. Microwave Heating

Microwave heating has many advantages over other methods like gamma radiation and ultraviolet radiation. Microwave ovens do not need a large space, do not heat the air, can be used continuously, save time and energy, and preserve more of the nutrients in the food (Wang et al., 2003). The practical problems in achieving uniform and reproducible heating have been addressed by carefully designed microwave cavities, by application of shielding or field directing films onto certain areas of the packages, by complicated power control schemes to direct controlled amounts of MW-power to defined areas on food trays or by the use of multi-mode cavity tunnels, where the scattered MW-fields are expected to give a fairly uniform coupling of the MW-energy into the food packages. Many studies have reported non-thermal effects of microwave energy on microorganisms in foods, and bacteria reported to be inactivated by microwave heating include *B. cereus, Campylobacter jejuni,C. perfringens,* pathogenic *E. coli, Enterococcus, L. monocytogenes, Staphylococcus aureus,* and *Salmonella* (Chipley, 1980; Heddleson and Doors, 1994; Knutson et al., 1987; Rosenberg and Bögel, 1987). Kozempel et al. (1998) summarized the theories that have been advanced to explain non-thermal effects of electromagnetic energy and investigated the effect of microwave energy on yeast and bacteria at $\leq 45\ ^{o}C$. They concluded that there were no non-thermal effects. Welt et al. (1994) also concluded that the effect of microwave energy on viability of *C. sporogenes* was indistinguishable from the effect of conventional heating. This is also supported by the fact that the inactivation kinetics and resistant microorganisms to microwave heating are identical to conventional heating. The final conclusion on a possible non-thermal effect of microwaves is however, not drawn. Recently Barnabas et al. (2010) claimed that microwaves was able to destroy bacterial cells at specific parameters without causing heating of the substrate, when using low dielectric substrates such as a glass slide.

4.3. Radio Frequency Energy

RadioFrequency (RF) heating is applicable to packaged foods and a candidate for rapid heating in cook-chill processes. For RF heating, an operating frequency of 27.12 MHz is most commonly used. The advantage of RF- compared to MW fields is the larger penetration depth that enables more uniform heating of large package volumes, the reduced presence of power intensity interference patterns with the much larger wavelength and the better availability of industrial power generators rated above 50 kW, which are needed in high volume food processes. However, RF heating also suffers from uneven heating, and so far is rarely used for the pasteurization of foods (McKenna et al., 2006). Using an oven power output of 450W a 4.3-fold reduction in cooking time compared to conventional steam oven cooking could be achieved on cased comminuted meat emulsion (white pudding) (Lyng et al., 2007). A pilot scale RF heater has been constructed and the thermal aspects of RF-heating have been modelled (Skipnes and Pfeiffer, 2005). The existence of a non-thermal effect of RF energy on microorganisms has been debated for more than 50 years. This would be attractive because fresh-like qualities and nutrients would be maintained in foods if RF energy could cause non thermal inactivation of microorganisms. However, no non-thermal effect of RF energy was detected on *E. coli* K-12, *L. innocua,* or yeast in apple cider, beer, deionised water, liquid whole egg, and tomato juice; nor were there any synergistic effects of RF energy with heat (Geveke et al., 2002).

4.4. Ohmic Heating

Ohmic heating is an advanced thermal processing method wherein the food material, which serves as an electrical resistor, is heated by passing electricity through it (Palaniappan and Sastry, 2002). Electrical energy is dissipated into heat and the rate of heat is directly proportional to the intensity of the electric field and to the electric conductivity of the sample. Ohmic heating can be used for heating liquid foods containing large particulates, such as soups, milk, liquid eggs, juice and yogurt, and heat sensitive liquids (Ho and Mittal, 2000).

The technology is useful for the treatment of proteinaceous foods, which tend to denature and coagulate when thermally processed. For example, liquid egg can be ohmically heated in a fraction of a second without coagulating it. Juices can be treated to inactivate enzymes without affecting the flavour. Other potential applications of ohmic heating include blanching, thawing, on-line detection of starch gelatinization, fermentation, peeling, dehydration, and extraction. The use of ohmic heating to other applications is now greatly expanding, e.g. inactivation kinetics of enzymes (milk and vegetable juices) (Jakob et al., 2010), quality effects of turkey meat (Zell et al., 2010b) and whole meat muscle (Bozkurt and Icier, 2010a; Bozkurt and Icier, 2010b; Zell et al., 2009; Zell et al., 2010a). The main inactivation mechanism in ohmic heating seems to be thermal in nature. However, Knirsch et al. (2010) reviewed the literature on ohmic heating and several reports show that ohmic heating may present mild non-thermal cellular damage due to the presence of the electric field (Cho et al., 1999; Pereira et al., 2007; Sun et al., 2008). The principal reason for the additional effect of ohmic treatment may be its low frequency (usually 50-60 Hz), which allows cell walls to build up charges and form pores (USA-FDA, 2000). Machado et al. (2010) even demonstrated the inactivation of *E. coli* using moderate electric fields at temperature below 25 °C. As a main consequence of this effect, the *D*-value observed for the microbial inactivation under ohmic heating is reduced when compared to traditional heating methods. This reduction has been observed for *B. licheniformis* and *E. coli* (Pereira et al., 2007), *B. subtilis* (Cho et al., 1999), and *Streptococcus thermophilus* (Sun et al., 2008). Compounds with poor conductivity, especially the fat in meat products, do not generate heat at the same rate as the muscle, thus creating a cold spot (Shirsat et al., 2004). In fact, conductivity of the products has to be in the range of 0.01-10 S/min (Piette et al., 2004). Another relevant factor to be considered is the heating rate (almost instant to reach target temperature) which may greatly reduce the impact of adaptation mechanisms of microorganisms.

4.5. High Pressure Processing Combined with Heat

High pressure (HP) is a novel method of food processing to achieve microbial inactivation or to alter the food attributes to get desired qualities. There are two physical consequences that affect the microbial spore destruction kinetics: i) pressure increase and ii) adiabatic compression heating resulting in a temperature increase, both having potential to increase the microbial kill rate. Microorganisms vary in their response to high pressure and bacterial spores are regarded as the most resistant group (Patterson, 2005; Shao et al., 2010). Microbial inactivation by HP does not follow a first order kinetics and a tail of inactivation is sometimes present (Garriga et al., 2002) and surviving cells can grow during storage (Chen and Hoover, 2003). Combination treatments using high pressure and heat have been proposed as a method of producing shelf stable low acid foods. Relatively low pressures (below 200 MPa) can trigger spore germination (Gould and Sale, 1970). This has led to the suggestion that spores could be killed by applying pressure in two stages. However, the extent of the inactivation can be highly variable and another approach to the problem is to combine high temperature along with pressure treatment. This approach is now being actively used for the commercial production of shelf-stable foods and is the subject of a number of patents designed to achieve the commercial sterilization of foods that have a pH greater than 4,5 (Reddy et al., 2006; Scurrah et al., 2006).

The temperature, pressure level, treatment time and time interval between the cycles can be varied depending on the product but are designed to give greater than the equivalent of a 12D process for *C. botulinum*. These treatments use an initial temperature of below 100°C and rely on the fact that adiabatic heating occurs when the product is pressure treated. The extent of the temperature increase varies with the composition of the food but is normally 3–9 °C/100 MPa, depending on the product temperature and composition (Patazca et al., 2007). Effects of HP combined with high temperature on spores, e.g. log reductions of *Clostridium* spp. and *Bacillus* spp., have been summarized in several reviews (Zhang and Mittal, 2008; Patterson, 2005; Wilson et al., 2008).

5. Sublethal Thermal Treatment of Spores

Dormant spores return to their vegetative growth cycle in a process including germination (Paidhungat and Setlow, 2002). Germinated spores have a lower heat resistance than dormant spores, and can be inactivated at lower temperatures (Setlow, 2003). Several physical and chemical factors can induce germination of spores, e.g. sub-lethal heat treatment, high pressure processing, or chemical germinants. From this, it has been questioned if an improved safety can be achieved by double heat treatment, often referred to as Tyndallisation. This involves two heat treatments separated by a period of lower temperature (e.g. 30 °C) storage. Induced germination before the second heat treatment, with subsequent inactivation of susceptible germinated spores, could be used as a method to improved spore control. In developing such processes, use of techniques known to enhance germination, e.g. heat activation (Berg and Sandine, 1970; Keynan and Evenchik, 1969) and addition of germinants (Moir et al., 2002; Moir, 2006), would be valuable. Heat treatment at 70 °C is one of the temperatures often used to heat activate different strains of *B. cereus* (Clements and Moir, 1998; Collado et al., 2003; Gould, 1966; Gounina-Allouane et al., 2008; Hornstra et al., 2005; Johnson et al., 1983). In addition, L-alanine and inosine have previously been shown to be powerful germinants for *B. cereus*, often generating stronger germination response used in combination than alone (Clements and Moir, 1998; Foerster and Foster, 1966; Gould et al., 1968; Hornstra et al., 2006).

A major problem in using germination as a spore control procedure in a Tyndallization process is the nonlog-linear nature of germination kinetics (Gould et al., 1968), which often displays long tails in long-term storage experiments. Clearly, the heterogeneity of spore populations, also observed by high pressure induction of germination, is reflected in a number of spore properties, including wet heat resistance, superdormancy and species specific heat activation temperature optima (Ghosh et al., 2009). Most experiments with heat induction of spore germination have experienced limited spore germination, compared to an initial inoculated spore numbers, and the effects of spore germination are therefore difficult to predict.

Conclusion

- Minimally processed, vacuum or modified atmosphere packaged food products are known to be of very high quality and increasingly used in gastronomy and food service. Their taste and texture in many cases compare to that of freshly prepared foods. The convenience aspect they offer has also led to increasing demand by quality conscious consumers in retail markets. These products, however, have a limited shelf life and require chilled distribution, and heat treatments designed to achieve a specific lethality for foodborne pathogens is one of the fundamentally important strategies used to assure the microbial safety of thermally processed foods.

- Several factors affect the heat resistance of the microorganisms; species and strain of bacteria, food composition, the physiological stage of microbial cells and spores, and recovery conditions (e.g. media and growth temperature). It is also documented that increased heat resistance of an organism can be caused by water activity and the presence of carbohydrates, lipids, proteins, salt, etc., and heat resistance of spores is altered by thermal adaptation, mineralization, and degree of dehydration. Research on the effects of these conditions and interaction among the variables is still needed for a better understanding of the heat inactivation kinetics and consideration of the products safety.
- Special attention must be given to analyses of emerging pathogens adapted to e.g. chilled temperatures or specific ingredients and to analyses of heat inactivation kinetics when novel processing technologies are introduced. Data from such experiments will be valuable to predict the heat resistance and survival of microorganisms in future food processing and distribution. Modelling of predictive thermal inactivation kinetics will most likely be an important aid in the future to analyse complex multifactorial experiments.
- A better understanding of the inactivation mechanisms in microorganisms has been suggested. These include i.e. denaturation of proteins and enzymes, disruption of cellular membranes. The manner in which a technology affects pathogen viability should be addressed in order to optimize pathogen destruction.

References

Ababouch, L. & Busta, F. F. (1987). Effect of thermal treatments in oils on bacterial spore survival. *J. Appl. Bacteriol.,* 62, 491-502.

Abeyta, C. (1983). Bacteriological quality of fresh seafood products from Seattle retail markets. *J. Food Prot.,* 46, 901-909.

ACMSF. Report on vacuum packaging and associated processes. (1992). London, UK, HMSO.

Afchain, A. L., Carlin, F., Nguyen-The, C., & Albert, I. (2008). Improving quantitative exposure assessment by considering genetic diversity of *B. cereus* in cooked, pasteurised and chilled foods. *Int. J. Food Microbiol.,* 128, 165-173.

Akhtar, S., Paredes-Sabja, D., Torres, J. A., & Sarker, M. R. (2009). Strategy to inactivate *Clostridium perfringens* spores in meat products. *Food Microbiology,* 26, 272-277.

Alderton,G., Chen,J.K., & Ito,K.A. (1974). Effect of lysozyne on the recovery of heated Clostridium botulinum spores. *Appl. Microbiol.* 27, 613-615.

Ando, Y., Tsuzuki, T., Sunagawa, H., & Oka, S. (1985). Heat-resistance, spore germination, and enterotoxigenicity of *Clostridium perfringens. Microbiology and Immunology,* 29, 317-326.

Anniballi, F., Fenicia, L., Franciosa, G., & Aureli, P. (2002). Influence of pH and temperature on the growth of and toxin production by neurotoxigenic strains of *Clostridium butyricum* type E. *J. Food Prot.,* 65, 1267-1270.

Augustin, J. C., Carlier, V., & Rozier, J. (1998). Mathematical modelling of the heat resistance of *Listeria monocytogenes*. *J. Appl. Microbiol.,* 84, 185-191.

Bae, Y. Y., Lee, H. J., Kim, S. A., & Rhee, M. S. (2009). Inactivation of *Alicyclobacillus acidoterrestris* spores in apple juice by supercritical carbon dioxide. *Int. J. Food Microbiol.*, 136, 95-100.

Banerjee, M. & Sarkar, P. K. (2004). Growth and enterotoxin production by sporeforming bacterial pathogens from spices. *Food Control,* 15, 491-496.

Barnabas, J., Siores, E., & Lamb, A. (2010). Non-thermal microwave reduction of pathogenic cellular population. *International Journal of Food Engineering, 6*, (issue 5), article 9, 1-20. DOI: 10.2202/1556-3758.1878.

Barrett, D. M., Somogyi, L., & Ramaswamy, H. S. (2004). *Processing fruits: Science and Technology*. Boca Raton, FL, CRC Press.

Ben Embarek, P. K. (1994). Thesis/Dissertation: *Microbial safety and spoilage of sous vide fish products*. Pp.1-168. Technological Laboratory of the Danish Ministry of Agriculture and Fisheries & the Royal Veterinary and Agricultural University, Denmark.

Ben Embarek, P. K. & Huss, H. H. (1993). Heat-resistance of *Listeria monocytogenes* in vacuum-packaged pasteurized fish fillets. *Int. J. Food Microbiol.,* 20, 85-95.

Berg, R. W. & Sandine, W. E. (1970). Activation of bacterial spores. A review. *Journal of milk and food technology,* 33, 435-441.

Bolton, D. J., McMahon, C. M., Doherty, A. M., Sheridan, J. J., McDowell, D. A., Blair, I. S., & Harrington, D. (2000). Thermal inactivation of *Listeria monocytogenes* and *Yersinia enterocolitica* in minced beef under laboratory conditions and in sous-vide prepared minced and solid beef cooked in a commercial retort. *J. Appl. Microbiol.,* 88, 626-632.

Boyle, D. L., Schmidt, G. R., & Sofos, J. N. (1990). Growth of *Listeria monocytogenes* inoculated in waste fluids from cleanup of a meat grinder. *J. Food Sci.,* 55, 277-278.

Bozkurt, H. & Icier, F. (2010a). Electrical conductivity changes of minced beef-fat blends during ohmic cooking. *Journal of Food Engineering,* 96, 86-92.

Bozkurt, H. & Icier, F. (2010b). Ohmic cooking of ground beef: Effects on quality. *Journal of Food Engineering,* 96, 481-490.

Breand, S., Fardel, G., Flandrois, J. P., Rosso, L., & Tomassone, R. (1997). A model describing the relationship between lag time and mild temperature increase duration. *Int. J. Food Microbiol.,* 38, 157-167.

Breand, S., Fardel, G., Flandrois, J. P., Rosso, L., & Tomassone, R. (1999). A model describing the relationship between regrowth lag time and mild temperature increase for *Listeria monocytogenes*. *Int. J. Food Microbiol.,* 46, 251-261.

Bremer, P. J. & Osborne, C. M. (1995). Thermal-death times of *Listeria monocytogenes* in green shell mussels (*Perna canaliculus*) prepared for hot smoking. *J. Food Prot.,* 58, 604-608.

Brown, K. L. & Martinez, A. (1992). The Heat-resistance of spores of *Clostridium botulinum* 213B heated at 121-130-degrees-C in acidified mushroom extract. *J. Food Prot.,* 55, 913-915.

Buchanan, R., Damert, W., Whiting, R., & van Schthorst, M. (1997). Use of epidemiological and food survey data to estimate a purposefully conservative dose-response relationship for *L. monocytogenes* levels and incidence of listeriosis. *J. Food Prot.,* 60, 918-922.

Buchrieser, C. & Rocourt, J. (2007). The genus *Listeria* and *Listeria monocytogenes*. In *Listeria, listeriosis, and food safety*. Ryser, E. T. & Marth, E. H. (eds). Boca Raton, CRC Press, 1-20.

Budu-Amoako, E., Toora, S., Walton, C., Ablett, R. F., & Smith, J. (1992). Thermal death times for *Listeria monocytogenes* in lobster meat. *J. Food Prot.,* 55, 211-213.

Cameron, M. S., Leonard, S. J., & Barrett, E. L. (1980). Effect of moderately acidic pH on heat resistance of *Clostridium sporogenes* spores in phosphate buffer and in buffered pea puree. *Appl. Environ. Microbiol.,* 39, 943-949.

Cann, D. C., Wilson, B. B., Shewan, J. M., & Hobbs, G. (1966). Incidence of *Clostridium botulinum* type E in fish products in the United Kingdom. *Nature,* 211, 205-206.

Carlin, F., Fricker, M., Pielaat, A., Heisterkamp, S., Shaheen, R., Salonen, M. S., Svensson, B., Nguyen-the, C., Ehling-Schulz, M. (2006). Emetic toxin-producing strains of *Bacillus cereus* show distinct characteristics within the *Bacillus cereus* group. *Int. J. Food Microbiol.,* 109, 132-138.

Casadei, M. A., de Matos, R. E., Harrison, S. T., & Gaze, J. E. (1998). Heat resistance of *Listeria monocytogenes* in dairy products as affected by the growth medium. *J. Appl. Microbiol.,* 84, 234-239.

Chen, H. & Hoover, D. G. (2003). Bacteriocins and their food applications. *Comprehensive Reviews in Food Science and Food Safety,* 2, 81-100.

Chipley, J. R. (1980). *Effects of microwave irradiation on microorganisms*. Academic Press Inc. Advances in Applied Microbiology, 26, 129-145.

Cho, H. Y., Yousef, A. E., & Sastry, S. K. (1999). Kinetics of inactivation of *Bacillus subtilis* spores by continuous or intermittent ohmic and conventional heating. *Biotechnology and Bioengineering,* 62, 368-372.

Choma, C., Guinebretiere, M. H., Carlin, F., Schmitt, P., Velge, P., Granum, P. E., & Nguyen-The, C. (2000). Prevalence, characterization and growth of *Bacillus cereus* in commercial cooked chilled foods containing vegetables. *J. Appl. Microbiol.,* 88, 617-625.

Christiansson, A., Naidu, A. S., Nilsson, I., Wadstrom, T., & Pettersson, H. E. (1989). Toxin production by *Bacillus cereus* dairy isolates in milk at low temperatures. *Appl. Environ. Microbiol.,* 55, 2595-2600.

Clements, M. O. & Moir, A. (1998). Role of the gerI operon of *Bacillus cereus* 569á in the response of spores to germinants. *J. Bacteriol.,* 180, 6729-6735.

Codex Alimentarius. (2002). Report on the thirty-fourth session of the Codex Committee on food hygiene (ALINORM 03/13). Rome, Italy, FAO/WHO.

Collado, J., Fernández, A., Rodrigo, M., Camats, J., & Lopez, A. M. (2003). Kinetics of deactivation of *Bacillus cereus* spores. *Food Microbiology,* 20, 545-548.

Condon, S. & Sala, F. J. (1992). Heat resistance of *Bacillus subtilis* in buffer and foods of different Ph. *J. Food Prot.,* 55, 605-608.

Conesa, R., Andreu, S., Fernandez, P. S., Esnoz, A., & Palop, A. (2009). Nonisothermal heat resistance determinations with the thermoresistometer Mastia. *J. Appl. Microbiol.,* 107, 506-513.

de Jong, A. E. I., Rombouts, F. M., & Beumer, R. R. (2004). Effect of cooling on *Clostridium perfringens* in Pea Soup. *J. Food Prot.,* 67, 352-356.

Dorsa, W. J., Marshall, D. L., Moody, M. W., & Hackney, C. R. (1993). Low-temperature growth and thermal inactivation of *Listeria monocytogenes* in precooked crawfish tail meat. *J. Food Prot.,* 56, 106-109.

Doyle, M. E., Mazzotta, A. S., Wang, T., Wiseman, D. W., & Scott, V. N. (2001). Heat resistance of *Listeria monocytogenes*. *J. Food Prot.,* 64, 410-429.

ECFF. (1996). Guideline for the hygienic manufacture of chilled foods. European chilled food federation.http://www.ecff.net. 03 October 2011.

ECFF. (2006). *Recommendations for the production of prepackaged chilled food.* European chilled food federation. http://www.ecff.net/ 03 October 2011.

EFSA (2005). Opinion of the scientific panel on biological hazards on *Bacillus cereus* and other *Bacillus* spp. in foodstuffs. Question No EFSA-Q_2004-010. *The EFSA Journal,* 175, 1-48.

Eklund, M. W., Peterson, M. E., Paranjpye, R., & Pelroy, G. A. (1988). Feasibility of a heat-pasteurization process for the inactivation of nonproteolytic *Clostridium botulinum* type-B and type-E in vacuum-packaged, hot-process (smoked) fish. *J. Food Prot.,* 51, 720-726.

Esty, J.R. & Meyer, K.F. (1922). The heat resistance of spores of *B. botulinus* and allied anaerobes. *Journal of Infectious Diseases* 31, 650-663.

European Commission. (1997). Harmonization of safety criteria for minimally processed foods. Martens, T. & Luchetti, A. 36-37. European Commission. FAIR Concerted action CT96-1020.

Fach, P., Perelle, S., Dilasser, F., Grout, J., Dargaignaratz, C., Botella, L., Gourreau, J-M., Carlin, F., Popoff, M. R., & Broussolle, V. (2002). Detection by PCR-enzyme-linked-immunosorbent assay of *Clostridium botulinum* in fish and environmental samples from a coastal area in Northern France. *Appl. Environ. Microbiol.*, 68, 5870-5876.

Farber, J. M. (1989). Thermal-resistance of *Listeria monocytogenes* in foods. *Int. J. Food Microbiol.,* 8, 285-291.

Fernández, A., López, M., Bernardo, A., Condòn, S., & Raso, J. (2007). Modelling thermal inactivation of *Listeria monocytogenes* in sucrose solutions of various water activities. *Food Microbiology,* 24, 372-379.

Fernandez, A., Ocio, M. J., Fernandez, P. S., & Martinez, A. (2001a). Effect of heat activation and inactivation conditions on germination and thermal resistance parameters of *Bacillus cereus* spores. *Int. J. Food Microbiol.,* 63, 257-264.

Fernandez, P. S., Baranyi, J., & Peck, M. W. (2001b). A predictive model of growth from spores of non-proteolytic *Clostridium botulinum* in the presence of different CO_2 concentrations as influenced by chill temperature, pH and NaCl. *Food Microbiology,* 18, 453-461.

Fernández, P. S., Gómez, F. J., Ocio, M. J., Rodrigo, M., Sánchez, T., & Martínez, A. (1995). *D* values of *Bacillus stearothermophilus* spores as a function of pH and recovery medium acidulant. *J. Food Prot.,* 58, 628-632.

Fernandez, P. S. & Peck, M. W. (1997). Predictive model describing the effect of prolonged heating at 70 to 80 degrees C and incubation at refrigeration temperatures on growth and toxigenesis by nonproteolytic *Clostridium botulinum*. *J. Food Prot.,* 60, 1064-1071.

Fernandez, P. S. & Peck, M. W. (1999). A predictive model that describes the effect of prolonged heating at 70 to 90 degrees C and subsequent incubation at refrigeration temperatures on growth from spores and toxigenesis by nonproteolytic *Clostridium botulinum* in the presence of lysozyme. *Appl. Environ. Microbiol.,* 65, 3449-3457.

Foegeding, P. M. & Stanley, N. W. (1991). *Listeria innocua* transformed with an antibiotic-resistance plasmid as a thermal-resistance indicator for *Listeria monocytogenes. J. Food Prot.,* 54, 519-523.

Foerster, H. F. & Foster, J. W. (1966). Response of *Bacillus* spores to combinations of germinative compounds. *J. Bacteriol.,* 91, 1168-1177.

Food Safety and Inspection Service. (2001). Performance standards for the production of certain meat and poultry products; final rule. *Fed. Regist.* 64, 732-749.

Garibaldi, J. A., Donovan, J. W., Davis, J. G., & Cimino, S. L. (1968). Heat denaturation of the ovomucin-lysozynie electrostatic complex-A source of damage to the whipping properties of pasteurized egg white. *J. Food Sci.,* 33, 514-524.

Garriga, M., Aymerich, M. T., Costa, S., Monfort, J. M., & Hugas, M. (2002). Bactericidal synergism through bacteriocins and high pressure in a meat model system during storage. *Food Microbiology,* 19, 509-518.

Gaze, J. E., Brown, D. E., Gaskell, D. E., & Banks, J. G. (1989). Heat resistance of *Listeria monocytogenes* in homogenates of chicken, beef steak and carrot. *Food Microbiology,* 6, 251-259.

Gaze, J. E. & Brown, G. D. (1990). Determination of the heat resistance of a strain of non-proteolytic *Clostridium botulinum* Type B and a strain of Type E, heated in cod and carrot homogenate over the temperature range 70 to 92 °C. *592,* 1-34. Campden Food & Drink Research Association. Tecnical Memorandum.

Geveke, D. J., Kozempel, M., Scullen, O. J., & Brunkhorst, C. (2002). Radio frequency energy effects on microorganisms in foods. *Innovative Food Science & Technologies,* 3, 133-138.

Ghosh, S., Zhang, P., Li,Y. Q., & Setlow, P. (2009). Superdormant spores of *Bacillus* species have elevated wet heat resistance and temperature requirements for heat activation. *J. Bacteriol.,* 191, 5584-5591.

Golden, D. A., Beuchat, L. R., & Brackett, R. E. (1988). Inactivation and injury of *Listeria monocytogenes* as affected by heating and freezing. *Food Microbiology,* 5, 17-23.

Gorris, L. G. M. & Peck, M. W. (1999). Microbiological safety consideration when using the hurdle technology with refrigerated processed foods of extended durability. In: *Sous vide and cook-chill processing for the food industry*. Ghazala, S. (ed). New York, Aspen Publishers, 206-233.

Gould, G. W. (1966). Stimulation of L-alanine-induced germination of *Bacillus cereus* spores by D-cycloserine and O-carbamyl-D-serine. *J. Bacteriol.,* 92, 1261-1262.

Gould, G. W., Jones, A., & Wrighton, C. (1968). Limitations of the initiation of germination of bacterial spores as a spore control procedure. *J. Appl. Bacteriol.,* 31, 357-366.

Gould, G. W. & Sale, A. J. H. (1970). Initiation of germination of bacterial spores by hydrostatic pressure. *J. General Microbiol.,* 60, 335-&.

Gould, W. G. (1989). Heat induced injury and inactivation. In: *Mechanisms of action of food preservation procedures*. Gould,G.W. (ed). London, Elsevier, 11-42.

Gounina-Allouane, R., Broussolle, V., & Carlin, F. (2008). Influence of the sporulation temperature on the impact of the nutrients inosine and L-alanine on *Bacillus cereus* spore germination. *Food Microbiology,* 25, 202-206.

Graham, A. F., Mason, D. R., Maxwell, F. J., & Peck, M. W. (1997). Effect of pH and NaCl on growth from spores of non-proteolytic *Clostridium botulinum* at chill temperature. *Lett. Appl. Microbiol.,* 24, 95-100.

Graham, A. F., Mason, D. R., & Peck, M. W. (1996). Predictive model of the effect of temperature, pH and sodium chloride on growth from spores of non-proteolytic *Clostridium botulinum. Int. J. Food Microbiol.,* 31, 69-85.

Granum, P. E. (2007). *Bacillus cereus.* In *Food microbiology, Fundamentals and frontiers.* Doyle, M. P. and Beuchat, L. R. (eds). pp. 445-455. ASM Press ,Washington DC.

Greig, J. D. & Ravel, A. (2009). Analysis of foodborne outbreak data reported internationally for source attribution. *Int. J. Food Microbiol.,* 130, 77-87.

Guinebretiere, M. H., Thompson, F. L., Sorokin, A., Normand, P., Dawyndt, P., Ehling-Schulz, M. et al. (2008). Ecological diversification in the *Bacillus cereus* group. *Environmental Microbiology,* 10, 851-865.

Harrison, M. A. & Huang, Y. W. (1990). Thermal death times for *Listeria monocytogenes* (Scott A) in crabmeat. *J. Food Prot.,* 53, 878-880.

Hauschild, A. W. H. (1989). *Clostridium botulinum.* In: *Foodborne bacterial pathogens.* Doyle, M. P. (ed). New York, Marcel Dekker, 112-189.

Heddleson, R. A. & Doors, S. (1994). Factors affecting microwave heating of foods and microweve induced destruction of foodborne pathogens - a review. *J. Food Prot.,* 57, 1025-1037.

Heredia, N. L. & Labbé, R. G. (2001). *Clostridium perfringens.* In: *Guide to foodborne pathogens.* García-Alvarado, J. S. & Labbé, R. G. (eds). New Yourk, USA, John Wiley & Sons, 133-142.

Hielm, S., Hyytiä, E., Andersin, A. B., & Korkeala, H. (1998). High prevalence of *Clostridium botulinum* type E in Finnish freshwater and Baltic Sea sediment samples. *J. Appl. Microbiol.* 84, 133-137.

Ho, S. & Mittal, G. S. (2000). High voltage pulsed electrical field for liquid food pasteurization. *Food Reviews International,* 16, 395-434.

Hong, H. A., To, E., Fakhry, S., Baccigalupi, L., Ricca, E., & Cutting, S. M. (2009). Defining the natural habitat of *Bacillus* spore-formers. *Research in Microbiology,* 160, 375-379.

Hornstra, L. M., de Vries, Y. P., de Vos, W. M., Abee, T., & Wells-Bennik, M. H. J. (2005). *gerR*, a novel *ger* operon involved in L-alanine- and inosine-initiated germination of *Bacillus cereus* ATCC 14579. *Appl. Environ. Microbiol.,* 71, 774-781.

Hornstra, L. M., de Vries, Y. P., Wells-Bennik, M. H. J., de Vos, W. M., & Abee, T. (2006). Characterization of Germination Receptors of *Bacillus cereus* ATCC 14579. *Appl. Environ. Microbiol.,* 72, 44-53.

Huss, H. H. (1980). Distribution of *Clostridium botulinum. Appl. Environ. Microbiol.,* 39, 764-769.

Huss, H. H. (1994). *Assurance of Seafood Quality.* 334, 43-44. Rome, Italy, FAO (United Nations).

Huss, H. H. (1997). Control of indigenous pathogenic bacteria in seafood. *Food Control,* 8, 91-98.

Hyytiä, E., Hielm, S., & Korkeala, H. (1998). Prevalence of *Clostridium botulinum* type E in Finnish fish and fishery products. *Epidemiol. Infect.,* 120, 245-250.

ICMSF (1996). *Microorganisms in food. Microbiological specifications of Food Pathogens.* London, Blackie Academic and Professional Publ.

ICMSF (2002). *Microorganisms in foods: 7. Microbiological testing in food safety management.* Kluwer Academic Pulishing, New York, NY, 285-309.

Jakob, A., Bryjak, J., Wojtowicz, H., Illeova, V., Annus, J., & Polakovic, M. (2010). Inactivation kinetics of food enzymes during ohmic heating. *Food Chemistry,* 123, 369-376.

Jensen, G. B., Hansen, B. M., Eilenberg, J., & Mahillon, J. (2003). The hidden lifestyles of *Bacillus cereus* and relatives. *Environmental Microbiology,* 5, 631-640.

Johnson, K. M., Nelson, C. L., & Busta, F. F. (1983). Influence of temperature on germination and growth of spores of emetic and diarrheal strains of *Bacillus cereus* in a broth medium and in rice. *J. Food Sci.,* 48, 286-287.

Johnston, M. D. & Brown, M. H. (2002). An investigation into the changed physiological state of *Vibrio* bacteria as a survival mechanism in response to cold temperatures and studies on their sensitivity to heating and freezing. *J. Appl. Microbiol.,* 92, 1066-1077.

Jorgensen, F., Hansen, T. B., & Knochel, S. (1999). Heat shock-induced thermotolerance in *Listeria monocytogenes* 13-249 is dependent on growth phase, pH and lactic acid. *Food Microbiology,* 16, 185-194.

Juneja, V. K. (2003). A comparative heat inactivation study of indigenous microflora in beef with that of *Listeria monocytogenes*, *Salmonella* serotypes and *Escherichia coli* O157 : H7. *Letters in Applied Microbiology,* 37, 292-298.

Juneja, V. K. (2002). Thermal inactivation of microorganisms. In: *Control of foodborne microorganisms.* Eds: Juneja, V. K. & Sofos, J. N. Marcel Decker, Inc. New York. 13 – 54.

Juneja, V. K., Novak, J. S., Huang, L., & Eblen, B. S. (2003). Increased thermotolerance of *Clostridium perfringens* spores following sublethal heat shock. *Food Control,* 14, 163-168.

Kennedy, J. & Wall, P. (2007). Food safety challenges. In: *Safety handbook: Microbiological challenges* Ed: Storrs, M., Devoluy,M. -C. & Cruveiller, P. 8-19. France, Bio-Mérieux Education

Keynan, A. & Evenchik, Z. (1969). Activation. In: *The Bacterial Spore.* Gould, G. W. & Hurst, A. (eds). New York, Academic press, 359-396.

Knirsch, M. C., Alves dos Santos, C., Martins de Oliveira Soares Vicente, A. A., & Vessoni Penna, T. C. (2010). Ohmic heating - a review. *Trends in Food Science & Technology,* 21, 436-441.

Knutson, K. M., Marth, E. H., & Wagner, M. K. (1987). Microwave heating of food. *Lebensm. Wiss. Technol.,* 20, 101-110.

Korkeala, H., Stengel, G., Hyytiä, E., Vogelsang, B., Bohl, A., Wihlman, H. et al. (1998). Type E botulism associated with vacuum-packaged hot-smoked whitefish. *Int. J. Food Microbiol.,* 43, 1-5.

Kozempel, M. F., Annous, B. A., Cook, R. D., Scullen, O. J., & Whiting, R. C. (1998). Inactivation of microorganisms with microwaves at reduced temperatures. *J. Food Prot.,* 61, 582-585.

Labbe, R. G. & Huang, T. H. (1995). Generation times and modelling of enterotoxin-positive and enterotoxin-negative strains of *Clostridium perfringens* in laboratory media and ground beef. *J. Food Prot.,* 58, 1303-1306.

Larousse, J. & Brown, B. (1997). Thermobacteriology. In: *Food Canning Technology*. Larousse, J. & Brown, B. (eds). New York, Wiley- CVH. Lynch, 150.

Lianou, A., Stopforth, J. D., Yoon, Y., Wiedmann, M., & Sofos, J. N. (2006). Growth and stress resistance variation in culture broth among *Listeria monocytogenes* strains of various serotypes and origins. *J. Food Prot.,* 69, 2640-2647.

Lie, Ø., Evensen, Ø., Sørensen, A., & Frøysdal, E. (1989). Study on lysozyme activity in some fish species. *Diseases of aquatic organisms,* 6, 1-5.

Lin, Y. D. & Chou, C. C. (2004). Effect of heat shock on thermal tolerance and susceptibility of *Listeria monocytogenes* to other environmental stresses. *Food Microbiology,* 21, 605-610.

Lindstrom, M., Kiviniemi, K., & Korkeala, H. (2006). Hazard and control of group II (non-proteolytic) *Clostridium botulinum* in modern food processing. *Int. J. Food Microbiol.,* 108, 92-104.

Lindstrom, M., Nevas, M., Hielm, S., Lahteenmaki, L., Peck, M. W., & Korkeala, H. (2003). Thermal inactivation of nonproteolytic *Clostridium botulinum* Type E spores in model fish media and in vacuum-packaged hot-smoked fish products. *Appl. Environ. Microbiol.,* 69, 4029-4036.

Lorentzen, G., Olsen, R. L., Bjorkevoll, I., Mikkelsen, H., & Skjerdal, T. (2010). Survival of *Listeria innocua* and *Listeria monocytogenes* in muscle of cod (*Gadus morhua* L.) during salt-curing and growth during chilled storage of rehydrated product. *Food Control,* 21, 292-297.

Lou, Y. & Yousef, A. E. (1999). Characterization of *Listeria monocytogenes* important to food processors. In: *Listeria, Listeriosis, and Food Safety*. Ryser, E. T. & Marth, E. H. (eds). New Yourk NY, Marcel Dekker, 131-225.

Lund, B. M. & Notermans, S. H. W. (1992). Potential hazards associated with REPFEDS. In *Clostridium botulinum: ecology and control in foods*. Hauser, D. & Dodds, K. L. (eds). New York, Marcel Dekker, 279-303.

Lund, B. M. & Peck, M. W. (1994). Heat resistance and recovery of spores of non-proteolytic *Clostridium botulinum* in relation to refrigerated, processed foods with an extended shelf-life. *Soc. Appl. Bacteriol. Symp. Ser.,* 23, 115-128.

Lund, B. M. & Peck, M. W. (2000). *Clostridium botulinum*. In: *The microbiological safety and quality of food*. Lund, B. M., Baired-Parker, T. C., & Gould, G. W. (eds). Gaithersburg, Aspen Published, 1057-1109.

Lunden, J., Tolvanen, R., & Korkeala, H. (2008). Acid and heat tolerance of persistent and nonpersistent *Listeria monocytogenes* food plant strains. *Letters in Applied Microbiology,* 46, 276-280.

Lynch, D. J. & Potter, N. N. (1989). Effects of acidification and processing variables on thermal inactivation of *Bacillus coagulans* spores in meat particulates. *J. Food Prot.,* 52, 320-328.

Lyng, J. G., Cronin, D. A., Brunton, N. P., Li, W., & Gu, X. (2007). An examination of factors affecting radio frequency heating of an encased meat emulsion. *Meat Science,* 75, 470-479.

Lynt, R. K., Solomon, H. M., Lilly, T., & Kautter, D. A. (1977). Thermal death time of *Clostridium botulinum* Type E in meat of the blue crab. *J. Food Sci.,* 42, 1022-1025.

Machado, L. F., Pereira, R. N., Martins, R. C., Teixeira, J. A., & Vicente, A. A. (2010). Moderate electric fields can inactivate *Escherichia coli* at room temperature. *Journal of Food Engineering,* 96, 520-527.

Mackey, B. M., Pritchet, C., Norris, A., & Mead, G. C. (1990). Heat-resistance of *Listeria* - strain differences and effects of meat type and curing salts. *Letters in Applied Microbiology,* 10, 251-255.

Mataragas, M., Skandamis, P. N., & Drosinos, E. H. (2008). Risk profiles of pork and poultry meat and risk ratings of various pathogen/product combinations. *Int. J. Food Microbiol.,* 126, 1-12.

Mazzotta, A. S. & Gombas, D. E. (2001). Heat resistance of an outbreak strain of *Listeria monocytogenes* in hot dog batter. *J. Food Prot.,* 64, 321-324.

McKenna, B. M., Lyng, J., Brunton, N., & Shirsat, N. (2006). Advances in radio frequency and ohmic heating of meats. *Journal of Food Engineering,* 77, 215-229.

Mcmahon, C. M. M., Byrne, C. M., Sheridan, J. J., McDowell, D. A., Blair, I. S., & Hegarty,T. (2000). The effect of culture growth phase on induction of the heat shock response in *Yersinia enterocolitica* and *Listeria monocytogenes*. *J. Appl. Microbiol.,* 89, 198-206.

Mead, P. S., Slutsker, L., Griffin, P. M., & Tauxe, R. V. (1999). Food-related illness and death in the United States - Reply. *Emerging Infectious Diseases,* 5, 841-842.

Miettinen, H., Arvola, A., & Wirtanen, G. (2005). Pasteurization of rainbow trout roe: *Listeria monocytogenes* and sensory analyses. *J. Food Prot.,* 68, 1641-1647.

Moir, A. (2006). How do spores germinate? *J. Appl. Microbiol.,* 101, 526-530.

Moir, A., Corfe, B. M., & Behravan, J. (2002). Spore germination. *Cellular and Molecular Life Sciences,* 59, 403-409.

Murphy, R. Y., Driscoll, K. H., Arnold, M. E., Marcy, J. A., & Wolfe, R. E. (2003a). Lethality of *Listeria monocytogenes* in fully cooked and vacuum packaged chicken leg quarters during steam pasteurization. *J. Food Sci.,* 68, 2780-2783.

Murphy, R. Y., Duncan, L. K., Beard, B. L., & Driscoll, K. H. (2003b). *D* and *z* values of *Salmonella*, *Listeria innocua*, and *Listeria monocytogenes* in fully cooked poultry products. *J. Food Sci.,* 68, 1443-1447.

Murphy, R. Y., Osaili, T., Duncan, L. K., & Marcy, J. A. (2004a). Effect of sodium lactate on thermal inactivation of *Listeria monocytogenes* and *Salmonella* in ground chicken thigh and leg meat. *J. Food Prot.,* 67, 1403-1407.

Murphy, R. Y., Osaili, T., Duncan, L. K., & Marcy, J. A. (2004b). Thermal inactivation of *Salmonella* and *Listeria monocytogenes* in ground chicken thigh/leg meat and skin. *Poultry Science,* 83, 1218-1225.

NACMCF (U.S.National Advisory Committee on Microbiological Criteria for Foods) (1992). Microbiological criteria for raw molluscan shellfish. *J. Food Prot.,* 55, 463-480.

Nelson, P. & Tressler, D. (1980). *Fruit and vegetable juice processing technology*. Westport, CT,AVI Pub. Co.

Nissen, H., Rosnes, J. T., Brendehaug, J., & Kleiberg, G. H. (2002). Safety evaluation of sous vide-processed ready meals. *Letters in Applied Microbiology,* 35, 433-438.

Nutsch, A. L., Phebus, R. K., Riemann, M. J., Kotrola, J. S., Wilson, J. S., Boyer, R. C., & Brown, T. L. (1998). Steam pasteurization of commercially slaughtered beef carcasses:

evaluation of bacterial populations at five anatomical locations. *J. Food Prot.,* 61, 571-577.

O'Bryan, C. A., Crandall, P. G., Martin, E. M., Griffis, C. L., & Johnson, M. G. (2006). Heat resistance of *Salmonella* spp., *Listeria monocytogenes*, *Escherichia coli* O157 : H7 and *Listeria innocua* M1, a potential surrogate for *Listeria monocytogenes*. in meat and poultry: A review. *J. Food Sci.,* 71, 23-30.

Ocio, M. J., Sanchez, T., Fernandez, P. S., Rodrigo, M., & Martinez, A. (1994). Thermal-resistance characteristics of PA-3679 in the temperature-range of 110-121-Degrees-C as affected by Ph, type of acidulant and substrate. *Int. J. Food Microbiol.,* 22, 239-247.

Odlaug, T. E. & Pflug, I. J. (1978). *Clostridium botulinum* and acid foods. *J. Food Prot.,* 41, 566-573.

Paidhungat, M. & Setlow, P. (2002). Spore germination and outgrowth. In: *Bacillus subtilis and its closest relatives: from genes to cells*. Sonenshein, A. L., Hoch, J. A., & Losick, R. (eds). Washington, DC, American Society for Microbiology, 537-548.

Palaniappan, S. & Sastry, S. (2002). Ohmic heating. In: *Control of foodborne microorganisms*. Juneja, V. K. & Sofos, J. N. (eds). New York, Marcel Decker, 451-460.

Palop, A., Raso, J., Condón, S., & Sala, F. J. (1996). Heat resistance of *Bacillus subtilis* and *Bacillus coagulans*: Effect of sporulation temperature in foods with various acidulants. *J. Food Prot.,* 59, 487-492.

Paoli, G. C., Bhunia, A. K., & Bayles, D. O. (2005). *Listeria monocytogenes*. In: *Foodborne pathogens*. Fratamico, P. M., Bhunia, A. K., & Smith, J. L. (eds). Caister Academic Press, UK, 295-324.

Patazca, E., Koutchma, T., & Balasubramaniam, V. M. (2007). Quasi-adiabatic temperature increase during high pressure processing of selected foods. *Journal of Food Engineering,* 80, 199-205.

Patchett, R. A., Watson, N., Fernandez, P. S., & Kroll, R. G. (1996). The effect of temperature and growth rate on the susceptibility of *Listeria monocytogenes* to environmental stress conditions. *Letters in Applied Microbiology,* 22, 121-124.

Patterson, M. F. (2005). Microbiology of pressure-treated foods. *J. Appl. Microbiol.,* 98, 1400-1409.

Peck, M. W. & Stringer, S. C. (2005). The safety of pasteurised in-pack chilled meat products with respect to the foodborne botulism hazard. *Meat Science,* 70, 461-475.

Pereira, R., Martins, J., Mateus, C., Teixeira, J., & Vicente, A. (2007). Death kinetics of *Escherichia coli* in goat milk and *Bacillus licheniformis* in cloudberry jam treated by ohmic heating. *Chemical Papers,* 61, 121-126.

Periago, P. M., Fernandez, P. S., Salmeron, M. C., & Martinez, A. (1998). Predictive model to describe the combined effect of pH and NaCl on apparent heat resistance of *Bacillus stearothermophilus*. *Int. J. Food Microbiol.,* 44, 21-30.

Peterson, M. E., Pelroy, G. A., Poysky, F. T., Paranjpye, R. N., Dong, F. M., Pigott, G. M., & Eklund, M. W. (1997). Heat-pasteurization process for inactivation of nonproteolytic types of *Clostridium botulinum* in picked dungeness crabmeat. *J. Food Prot.,* 60, 928-934.

Piette, G., Buteau, M. L., de Halleux, D., Chiu, L., Raymond, Y., Ramaswamy, H. S., & Dostie, M. (2004). Ohmic cooking of processed meats and its effects on product quality. *J. Food Sci.,* 69, 71-78.

Quintavalla, S. & Campanini, M. (1991). Effect of rising temperature on the heat-resistance of *Listeria monocytogenes* in meat emulsion. *Letters in Applied Microbiology.*, 12, 184-187.

Rahmati, T. & Labbe, R. (2008). Levels and toxigenicity of *Bacillus cereus* and *Clostridium perfringens* from retail seafood. *J. Food Prot.,* 71, 1178-1185.

Raso, J., Condon, S., & Pagan, R. (2005). Nonthermal technologies in combination with other preservation factors. In: *Novel food processing technologies*. Barbosa-Canovas, G. V., Tapia, M. S. and Cano, M. P. (eds). New York, Marcel Dekker, 453-475.

Reddy, N. R., Tetzloff, R. C., Solomon, H. M., & Larkin, J. W. (2006). Inactivation of *Clostridium botulinum* nonproteolytic type B spores by high pressure processing at moderate to elevated high temperatures. *Innovative Food Science & Emerging Technologies,* 7, 169-175.

Reichart, O. & Mohácsi-Farkas, C. (1994). Mathematical modelling of the combined effect of water activity, pH and redox potential on the heat destruction. *Int. J. Food Microbiol.,* 24, 103-112.

Retzlaff, D., Phebus, R., Nutsch, A., Riemann, J., Kastner, C., & Marsden, J. (2004). Effectiveness of a laboratory-scale vertical tower static chamber steam pasteurization unit against *Escherichia coli* O157 : H7, *Salmonella* Typhimurium, and *Listeria innocua* on prerigor beef tissue. *J. Food Prot.,* 67, 1630-1633.

Roberts, T. A., Gibson, A. M., & Robinson, A. (1981a). Factors controlling the growth of *Clostridium botulinum* Type-A and Type-B in pasteurized, cured meats.1. Growth in pork slurries prepared from low pH meat (pH range 5.5-6.3). *Journal of Food Technology,* 16, 239-266.

Roberts, T. A., Gibson, A. M., & Robinson, A. (1981b). Factors controlling the growth of *Clostridium botulinum* Types-A and Type-B in pasteurized, cured meats.2. Growth in pork slurries prepared from high pH meat (range 6.3-6.8). *Journal of Food Technology,* 16, 267-281.

Rodrigo, F., Fernandez, P. S., Rodrigo, M., Ocio, M. J., & Martinez, A. (1997). Thermal resistance of *Bacillus stearothermophilus* heated at high temperatures in different substrates. *J. Food Prot.,* 60, 144-147.

Rodrigo, M. & Martinez, A. (1988). Determination of a process time for a new product: canned low acid artichoke hearts. *International Journal of Food Science & Technology,* 23, 31-41.

Rosenberg, U. & Bögel, W. (1987). Microvawe pasteurization, sterilization, blanching, and pest control in the food industry. *Food Technology,* 41, 92-99.

Sagoo, S. K., Little, C. L., Greenwood, M., Mithani, V., Grant, K. A., McLauchlin, J. de Pinnaa, E. & Threlfall, E. J. (2009). Assessment of the microbiological safety of dried spices and herbs from production and retail premises in the United Kingdom. *Food Microbiology,* 26, 39-43.

Sanchez, T., Rodrigo, M., Ocio, M. J., Fernandez, P., & Martinez, A. (1995). Growth and heat-resistance of *Clostridium sporogenes* PA-3679 spores heated and recovered in acidified media. *J. Food Prot.,* 58, 656-660.

Schellekens, W. & Martens, T. (1993). FLAIR. "Sous vide" coking. Part I: Scientific literature review. Publication NO EUR 15018 EN of the commission of the European Communities, Directorate General XII Research and Development. Brussels-Luxenburg.

Scurrah, K. J., Robertson, R. E., Craven, H. M., Pearce, L. E., & Szabo, E. A. (2006). Inactivation of *Bacillus spores* in reconstituted skim milk by combined high pressure and heat treatment. *J. Appl. Microbiol.,* 101, 172-180.

Selby, T. L., Berzins, A., Gerrard, D. E., Corvalan, C. M., Grant, A. L., & Linton, R. H. (2006). Microbial heat resistance of *Listeria monocytogenes* and the impact on ready-to-eat meat quality after post-package pasteurization. *Meat Science,* 74, 425-434.

Sergelidis, D. & Abrahim, A. (2009). Adaptive response of *Listeria monocytogenes* to heat and its impact on food safety. *Food Control,* 20, 1-10.

Setlow, P. (2003). Spore germination. *Current Opinion in Microbiology,* 6, 550-556.

Shao, Y. W., Zhu, S. M., Ramaswamy, H., & Marcotte, M. (2010). Compression heating and temperature control for high-pressure destruction of bacterial spores: An experimental method for kinetics evaluation. *Food and Bioprocess Technology,* 3, 71-78.

Shirsat, N., Brunton, N. P., Lyng, J. G., McKenna, B., & Scannell, A. (2004). Texture, colour and sensory evaluation of a conventionally and ohmically cooked meat emulsion batter. *J. Sci. Food Agric.,* 84, 1861-1870.

Sikorski, Z. E. & Kolodziejska, I. (2002). Microbial risks in mild hot smoking of fish. *Critical Reviews in Food Science and Nutrition,* 42, 35-51.

Silla Santos, M. H., Nunez, K. H., Casado, G.A., & Rodrigo, E. M. (1992). The effect of pH on the thermal resistance of *Clostridium sporogenes* (PA 3679) in asparagus puree acidified with citric acid and glucono-delta-lactone. *Int. J. Food Microbiol ,* 16, 275-281.

Silva, F. V. M. & Gibbs, P. (2004). Target selection in designing pasteurization processes for shelf-stable high-acid fruit products. *Critical Reviews in Food Science and Nutrition,* 44, 353-360.

Skandamis, P. N., Yoon, Y., Stopforth, J. D., Kendall, P. A., & Sofos, J. N. (2008). Heat and acid tolerance of *Listeria monocytogenes* after exposure to single and multiple sublethal stresses. *Food Microbiology,* 25, 294-303.

Skipnes, D. & Pfeiffer, T. (2005) Radio-Frequency heating of food in a water bath. Building a FEM model of the process. Leuven, Belgium. The third international symposium on: Applications of modelling as an innovative technology in the agri-food chain. Conference Proceeding.

Sofos, J. N. & Geornaras, I. (2010). Overview of current meat hygiene and safety risks and summary of recent studies on biofilms, and control of *Escherichia coli* O157:H7 in nonintact, and *Listeria monocytogenes* in ready-to-eat, meat products. *Meat Science,* 86, 2-14.

Sofos, J. N. (2008). Challenges to meat safety in the 21st century. *Meat Science,* 78, 3-13.

Sörqvist, S. (1993). Heat-resistance of *Listeria monocytogenes* by 2 recovery media used with and without cold preincubation. *J. Appl. Bacteriol.,* 74, 428-432.

Sörqvist, S. (1994). Heat-resistance of different serovars of *Listeria monocytogenes. J. Appl. Bacteriol.,* 76, 383-388.

Sörqvist, S. (2003). Heat resistance in liquids of *Enterococcus* spp., *Listeria* spp., *Escherichia coli*, *Yersinia enterocolitica*, *Salmonella* spp. and *Campylobacter* spp. *Acta Veterinaria Scandinavica,* 44, 1-19.

Stephens, P. J., Cole, M. B., & Jones, M. V. (1994). Effect of heating rate on the thermal inactivation of *Listeria monocytogenes. J. Appl. Bacteriol.,* 77, 702-708.

Stringer, S. C. & Peck, M. W. (2008). Foodborne *Clostridia* and the safety of in-pack preserved foods. In: *In-pack processed foods*. Richardson, P. (ed). Boca Raton, Woodhead Publishing Limited, 251-276.

Stumbo, C. R. (1973). *Thermobacteriology in food processing*. Academic Press Inc.

Stumbo, C. R., Purohit, K.S., & Ramakrishna, T. V. (1975). Thermal process lethality guide for low-acid foods in metal containers. *J. Food Sci.,* 40, 1316-1323.

Sun, H. X., Kawamura, S., Himoto, J. I., Itoh, K., Wada, T., & Kimura, T. (2008). Effects of ohmic heating on microbial counts and denaturation of proteins in milk. *Food Science and Technology Research,* 14, 117-123.

Svensson, B., Monthán, A., Guinebretière, M. H., Nguyen-Thé, C., & Christiansson, A. (2007). Toxin production potential and the detection of toxin genes among strains of the *Bacillus cereus* group isolated along the dairy production chain. *International Dairy Journal,* 17, 1201-1208.

Swiecicka, I. (2008). Natural occurrence of *Bacillus thuringiensis* and *Bacillus cereus* in eukaryotic organisms: a case for symbiosis. *Biocontrol Science and Technology,* 18, 221-239.

Taormina, P. J., Bartholomew, G. W., & Dorsa, W. J. (2003). Incidence of *Clostridium perfringens* in commercially produced cured raw meat product mixtures and behaviour in cooked products during chilling and refrigerated storage. *J. Food Prot.,* 66, 72-81.

Thevenot, D., Dernburg, A., & Vernozy-Rozand, C. (2006). An updated review of *Listeria monocytogenes* in the pork meat industry and its products. *J. Appl. Microbiol.,* 101, 7-17.

Tompkin, R. B. (2002). Control of *Listeria monocytogenes* in the food-processing environment. *J. Food Prot.,* 65, 709-725.

Trivedi, S., Reynolds, A. E., & Chen, J. (2008). Effectiveness of commercial household steam cleaning systems in reducing the populations of *Listeria monocytogenes* and spoilage bacteria on inoculated pork skin surfaces. *LWT - Food Science and Technology,* 41, 295-302.

USA-FDA. (2000). United States of America,F.a.D.A.C.f.F.S.a.A.N. Kinetics of microbial inactivation for alternative food processing technologies: ohmic and inductive heating Available from URL: http://www.fda.gov/Food/ScienceResearch/ResearchAreas/SafePracticesforFoodProcesses/ucm101246.htm.

Van Netten, P., Vandemoosdijk, A., Vanhoensel, P., Mossel, D. A. A., & Perales, I. (1990). Psychrotrophic strains of *Bacillus cereus* producing enterotoxin. *J. Appl. Bacteriol.,* 69, 73-79.

van Zuijlen, A., Periago, P. M., Amézquita, A., Palop, A., Brul, S., & Fernández, P. S. (2010). Characterization of *Bacillus sporothermodurans* IC4 spores; putative indicator microorganism for optimisation of thermal processes in food sterilisation. *Food Research International,* 43, 1895-1901.

Wang, J. C., Hu, S. H., & Lin, C. Y. (2003). Lethal effect of microwaves on spores of *Bacillus* spp. *J. Food Prot.,* 66, 604-609.

Welt, B. A., Tong, C. H., Rossen, J. L., & Lund, D. B. (1994). Effect of microwave radiation on inactivation of *Clostridium sporogenes* (PA 3679) spores. *Appl. Environ. Microbiol.,* 60, 482-488.

Wilson, D. R., Dabrowski, L., Stringer, S., Moezelaar, R., & Brocklehurst, T. F. (2008). High pressure in combination with elevated temperature as a method for the sterilisation of food. *Trends in Food Science & Technology,* 19, 289-299.

Wong, H. C., Peng, P. Y., Lan, S. L., Chen, Y. C., Lu, K. H., Shen, C. T., & Lan, S. F. (2002). Effects of heat shock on the thermotolerance, protein composition, and toxin production of *Vibrio parahaemolyticus. J. Food Prot.,* 65, 499-507.

Xezones, H. & Hutching, I. J. (1965). Thermal Resistance of *Clostridium botulinum* (62A) Spores as affected by fundamental food constituents. I. Effect of pH. *Food Technology,* 19, 1003-1005.

Zell, M., Lyng, J. G., Cronin, D. A., & Morgan, D. J. (2009). Ohmic cooking of whole beef muscle - Optimisation of meat preparation. *Meat Science,* 81, 693-698.

Zell, M., Lyng, J. G., Cronin, D.A., & Morgan, D. J. (2010a). Ohmic cooking of whole beef muscle - Evaluation of the impact of a novel rapid ohmic cooking method on product quality. *Meat Science,* 86, 258-263.

Zell, M., Lyng, J. G., Cronin, D. A., & Morgan, D. J. (2010b). Ohmic cooking of whole turkey meat - Effect of rapid ohmic heating on selected product parameters. *Food Chemistry,* 120, 724-729.

Zhang, H. & Mittal, G. S. (2008). Effects of high-pressure processing (HPP) on bacterial spores: An overview. *Food Reviews International,* 24, 330-351.

In: Progress on Quantitative Approaches in Thermal ... ISBN: 978-1-62100-842-2
Editors: V. P. Valdramidis and J. F. M. Van Impe © 2012 Nova Science Publishers, Inc.

Chapter 2

THERMORESISTOMETERS FOR ASSESSING MICROBIAL HEAT RESISTANCE

Alfredo Palop[1], Francisco J. Sala[2] and Santiago Condón[2]
[1]Universidad Politecnica de Cartagena,
Departamento de Ingeniería de Alimentos y Equipamiento Agrícola, Spain
[2] Universidad de Zaragoza, Tecnología de los Alimentos, Facultad de Veterinaria de Zaragoza, Spain

Abstract

This chapter deals with the methodology to characterise the heat resistance of microorganisms. Microbial heat resistance is affected by many different factors and a proper method should be able to take into account each one of these factors. After a detailed revision of the different devices used for this purpose, the chapter focuses on the equipment with which most articles have been published, the thermoresistometer TR-SC, an improved version of this one, the thermoresistometer Mastia, and other modifications that have been also developed. Both thermoresistometers can be helpful tools for food processors in order to estimate the level of safety of the treatments they apply and, specially, for researchers working on heat inactivation, damage and adaptation mechanisms and on heat inactivation kinetics.

Introduction

Bigelow and Esty showed in 1920 that the inactivation of microorganisms by heat followed a first order kinetics nature (Bigelow and Esty, 1920). For those early experiments, carried out at the facilities of the National Canners Association, Bigelow and Esty used pyrex tubes, in which they placed the spore suspensions. The tubes were sealed by fusing the ends and heated in an oil bath. At preset time intervals the tubes were withdrawn from the oil bath and incubated.

The important discovery of Bigelow and Esty opened the possibility of setting thermal treatments according to the microbial load of the food product under process. Microbial heat resistance studies have been performed since then, thus laying the basis of Thermobacteriology. Although many researchers have reported deviations from log linearity

(Peleg and Cole, 1998), the assumption of a logarithmic order of death has served very much the food industry since then. The knowledge provided by all these studies on microbial destruction kinetics and on the mechanisms of this destruction has allowed the design and development of safe processes, reducing the risk of food borne diseases as well as of food spoilage.

Many researchers agree that the heat resistance determination method may influence the results obtained (Schmidt, 1957; Reichart, 1983; Cerf, 1987; van Zuijlen et al., 2010). A proper method should have, at least, the following features:

- be able to keep the temperature stable and allow for a proper control of time and temperature of treatment
- allow for the instant (or monitored) heating and cooling of the samples
- allow for taking D_T values as low as possible in a wide range of temperatures
- enable to perform heat treatments in foods
- make it possible to know and monitor the parameters that influence microbial heat resistance
- reduce handling during the thermal treatment
- have a simple design and low cost

Some heat resistance determination methods expose the microbial suspension to a temperature gradient before achieving thermal stability. This lag phase contributes to the thermal inactivation and may hide the actual effect of heat at the treatment temperature. Although several mathematical models have been proposed in order to amend the lethal effects of such lag phases, the different heating rates obtained in different heating media, make it difficult to estimate the real temperature profile. Moreover, it is necessary to know the exact *z* value of the microbial species under study to perform such amendments. Hence, the solution of this problem is not easy (Cerf, 1987). According to Stumbo the heat resistance of a microorganism cannot be properly estimated if more than 5% of the whole thermal treatment is applied at temperatures lower than 1ºC of the set temperature (Stumbo, 1973). Brown and Ayres agreed and added that the problem is worst the higher the treatment temperature (Brown and Ayres, 1982). Moreover, when exposing the microorganisms to increasing temperatures before reaching the treatment temperature, they have the chance of increasing their heat resistance (Pagán et al., 1997; Yura et al., 2000; Periago et al., 2002a; 2002b).

Since inactivation rate is exponentially related to temperature, minor temperature variations may lead to important changes in the slope of the survival curve, especially when *z* value is small (for example for most of the vegetative bacteria). The problem is bigger the higher the treatment temperature. Inaccuracies in the treatment time may also lead to changes in the slope of the survival curve.

Heat resistance determination temperatures should be set accurately in a wide range of temperatures, especially when it is intended to work with bacterial spores as well as with vegetative cells. It has been shown that food quality attribute losses are less dependent on treatment temperature than microbial inactivation kinetics (Pflug and Esselen, 1953). This finding led to the application of high temperature short time and ultra high temperature treatments several decades ago. Although there is a general agreement in the exponential relation of the inactivation rate and temperature, it has also been shown that estimation of *D* values at ultra high temperatures by extrapolation of thermal death time (TDT) lines may lead to important errors (Brown et al., 1981; Reichart, 1983). The temperature range at which a heat resistance determination equipment can be used does not depend on the heating capability or on the thermal stability, but on other factors, such as the heating lag time duration, the sampling speed, etc. Hence, this temperature range depends on the minimum D value that such equipment is able to accurately provide. Most of the thermal resistance determination methods are not able to provide accurate D values lower than 0.1 min.

Sample handling during and after the experiment should be reduced to a minimum in order to lessen errors, contaminations and costs. The equipment should also be of simple design and easy use. In this way, variability of results among researchers using it would be reduced. According to Schmidt, this is the main cause for discrepancies found in literature (Schmidt, 1957).

An appropriate thermal resistance determination method should take into account the effect on heat resistance of any influencing factor. Heat resistance of microorganisms is mostly determined by time and temperature of treatment, and varies, not only among different species, but also among different strains of the same species (Cerf, 1987). Upon the genetic component, environmental factors may drastically influence heat resistance, even during the heat treatment (Murrell, 1964). The chemical composition of the heating medium is among the most important factors affecting microbial heat resistance (Murrell, 1964; Stumbo, 1973; Pflug and Odlaug, 1978). Hence, the thermal resistance determinations designed in order to set sterilisation treatments should be performed in the same food where the treatment will be applied (Busta and Ordal, 1964; Cerf, 1987). Apart from the chemical composition of the heating medium, other chemical and physical characteristics of the medium may influence microbial heat resistance. The most influencing chemical and physical characteristics are pH (Blocher and Busta, 1983; Condón and Sala, 1992; Palop et al., 1996b; Palop et al., 1999) and water activity (Murrell, 1964; Alderton et al., 1976; Gaillard et al., 1998). These factors vary with temperature (Condón and Sala, 1992), hence results concerning their effect on heat resistance may also be affected. For instance the effect of pH on *z* values (Sala et al., 1995; Palop et al., 1996b; 1999). Hence, a proper thermal resistance determination method should also enable to know the value of these parameters along the thermal treatment.

Methods for Heat Resistance Determination

Many different heat resistance determination methods have been used (Stumbo, 1973; Brown and Ayres, 1982), each of them having their own advantages and drawbacks. The methods for heat resistance determination available today may be grouped into: (1) indirect methods, (2) direct methods, (3) particle based methods and (4) mixing methods (Brown and Ayres, 1982).

Indirect Methods

Indirect methods include all those in which the heating fluid is not in contact with the heating medium in which microorganisms are suspended. A common defect of all these methods is their inherent long heating and cooling lags. The single tube method (Bigelow and Esty, 1920) was the first one in this group to be used. It consisted in sealing under the flame several glass tubes containing the spore suspension and placing them in a thermostat bath. At preset time intervals, tubes were taken out from the bath and incubated immediately. The time necessary to attain sterilization was set by the treatment times corresponding to the first unspoiled tube and the preceding one. This end point method was very practical, since it allowed drawing TDT lines by plotting log of sterilization times in front of temperature. Hence, it allowed knowing *z* values and sterilization times at any temperature. However, it did not allow describing complete microbial inactivation kinetics.

When using this method it was quite frequent to observe series in which after a sterile tube, a spoiled tube corresponding to a further treatment time was found (Williams et al., 1937). In order to solve this problem, Esty and Williams (1924) proposed a modification, which they called the multiple tube method. This method consisted in placing the tubes in the bath and taking them out in batches of 25 – 30 at preset time intervals. Plotting the proportion of spoiled tubes in front of exposure time, a straight line was obtained and, extrapolating, the time needed to achieve the sterilization of all the tubes could be calculated. Later on, the National Canners Association reduced the size of the batches to 3 – 4 tubes per treatment time (NCARL, 1968).

A modification of this method (Stumbo et al., 1945), widely used (Odlaug and Pflug, 1977; Montville and Sapers, 1981; Westhoff and Dougherty, 1981), consisted in evaluating the number of survivors in each tube, after opening and sampling their content. The plot of the log of survivors in front of time is the survival curve, which describes the inactivation kinetics and allows estimating D values.

These methods are simple, cheap and allow working with food homogenates. However, they are time consuming and have long heating and cooling lags. According to Stumbo, the accuracy of these methods is only acceptable at temperatures up to 115°C, although this will depend on the heat resistance of the microorganism (Stumbo, 1973).

In 1938 Twonsend et al. developed a modification of the method of the tubes, which they called the method of the cans (Twonsend et al., 1938). This method was recommended by the National Canners Association in 1943 (NCARL, 1968). The procedure is similar to that of the tubes, but replacing the tubes by cans and the thermostat bath by retorts. The main advantage is that microorganisms can be inoculated even in solid foods and, in case of gas-producing microorganisms, spoilage can be detected in incubated cans by the swelling they lead to. The main drawback is that the heating and cooling lags are even longer than in tubes.

Advances in the knowledge of nutrient and microorganism inactivation kinetics, together with the need of extending previous research towards higher temperatures, led several authors to minimize heating and cooling lags. This was achieved by reducing the inner diameter of the tubes and the wall thickness. This is the bases of the method of the capillary tubes, first proposed by Stern and Proctor (Stern and Proctor, 1954). In this method, thin capillary tubes (0.8 mm inner diameter) were filled with 0.01 mL of the heating medium, inoculated with the microbial suspension, by means of a microsyringe. The tubes were then sealed over the flame and put in a special rack, handled by an articulated arm, placed between two thermostat baths, one at the treatment temperature and the other refrigerated. When all the capillary tubes were in the rack, it was immersed in the treatment temperature bath and, once the treatment time was past, a solenoid valve acted on the articulated arm, which moved the rack to the refrigerated bath. This operation took 0.37 – 0.40 seconds. After cooling, the tubes were washed and broken, the content was flushed and the samples diluted, plated and incubated. According to the authors, the content of the capillary tube immersed in a bath at 131.5°C, spent 6.4 seconds to reach 131°C and 8.31 seconds to reach 131.4° C. Although the authors stated that their method could be used up to 140°C, this will obviously depend on the heat resistance of the microorganism. The lower *D* value they determined was of 0.153 min. Perkin et al. placed a thermocouple inside a thin wall capillary tube (0.15 mm), estimating that for heating times longer than 5 seconds, the lag time was negligible (Perkin et al., 1977). However, for shorter heating times it was necessary to make corrections. A modification of the original method consists in taking out manually several capillary tubes per treatment time, allowing then for obtaining the survival curve in one single experiment. This reduction in time and handling has made the modification even more accepted than the original method, being even more widely used (Busta, 1967; Burgos et al., 1972; Goldoni et al., 1980; Ordoñez et al., 1987; Leguerinel & Mafart, 2001; Lekogo et al., 2010; van Zuijlen et al., 2010). Moreover, heating and cooling lag times are considerably reduced, making this method relatively accurate. It has been used frequently as a reference method (Davies, 1975; Burton et al., 1977; Mikolajcik and Rajkowski, 1980; Brown et al., 1981; Reichart, 1983; Leguerinel and Mafart, 2001).

In Reichart's opinion, a proper system should allow performing all the thermal resistance determinations necessary to obtain the *z* value in a single experiment, in order to avoid the influence of different factors on the heat resistance (Reichart, 1979). According to Reichart, since lethality coefficient increases exponentially with temperature and temperature increments decrease logarithmically with time, it can be concluded that lethality coefficient increases linearly with time. Reichart (1979) placed thin capillary tubes (0.5 mm outer diameter and 0.2 mm wall thickness) in a bath which temperature followed a logarithmic profile. At preset intervals a sample was taken and the temperature was recorded. Plotting the survival counts in front of time, a downward concavity curve was obtained. Appropriate adjustments in function of temperature at each sampling time allow obtaining the *z* value directly. Drawbacks of this method are those already mentioned for capillary tubes at high temperatures. The lower *D* value obtained by this author was 0.3 minutes.

However, the handling of the content of the capillary tubes is a tedious task, especially when working with viscous media. Another drawback is that it does not allow working with particulate foods. Moreover, Brown et al. checked that the system used for holding the capillary tubes in the bath during the treatment can significantly influence heating lags and, consequently, *D* values obtained (Brown et al., 1981). Over the years, capillary tubes have been replaced by glass spheres (Brown and Thorpe, 1978), freeze dry vials (Cook and Gilbert, 1968), aluminium tubes (Odlaug and Pflug, 1977), etc. However, all these modifications do not make these methods significantly different in their advantages and drawbacks to that of Stern and Proctor.

An objection to all the above mentioned methods is the chance of finding physical and/or chemical differences between the tubes/cans/capillaries of a single experiment, which would interfere with the results. Williams et al. tried to solve this trouble by designing a piece of equipment in which heating was performed in a single container, from which samples were obtained along treatment time (Williams et al., 1937). This method was called the tank. The instrument is a small retort (900 mL volume) consisting of a main vessel surrounded by a jacket through which steam flows at the preset temperature. The vessel is sealed by a lid, crossed by a thermometer, an agitation shaft and a vent valve. At the bottom of the main vessel, several valves allow for sampling. According to the authors, the operation of the instrument consisted on filling the vessel with the heating medium, adding the microbial suspension, closing the vessel, shaking its content until achieving a homogeneous distribution of the inoculum and flowing the steam through the jacket to increase the temperature. When the preset temperature was reached, samples were obtained at preset times. The main drawbacks of this method are the long heating lags (two to three minutes) and the sophistication of the instrument (a steam generator was necessary), which did not work at temperature under 100ºC.

Ultrahigh temperature treatments are now widely used the food industry. Some authors have used in flow indirect heating systems in order to mimic the conditions existing in these treatments and to calculate heat resistance at these high temperatures. Such systems include plate and tube heat exchangers at a food plant and laboratory scale. Detailed descriptions of these devices can be found in the reviews by Brown and Ayres (1982) and Cerf (1987). The main drawbacks of these methods are the sophistication of the equipments, the big amount of microbial suspension needed to run each experiment (Stumbo, 1973) and the poor accuracy in the determination of heating times (Cerf, 1987). Also the extent of heating lags is difficult to measure due to the differences in viscosity and heat transfer coefficients among the heating media (Brown and Ayres, 1982). Another drawback of these methods is the difficulties found to reach adequate aseptic conditions (Cerf, 1987). When working with these equipments, the end point method is often used. Hence they are mainly used to set sterilization times, more than to know microbial inactivation kinetics or to determine accurate microbial heat resistance parameters.

More recent is the submerged-coil heating apparatus designed by Cole and Jones (1990). It lacks most of the drawbacks mentioned above: it needs a small inoculum volume (maximum 9.5 mL), heating lags are quite short (less than 1 s), multiple samples can be taken in each experiment with time intervals as short as 6 seconds and it is simple to operate since it is just a stainless-steel coil fully submerged in a thermostatically-controlled water bath. However it does not work at temperatures over 90ºC. The continuous microflow submerged microcoil apparatus of Loss and Hotchkiss (2004) increased the temperature range to 140ºC by submerging the microcoil in a heated oil bath and by regulating the back pressure with and adjustable screw clamp valve fixed to a section of flexible tubing at the outlet end.

Direct Methods

Brown and Ayres consider direct methods those in which the microorganisms are heated directly with steam, either by injecting the heating medium in a steam environment or steam into the heating medium (Brown and Ayres, 1982). The latent heat of steam condensation provides in both cases very fast heating, reducing to a minimum heating lags. The main advantage of these methods is that they enable to know microbial heat resistance at usual temperatures in UHT processes.

Current direct methods are complex, do not allow working at low temperatures (below 100ºC), and, consequently, are restricted to spore heat resistance determinations. Stumbo designed a thermoresistometer able to determine heat resistance in the interval 104 to 150ºC (Stumbo, 1948). This thermoresistometer consisted of three adjoining chambers, which can keep steam pressure separately, through tight closures. Steam was used to sterilise the chambers, besides to treat the samples. This steam came from a big tank, avoiding temperature fluctuations. Each chamber was provided with separate inlet and outlet steam pipelines, as well as with an air vent and a condensed water drain. A metallic tray carried the samples and passed consecutively through all three chambers by means of a chain system. Samples went into the waiting chamber, at room temperature, then into the treatment chamber, which was at high temperature, with steam, and then to the discharge chamber, also at room temperature. Thermal stability of this equipment was of +/- 0.167ºC and heating and cooling lags were considerably reduced. In this way, heating times longer than 6 seconds did not need for any correction and shorter times only needed a correction of 3.3 seconds as maximum (Stumbo, 1973).

Pflug and Esselen improved this thermoresistometer in a way that enabled to measure temperature with a smaller error (Pflug and Esselen, 1953). The metallic try was changed by a set of cylindrical pistons, which carried the samples. Such pistons closed tightly the treatment chambers by means of O-rings. These treatment chambers had a smaller volume. Pistons were switched on by a pneumatic cylinder, controlled by a solenoid valve and an automatic chronometer. The discharge chamber was removed, so samples were charged and discharged in the pre-treatment chamber, which was modified according to this new purpose. Treated samples were collected in test tubes, pushed by sterile compressed air, which also helped to avoid the evaporation of the sample when it was coming out from the treatment chamber at high temperature. The authors claimed better precision, which enabled to apply very short treatment times. However, according to Stumbo, this device did not improve significantly his own thermoresistometer, had the same limitations and was susceptible of more mechanical failures (Stumbo, 1973).

David and Merson developed a new thermoresistometer, computer controlled, able to determine heat resistance parameters up to 155ºC (David and Merson, 1990). It consisted only of one cylinder that carried the sample, which was punched on its longitudinal axis to put the thermocouples in the sample. The cylinder was coupled to a pneumatic piston, which was driven by a solenoid valve. A liquid sample was placed in a hydrophilic filter paper, which was placed between other two hydrophobic discs. Then the sample was exposed to the steam in the chamber through both sides, enabling almost instantaneous heating and cooling. The preheating chamber was provided with a water coil, to avoid heating of the sample prior to treatment. After the treatment, a flow of sterile air pushed the sample and also helped to cool it down quickly and to avoid its evaporation.

Based on the previous ideas, Brown et al. developed another thermoresistometer at the Campdem Food and Drink Research Association, with interesting improvements regarding the above mentioned designs (Brown et al., 1988). In this equipment, the steam tank was a vertical pipeline, which saved room in the lab. This steam tank was, at the same time, the treatment chamber, which helped very much to keep the temperature stable during all the thermal treatment. Charging and discharging the sample was done through the pre-treatment chamber, which was also refrigerated and provided by an injection system for pressurised air. The sample was placed in a paper filter, following David and Merson (1990), but was not located between the two hydrophobic discs. A T-type thermocouple provided the sample temperature profile. In order to control the temperature inside the treatment chamber, four Pt-100 were located besides the sample. The temperature in the chamber was then computer controlled, by opening or closing electro valves switched to the steam tank. This last model has been widely used by different researchers (Rodrigo et al., 1993; Silla et al., 1993; Fernández et al., 1994; Ocio et al., 1994; Rodrigo et al., 1998; Blasco et al., 2004).

In flow direct heating systems have also been designed. Perkin et al. designed a laboratory scale UHT plant able to treat small volumes of liquids at temperatures between 130 and 150ºC, with negligible heating lags (Perkin et al., 1977). In this method, the sample was heated up when it was sprayed in a steam chamber preheated at the treatment temperature. Afterwards, the medium was instantaneously cooled down when it passed to a low pressure chamber. Different treatment times were obtained by changing the flow rate. Other instruments, such as those of Burton and Perkin (1970) and Edwards et al. (1965) heated up the in flow heating medium injecting steam in it.

The main advantage of all these systems is that they enable to determine heat resistance values at very high heating temperatures. Common drawbacks are their high cost and complexity and that they can only be used in the range of high temperatures for just a few types of food products. Moreover, treatment time cannot be accurately known. Burton et al. found that an exposure time of 3.26 s at 145ºC was equivalent to less than 2.32 s of real heating time (Burton et al., 1977). On the other hand, Daudin and Cerf found that the heat resistance values shown for several microorganisms in the UHT temperature range was lower than that calculated by extrapolation of the TDT lines obtained at lower temperatures (Daudin and Cerf, 1977). Their *z* values were also smaller. These authors proposed that this effect could be explained by the thermal shock that the dramatic change in temperature caused. Burton et al. compared the data obtained with their equipment to those obtained using the capillary method, showing that *z* values were different and that TDT lines crossed at temperatures close to 145ºC (Burton et al., 1977).

Particle Based Methods

Particle based methods are the so-called Time-Temperature Integrators (TTIs), which are small beads consisting of a carrier system and a sensitive element. The sensitive element is dissolved in the carrier system, which is at first liquid and hardens later on, immobilising the sensitive element. Both components are then immobilised and the result is the TTI. They are included in the food product to be heat treated and, after the thermal treatment, the TTIs are recovered from the heating medium, physically or chemically broken and the survivors are counted. The carrier system should be stable to the thermal treatment and easy to handle. The sensitive element can be any chemical or microorganisms of known and adequate heat resistance (Dallyn et al., 1977; Ocio et al.,, 1997). Polymethyl-methacrylate beads and alginate beads designed by Hunter (1972) and Brown et al. (1984) respectively were among the first methods used within this group.

Apart from the behaviour of microorganisms when exposed to heat, these methods enable to research on the kinetics of heat penetration in these solid particles. Complexity of both phenomena when analysed separately is significantly increased when trying to analyse them together, hampering the discussion of the results. Moreover, the accuracy of the heat resistance values obtained is questionable. Hence, the usefulness of these methods to determine heat resistance values is also questionable. More than heat resistance determination methods, TTIs are used to evaluate the intensity of the heat treatment to which foods are exposed. For more details please refer to Chapter 'Time-temperature integrators in thermal processing' of this book.

Mixing Methods

Brown and Ayres (1982) included within the group of mixing methods, those in which the suspension of microorganisms is inoculated directly into the medium, once it has reached the temperature of treatment (Levine, 1956). In this way, heating and cooling lag times are dramatically reduced, since a very small volume of microbial suspension is mixed with a much bigger volume of preheated medium. Heat resistance values are calculated from the survival curves obtained when plotting the survival counts of samples taken at different treatment times in front of their corresponding treatment times.

The first mixing device corresponds to Levine et al. (Stumbo, 1973), who designed the flask method in 1927. This method consisted in putting a flask (formerly a Woulf bottle) which contained the preheated heating medium in a thermostat bath. A thermocouple was placed into the flask through one of its openings and the agitation axis through the central opening. The remaining opening was used to inject the microbial suspension, once the treatment temperature was reached and stable, and to pipette the samples. In 1967 Aiba and Toda modified this method by injecting 5 mL of the microbial suspension into 15 L of preheated buffer (Aiba and Toda, 1967). Later on, Kooiman and Geers reduced the inoculum volume to 0.1 mL, which was injected in 10 mL preheated tubes of treatment medium (Kooiman and Geers, 1975). Menegazzi and Ingledew injected 1 mL of microbial suspension in a preheated flask with 99 mL of heating medium (Menegazzi and Ingledew, 1980). These authors used two flasks for each experiment, one of them for the samples and the other to register the temperature. This method was called the tempered dilution blank method.

All these mixing methods were inexpensive and easy to use, and avoided heating lag times. However, temperatures well below the boiling temperature of water were only affordable and thick media should not be used. Hence, these methods have only been used to determine the heat resistance of vegetative bacteria or low heat-resistant spore-forming bacteria. Moreover, sampling is manual, usually pipetting, and this fact limits the minimum D_T values to be obtained.

Milolajcik and Rajkowski described the serum bottle technique (Milolajcik and Rajkowski, 1980). It consisted in inoculating the microbial suspension in injectable flasks containing the treatment medium already sterilised and preheated. Samples were taken along heating with hypodermic syringes. The main advantage of this method was that it enabled to perform experiments at temperatures above the boiling temperature. In the former manuscript, the authors were not able to make the heating lags disappear, since the medium temperature decreased when the microbial suspension was inoculated and it lasted 1 minute to stabilise again. This problem could be easily solved by reducing the inoculum volume or increasing that of the heating medium. However, shaking of flasks was performed manually, and sampling was performed with syringes, making sampling a very slow process, which limited sampling intervals to 1 min.

Brown et al. designed a new equipment, called flash off, which can be included in the mixing method group (Brown et al., 1981). The device consisted on a metallic vessel, tightly closed with a screw cap, in which several ports went across: one for the thermocouple, another for the injection of the microbial suspension and another for a sampling tube, closed at the end by a valve. When the valve was opened, a sample of the content of the vessel came out, just pushed by the own pressure generated inside the vessel. Heating of the vessel was achieved by immersing it in an oil thermostat bath. The main advantage of this system was its simplicity, and major drawbacks were that it only enabled to work at temperatures above the boiling temperature of low viscosity liquids. According to the authors, five seconds were needed to take a sample, which set the minimum D_T value to be obtained. Still, D_T values obtained using this device were lower than with the capillary method.

A common drawback to all the mixing methods described until now was that they used thermostat baths to heat up the heating medium. This makes temperature changes and stabilisation times very slow, which is an important limitation according to several authors (Reichart, 1979). Trying to avoid this problem, Cerny put an electric heating element inside the metallic vessel (Cerny, 1980). This vessel was tightly closed by a cap with a manometer, a vent valve and two thermocouples, one of them connected to the electric heating element. Inoculum distribution was achieved by a magnetic shaker. Inoculation and sampling was performed with a syringe through two septa placed at the wall of the vessel. Shaking and sampling systems did not enable to use thick media. Moreover, taking samples lasted for approx. 10 seconds, which limited D_T minimum values to 0.5 to 1 min. According to Reichart, at that moment there was no method able to determine D_T values lower than 0.06 min (Reichart, 1983). This author developed a method able to determine D values as low as 0.003 min, but only enabled to apply thermal treatments at temperatures lower than 100ºC.

Advancements on Resistometers

At the beginning of the decade of 1980s the research group of Food Technology of the University of Zaragoza intended to design a heat resistance determination method with all the advantages of the existing methods and avoiding their limitations. After significant efforts, the thermoresistometer TR-SC was developed (Condón et al., 1989), which was further improved in several aspects until the present design was reached, the thermoresistometer Mastia (Conesa et al., 2009). Other instruments with different purposes have also derived from the former thermoresistometer TR-SC. The thermoresistometers TR-SC and Mastia are also included in the group of the mixing methods, but will be explained in detail in the following sections.

The Thermoresistometer TR-SC

The former thermoresistometer TR-SC (Condón et al., 1989) incorporated most of the advantages of the existing methods. The instrument consisted of a small stainless steel insulated and pressurised vessel (8.5 × 12 cm outer diameter; approx. 400 mL) with a screw cap with O-ring. This cap had an agitation shaft with a teflon (PTFE) friction bearing with a propeller, and 7 ports with screw caps to hold: the pressure source (N_2); the inoculum injection port with a gas chromatography septum; the sampling tube; a thermocouple to monitor the temperature during heat treatment; the pH-meter; and two ports to hold the two branches of the electric heating element. Inside the vessel, attached to the removable cap there was also a deflector screen to improve turbulence. Figure 1 shows a diagram and a picture of this equipment.

The heating medium was pre-heated in the vessel by means of the heating element and it was homogenised by the propeller and the deflector screen. The main vessel could be pressurised through a manometer connected to the pressure source. To overcome the pressure when the microorganisms were injected inside the vessel, a Hamilton-type syringe was used.

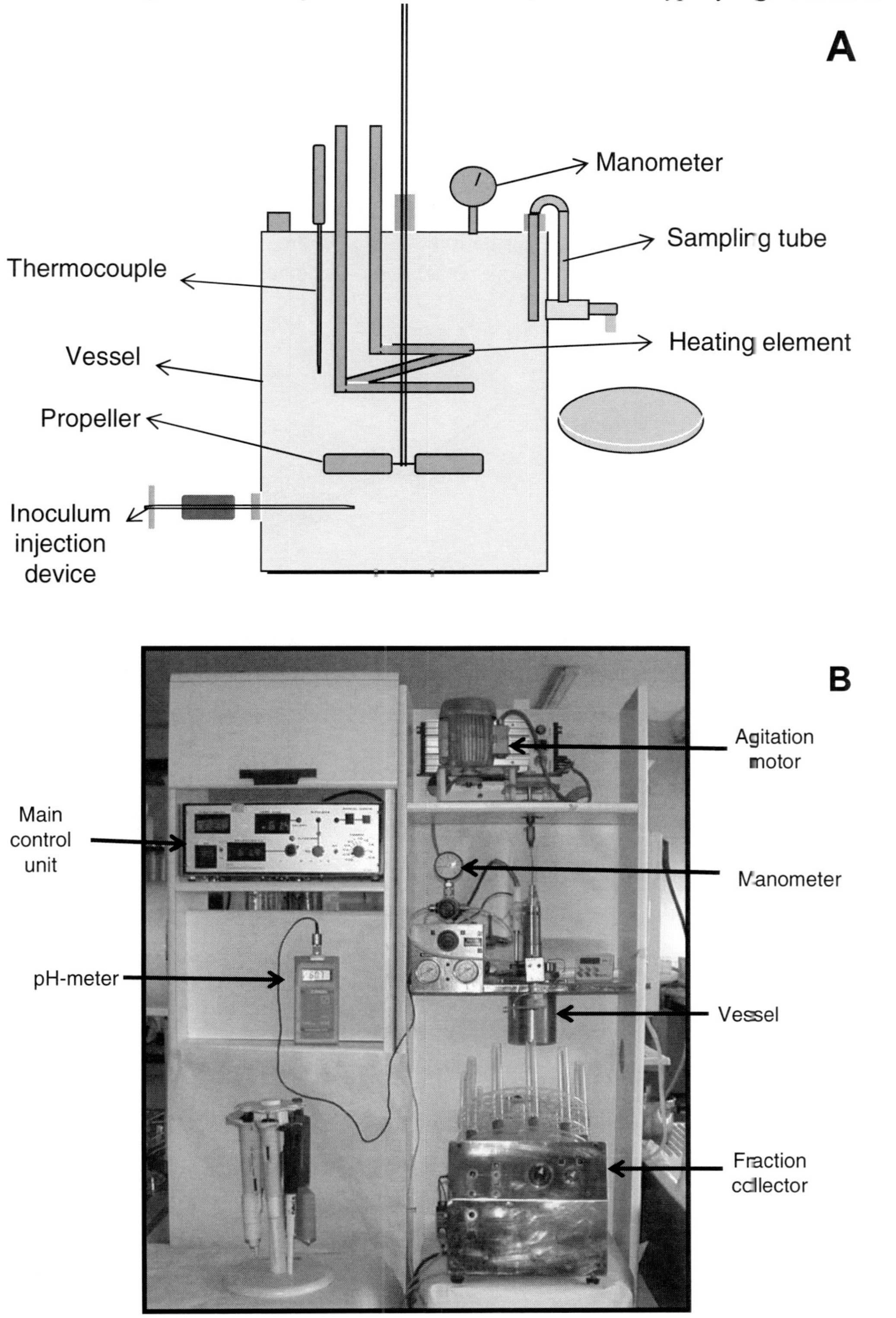

Figure 1. Diagram (A) and picture (B) of the Thermoresistometer TR-SC.

The agitation shaft was powered by a variable speed agitation motor. The sampling tube was closed in the outer extreme by means of a clamp. Sampling could be performed at time intervals as short as 2 seconds, or continuously by using a microswitch, to operate the sampling tube clamp which turned on a continuous turning regulable speed, custom built, fraction collector.

This thermoresistometer showed a high heating rate (5.7°C/min) and a significant thermal stability at the set temperature (+/- 0.1°C). On the other hand, pressurisation of the equipment enabled to keep a constant pressure inside the vessel, whatever the temperature. Hence, sample volume was constant, even at temperatures below the boiling temperature. The shaking system enabled a very fast distribution of the inoculum inside the vessel in buffer and milk (less than 0.13 seconds) as well as in potato purée and other food homogenates (less than two seconds).

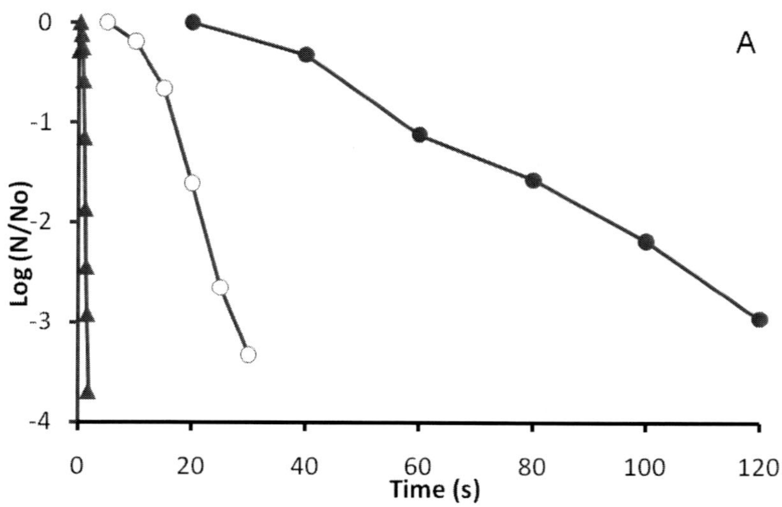

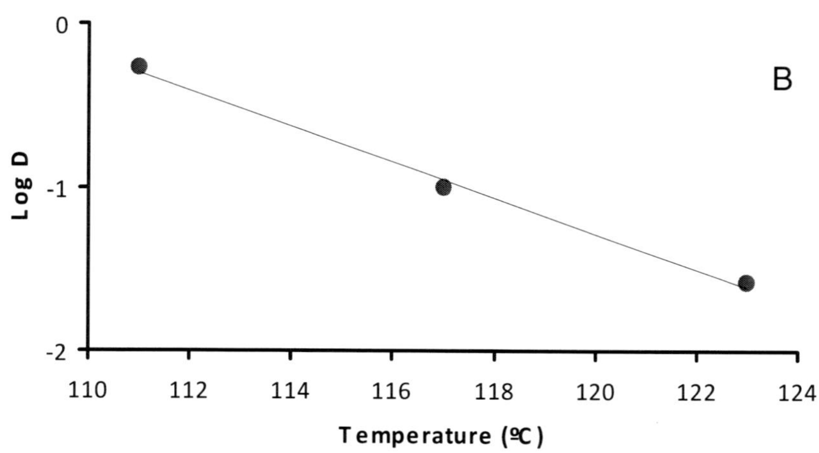

Figure 2. Survival curves (A) and thermal death time curve (B) of *Bacillus coagulans* spores heated in pH 5.3 asparagus purée at 111°C (●), 117°C (○) and 123°C (▲).

The thermoresistometer TR-SC was easy to handle and saved a lot of time and effort. The time to obtain the experimental data needed to build a TDT line was reduced from 20 hours, when using the capillary method, to 5 hours with the thermoresistometer TR-SC. No statistically significant differences were found between neither D_T nor z values obtained with both methods. However, data accuracy with the thermoresistometer was significantly higher ($p \geq 0.01$).

Particulate media, such as potato purée, or homogenized canned foods (peas, asparagus, tomatoes, garbanzo beans, lentils, mushrooms and meat balls) could be used as heating media in the thermoresistometer TR-SC. Figure 2 shows, as examples, survival and TDT lines obtained with this equipment when using homogenized asparagus. Survival curves at 111 and 117°C correspond to experiments in which samples were taken manually, while that at 123°C corresponds to an experiment performed with the fraction collector.

The thermal stability of the thermoresistometer, the fast distribution of the inoculum in liquid heating media and the sampling times as short as one tenth of second allowed obtaining survival curves at UHT temperatures. Figure 3 shows survival curves of *Bacillus coagulans* in milk. All three survival curves were obtained under continuous sampling at a sampling speed of 15 samples per second. Condón et al. reported the lowest D value obtained with this instrument, of 0.009 min (for *Bacillus coagulans* at 135.8°C; Condón et al., 1989). This was the lowest D value obtained until then with the multipoint system (*i.e.* sampling at different preset time intervals) and ten times lower than those usually reported with the capillary method.

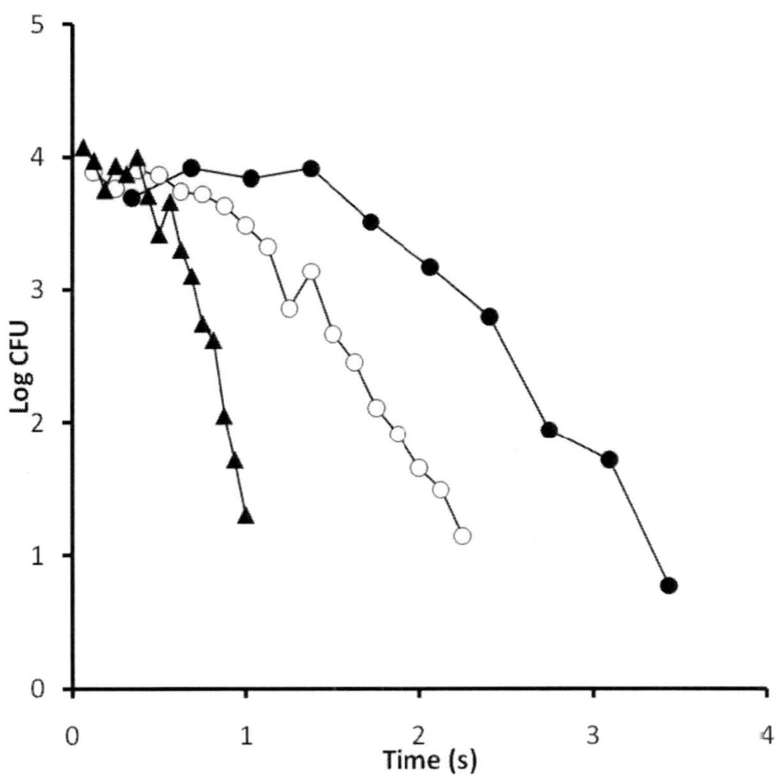

Figure 3. Survival curves of *Bacillus coagulans* in milk at 128°C (•), 130°C (○) and 132°C (▲).

Afterwards, the equipment was further improved in order to obtain lower *D* values. In 1993, Condón et al. reported the improvements performed (Condón et al., 1993). The most important ones was to synchronize the inoculation of the microbial suspension into the heating medium with the start of sampling and to use a faster and more accurate fraction collector. The improved thermoresistometer enabled to determine a D_T value as low as 0.0009 min (for *Bacillus coagulans* at 138°C), also with the multipoint system. This is the lowest D_T value reported to date.

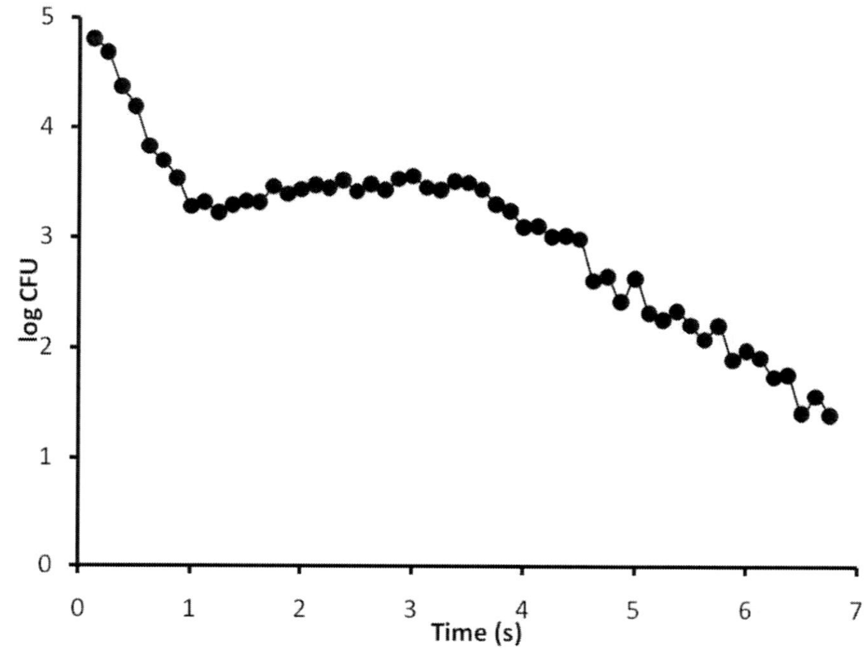

Figure 4. Survival curve of *Bacillus coagulans* in pH 7 McIlvaine buffer at 126°C.

Obtaining many samples in fractions of second enables to determine microbial heat resistance at UHT temperatures. Moreover, the instrument is very useful to describe non linear inactivation kinetics, since lag times are negligible. Figure 4 shows a non-linear survival curve obtained with the thermoresistometer TR-SC. The presence of different subpopulations with different heat resistance in a suspension of *B. coagulans* spores and the transformation of the heat sensitive subpopulation in the heat resistant one as a consequence of heating (Palop et al., 1997), and the modelling of the heat inactivation/adaptation kinetics in vegetative bacteria (Pagán et al., 1997) have also been reported with this instrument.

The behaviour of the thermoresistometer TR-SC at pasteurisation temperatures is also quite satisfactory. In the decade of 1990s, efforts were invested in vegetative bacteria and yeast heat inactivation kinetics (Condón et al., 1992; Garza et al., 1994; Pagán et al., 1997; 1998; 1999a; Mañas et al., 2003). More recently, efforts have been focused on vegetative emerging pathogens (Arroyo et al., 2009; Sagarzazu et al., 2010). Figure 5 shows, as an example, *Salmonella* enteritidis survival curves in pH 7 McIlvaine buffer. The instrument has also been successfully used to measure the thermal resistance of enzymes (López et al., 1997), antibiotics (Hassani et al., 2008) and the heat denaturation kinetics of proteins (Sánchez et al., 1992).

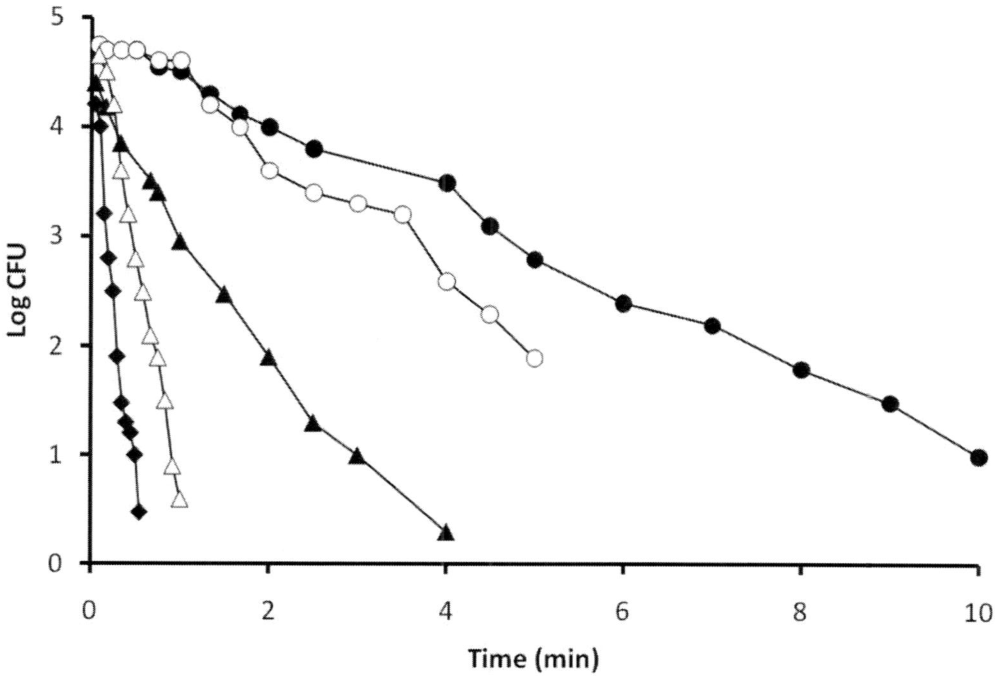

Figure 5. Survival curves of *Salmonella* enteritidis in pH 7 McIlvaine buffer at 53ºC (●), 54ºC (○), 56ºC (▲), 58ºC (Δ) and 60ºC (♦).

Another research field of interest that can be easily afforded with the thermoresistometer is the combined effect of heat and antimicrobials. Proper concentration of the antimicrobial can be added to the heating medium before inoculation of the microorganisms. Several antimicrobials have been tested with promising results since they achieve in general an important reduction of the heat resistance of microorganisms. Results have been published on the effect of organic acids on the heat resistance of vegetative bacteria of *Enterobacter faecium* (Fernández et al., 2009), and on bacterial spores (Palop et al., 1996a). Natural antimicrobials, such as those present in the essential oil fraction of aromatic herbs and spices, have also been tested (Espina et al., 2010). Figure 6 shows, as an example, the survival curves of *Escherichia coli* heated in presence of different concentrations of thymol. The inactivation of the microorganism was achieved at faster rates at higher concentrations of the antimicrobial.

This thermoresistometer was purchased by several research centres, which made this thermoresistometer one of the most widely used by researchers. It is probably the instrument with which more thermobacteriology data have been reported in the last 20 years, with more than eighty cites referenced up to date. The following are just examples of research articles performed with the thermoresistometer TR-SC in these centres: Condón and Sala, 1992; González et al., 1995; Palop et al., 1996a; Pagán et al., 1999a; Conesa et al., 2003; Arroyo et al., 2009; Fernández et al., 2009; Sagarzazu et al., 2010; van Zuijlen et al., 2010.

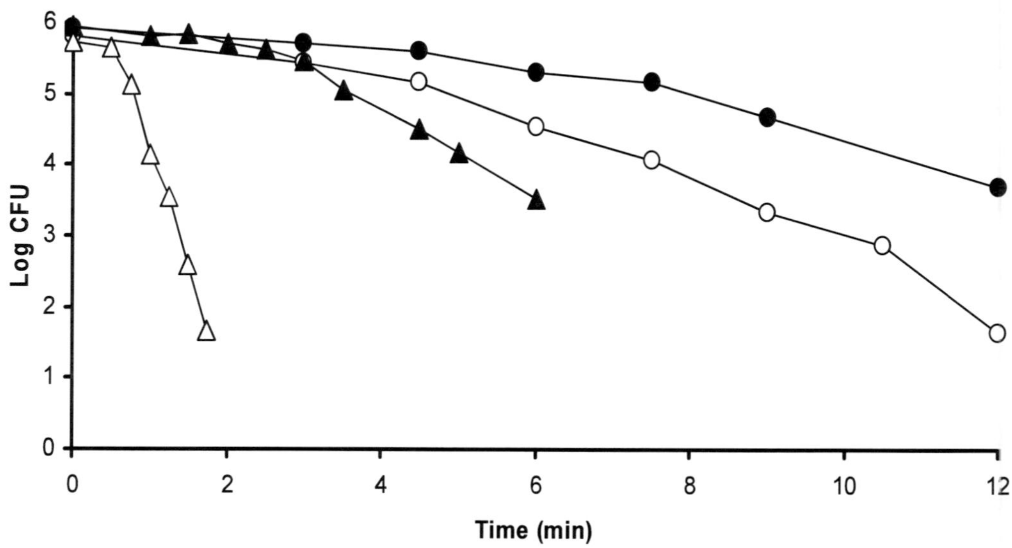

Figure 6. Survival curves of *Escherichia coli* in pH 7 McIlvaine buffer at 55°C without thymol (•) and in presence of 0.1 mM (○), 0.3 mM (▲) and 0.5 mM thymol (Δ).

Almost the only drawback of the former thermoresistometer TR-SC was that it was difficult to perform non-isothermal experiments with it. Most of the heat resistance determinations have been performed under isothermal conditions and, hence, omitting the effect on heat resistance of the heating and cooling rates. However, heat treatments applied in the food industry comprise stages of non-isothermal heating (corresponding to heating and cooling phases), which may be even more important than those of isothermal heating (holding phase) in food canning. This fact has lead to some authors to use non-isothermal methods as an alternative to the study of microbial inactivation kinetics (Reichart, 1979; Periago et al., 1998; Fernández et al., 1999; Conesa et al., 2003; Hassani et al., 2005; 2007) and differences between the heat resistance values obtained from isothermal and non-isothermal heating experiments have been found (De Cordt et al., 1992; Periago et al., 1998). Furthermore, during heating lags, bacterial cells may undergo sublethal heating and develop adaptative responses through the synthesis of heat shock proteins (Pagán et al., 1997; Yura et al., 2000; Periago et al., 2002a; 2002b; Hassani et al., 2006; Valdramidis et al., 2006; 2007; van Derlinden et al., 2010).

The Thermoresistometer Mastia

The former thermoresistometer TR-SC was modified by fitting it with a coil through which cold water could circulate, acting as a cooling system. The instrument was also implemented with a programmable logic controller (PLC) which powered the heating element and the cooling system to control the temperature inside the main vessel. The PLC was connected to a tactile screen, which allowed communication with the PLC and to a computer provided with software that enabled to program and/or register temperature profiles.

The vessel of the thermoresistometer Mastia (Conesa et al., 2009) was quite similar to that of the TR-SC, but the cap has 8 ports. The one for the pH-meter was removed and two ports that hold the two branches of the coil of the cooling system were included. A picture of the instrument can be seen in Figure 7.

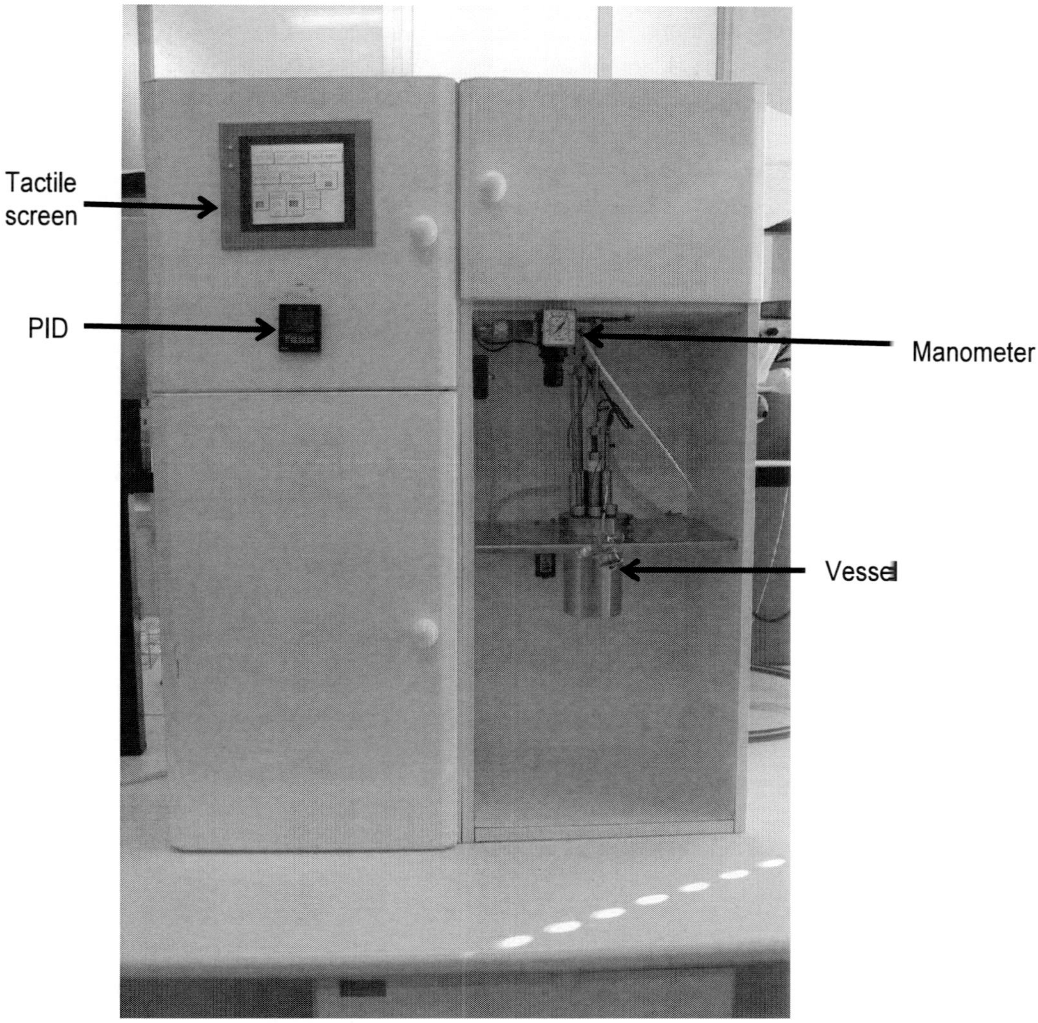

Figure 7. Picture of the Thermoresistometer Mastia.

The control of the temperature inside the thermoresistometer Mastia was done by the PLC, by means of a PID (Proportional Integral Derivative) connected to the electric heating element, to the solenoid valve that regulates the flow of water through the cooling system, and to the thermocouple. When the PID detects that the temperature is below the set temperature, the PLC powers the electric heating element; when the temperature is above the set temperature, the PLC acts on the solenoid valve fitted to the cooling system to control the flow of cold water through the coil.

The instrument can be programmed to perform isothermal treatments, non-isothermal treatments at a linearly increasing or decreasing temperature and more complex, industry-like heat treatment profiles, composed of several steps combining isothermal and non-isothermal periods. For isothermal treatments the temperature is kept constant through all the experiment, and for non-isothermal treatments, the thermoresistometer is programmed to perform the corresponding temperature profile. The use of a 2 kW electric heating element enabled the instrument to provide heating rates of up to 35°C/min. Similar cooling rates can also be achieved. The thermoresistometer Mastia enables, thus, to mimic heat treatments and makes it possible to know the number of survivors at any time of the treatment. This could help food processors to estimate if the heat treatment they apply to the foods they process is enough to inactivate all the microorganisms, with an adequate degree of safety or, maybe, it is too severe or too risky.

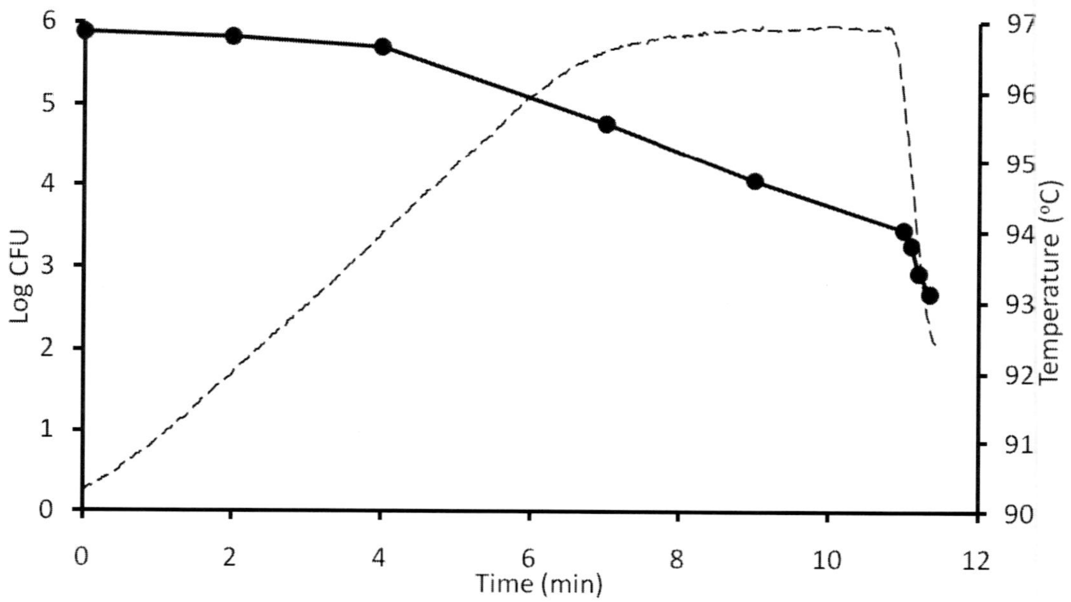

Figure 8. Survival curve of *Alicyclobacillus acidoterrestris* in orange juice when exposed to a non-isothermal treatment consisting of a heating phase at 1°C/min, a holding phase at 97°C for 4 min and a cooling phase at 20°C/min.

Figure 8 shows the effect of a non-isothermal treatment consisting of a heating phase at a slow heating rate (1°C/min), a holding phase and a cooling phase at a fast cooling rate (20°C/min), applied to *Alicyclobacillus acidoterrestris* spores in orange juice. Interestingly, the cooling phase contributed greatly to the inactivation of the microorganism (Conesa et al., 2009).

The instrument could also be operated in continuous mode. This operating mode consists in preheating the thermoresistometer with water at a preset temperature and circulating the microbial suspension through the coil, by means of a peristaltic pump at a controlled flow. In this way, the thermoresistometer works as an indirect method, similar to those reported by Cole and Jones (1990) or Loss and Hotchkiss (2004), with the advantage that changes in the set temperature can be performed much faster.

MTS-Resistometer

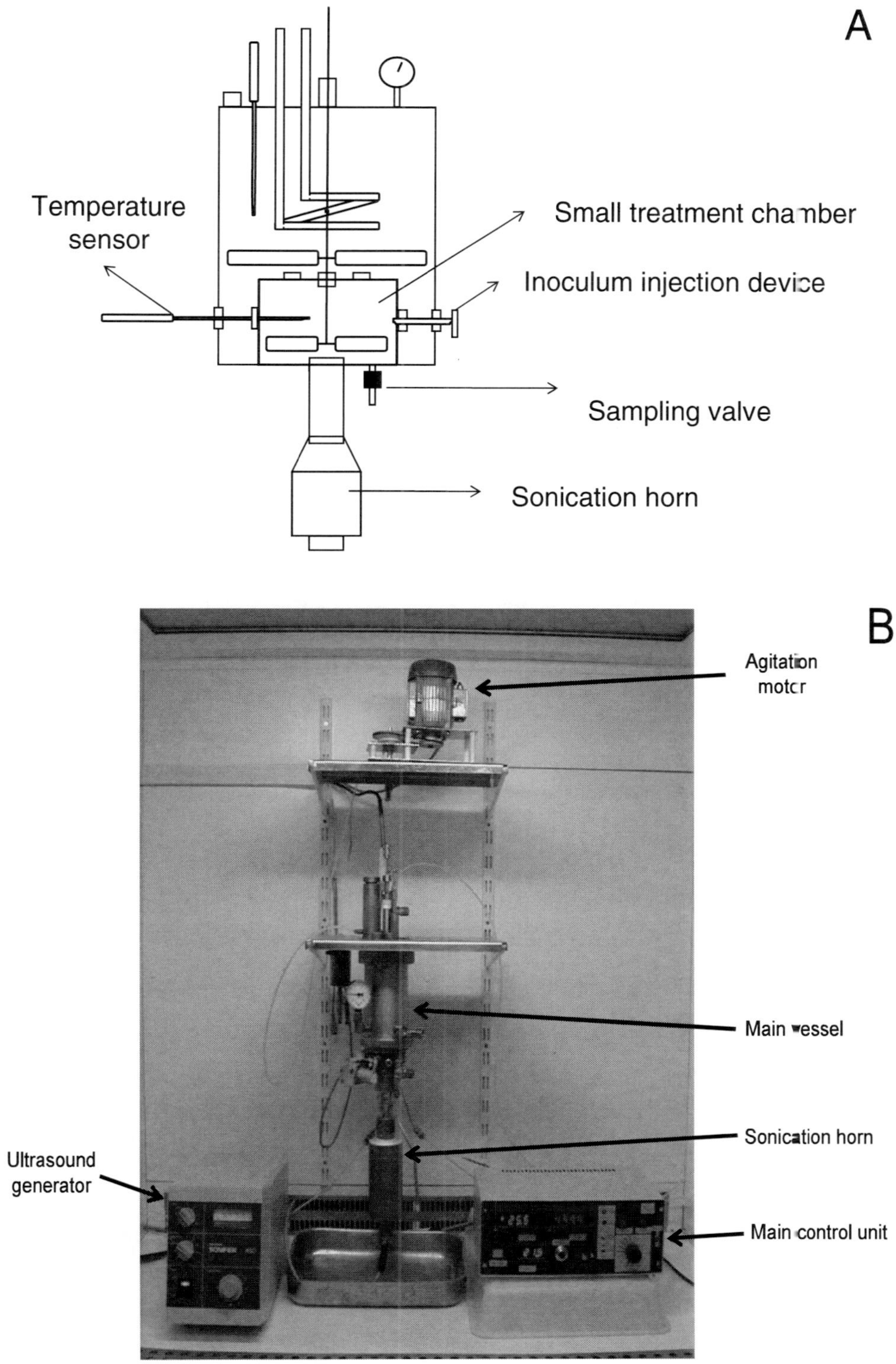

Figure 9. Diagram (A) and picture (B) of the MTS-resistometer.

At the end of the decade of 1980s, there was a remarkable interest on the research on non thermal technologies and on the development of combined processes. One research topic was focussed on the inactivation of microorganisms by ultrasound and its combinations with moderate pressure and heat. Based on these needs, the former thermoresistometer TR-SC was re-designed at the University of Zaragoza, leading to an MTS-resistometer, which description was published in 1998 by Raso et al. (1998a). The MTS-resistometer (Figure 9), built of stainless steel and teflon, consisted of a main vessel (650 mL capability) with removable top and bottom caps with o rings. The top cap was similar to that described for the former instrument. The bottom cap had a filling and emptying tube with a valve and a cooling coil to dissipate heat generated by ultrasound. A small treatment chamber (23 mL) was screwed onto this cap. This chamber also had a small agitation shaft and blade, coupled to the agitation shaft of the main vessel through a Teflon friction bearing to prevent leakage. The treatment chamber was connected to a capillary sampling tube that had a solenoid sampling valve activated by a timer. The chamber also had two one-way inverted sense valves in the top and three hermetical ports with o rings in the wall to hold a temperature sensor, a tube connected to a manometer and an automatic injection syringe holder. The top of the chamber was reached by the tip of a sonication horn of an ultrasonic generator held in place by a housing screwed on to the bottom cap of the main vessel. As in the former instrument, during heat treatments, samples could be taken manually (up to 30 samples/min) or continuously (as many as 32 samples/second) with a fraction collector. The fraction collector was fitted with a photocell detector that opened the solenoid sampling valve for the preset time every time a tube passing under the sampling tube was detected.

The temperature stability of the MTS-resistometer during heat treatments was +/- 0.05 ºC and during experiments including ultrasound was +/- 0.2 ºC. The heating-up rate was 10 ºC/min. In the range of 40 to 140 ºC, the instrument stabilized to a temperature 10 ºC higher in approximately 3 min. During MTS treatments the maximum sampling speed used was five samples/second. Reproducibility obtained in five consecutive MS treatments at room temperature, at 200 kPa, and at 117 microns of amplitude was very satisfactory (coefficient of variation lower than 7%).

This instrument enabled to research on the lethal efficacy of ultrasound in different environmental conditions on bacterial spores (Raso et al., 1998b) and on vegetative cells (Pagán et al., 1999b; 1999c), as well as on enzymes (Vercet et al., 1997; Vercet and López, 2002). Further information can be found in the reviews published in *Novel Food Processing Technologies* (Condón et al., 2005) and in *Ultrasound technologies for food and Bioprocessing* (Condón et al., 2010).

Thermoresistometer for TTI Applications

In the last 20-30 years there has been a dramatic change in food safety and food quality control procedures. The Hazard Analysis and Critical Control Points system pointed out that in order to ensure food safety it was necessary the control at the whole food processing chain rather than the control of the processed foods. More efficient methods for process control were required to achieve this goal. Wireless probes were designed in order to record temperature profiles, even in continuous retorts. Biological time-temperature integrators (TTIs) were developed in parallel, so the lethal effect of thermal treatments could be quantified, even those applied in heat exchangers. As already mentioned, one of the drawbacks of these TTIs was to quantify heat penetration during the thermal treatment. In order to solve this inconvenient and to validate their behaviour at a laboratory scale, a modification of the thermoresistometer TR-SC was performed, as depicted in Figure 10.

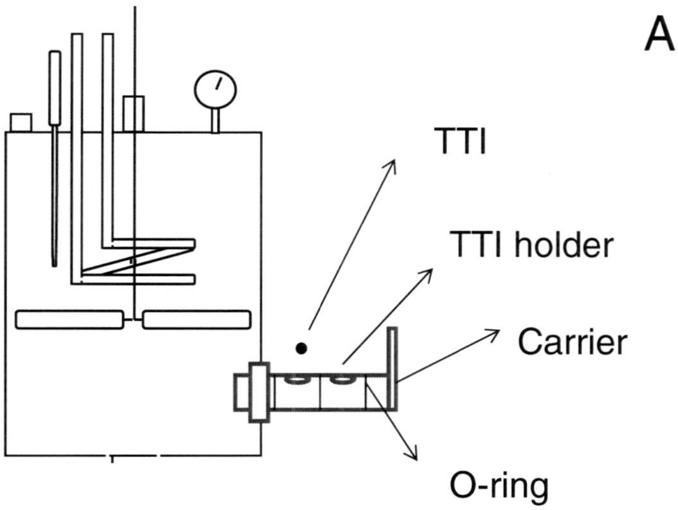

Figure 10. Diagram of the thermoresistometer adapted for TTI treatment.

This equipment, modified to allow for the thermal treatment of TTIs, is similar to the previous devices, with the peculiarity that has several carriers located at the wall of the vessel. Each carrier has holes to hold the TTIs. Carriers are pushed into the vessel when appropriate and, after the thermal treatment has been applied, at preset time intervals, they are pulled from the vessel. TTIs are then broken and the survivors are counted. The design of the sample injection and extraction is similar to that of some previous direct methods previously described. With this equipment, research on the suitability of different microbial species for the development of TTIs, or on the effect of the pH of the heating medium on the behaviour of TTIs, has been possible.

Conclusion

As soon as the importance of microbial inactivation kinetics was revealed, different devices were designed in order to estimate the heat resistance of microorganisms. The thermoresistometers TR-SC and Mastia are very suitable for estimating the heat resistance of microorganisms and can be helpful tools for food processors and researchers in order to estimate the level of safety of the treatments they apply. Not only microbial heat resistance, but also enzyme, nutrient or quality attribute thermal inactivation can be analysed with these instruments. Both are versatile instruments, custom built, in which modifications can be developed in order to focus on specific research areas, such as the application of combined processes including heat.

References

Aiba, S., & Toda, K. L. (1967). Thermal death rate of bacterial spores. *Process Biochemistry,* 36, 38-41.

Alderton, G., Chen, J. K., & Ito, K. A. (1976). Heat resistance of the chemical resistance forms of *Clostridium botulinum* 62A spores over the water activity range 0 to 0.9. *Applied and Environmental Microbiology,* 40, 511-515.

Arroyo, C., Condón, S., Pagán, R. (2009). Thermobacteriological characterization of *Enterobacter sakazakii. International Journal of Food Microbiology,* 136 (1), 110-118.

Bigelow, W. D., & Esty. J. R. (1920). Thermal death point in relation to time of typical thermophilic organisms. *Journal of Infectious Diseases* 27, 602-617.

Blasco, R., Esteve, M. J., Frigola, A., & Rodrigo, M. (2004). Ascorbic acid degradation kinetics in mushrooms in a high-temperature short-time process controlled by a thermoresistometer. *LWT-Food Science and Technology* 37, 171-175.

Blocher, J. C., & Busta, F. F. (1983). Bacterial spore resistance to acid. *Food Technology, 37,* 87-99.

Brown , K. L., & Thorpe, R.H. (1978) The effect of pH on the heat resistance and recovery of bacterial spores. CFPRA. *Technical Memorandum* 185. Campden Food Preservation Research Association, Chipping, Campden, Glos. England.

Brown, K.L., Pitcher, J.E., & Mottishaw, J. (1981). Microbiological aspects of aseptic packaging. *Technical Memorandum* 281. Campden Food Preservation Research Association, Chipping, Campden, Glos. England.

Brown, K. L., & Ayres, C. A. (1982). Thermobacteriology of UHT processed foods. In R. Davies (Ed.), *Developments in food microbiology.* London: Applied Science Publishers.

Brown, K. L., Ayres, C. A., Gaze, J. E. & Newman, M. E. (1984). Thermal destruction of bacterial spores immobilized in food alginate particles. *Food Microbiology,* 1, 187-198.

Brown, K. L., Gaze, J. E., McClement, R. H., & Withers, P. (1988). Construction of a computer-controlled thermoresistometer for the determination of the heat resistance of bacterial spores over the temperature range 100 to 150°C. *International Journal of Food Science and Technology*, 23 (4), 361-371.

Burgos, J., Ordoñez, J. A., & Sala, F. (1972). Effect of ultrasonic waves on the heat resistance of *Bacillus cereus* and *Bacillus licheniformis* spores. *Applied Microbiology* 24, 497-498.

Burton, H., & Perkin, A. G. (1970). Comparison of milks processed by the direct and indirect methods of ultra-high-temperature sterilization. I. An experimental ultra-high-temperature sterilizer and its characteristics. *Journal of Dairy Research* 37, 209-218

Burton, H., Perkins, A. G., Davies, F. L., & Underwood, H. H. (1977). Thermal death kinetics of *Bacillus stearothermophilus* spores at ultra-high temperatures. III. Relationship between data from capillary tube experiments and from UHT sterilizers. *Journal of Food Technology,* 12, 149-161

Busta, F. F., & Ordal, Z. J. (1964). Heat activation kinetics of endospores of *Bacillus subtilis.* Journal of Food Science, 29, 345-348.

Busta, F. F. (1967). Thermal inactivation characteristics of bacterial spores at ultra high temperatures. *Applied Microbiology,* 15, 640-645.

Cerf. O. (1987). Revue bibliographique: caractérisation de la thermorésistance des spores bactériennes pour l'optimisation des traitments UHT. *Lait*, 67 (1), 97-109.

Cerny, G. (1980). Dependence of thermal inactivation of microorganisms on the pH-value of media. II. Bacteria and bacterial spores. *Zeitschrift fuer Lebensmittel Untersuchung und Forschung,* 170, 180-186.

Cole, M. B., & Jones, M. V. (1990). A submerged-coil heating apparatus for investigating thermal inactivation of microorgansims. *Lettes in Applied Microbiology,* 11, 233-235.

Condón, S., López, P., Oria, R., & Sala, F. J. (1989). Thermal death determination: design and evaluation of a thermoresistometer. *Journal of Food Science,* 54 (2), 451-457.

Condón, S., García, M. L., Otero, A., & Sala, F. J. (1992). Effect of culture age, pre-incubation at low temperature and pH on the thermal resistance of *Aeromonas hydrophila*. *Journal of Applied Bacteriology,* 72, 322-326.

Condón, S., & Sala, F. J. (1992). Heat resistance of *Bacillus subtilis* in buffer and foods of different pHs. *Journal of Food Protection,* 55(8), 605-608.

Condón, S., Arrizubieta, M. J., & Sala, F. J. (1993). Microbial heat resistance determinations by the multipoint system with the thermoresistometer TR-SC. *Journal of Microbiological Methods*, 18 (4), 357-366.

Condón, S., Raso, J., & Pagán, R. (2005). Microbial inactivation by ultrasound. In G. V. Barbosa-Cánovas, M. S. Tapia & P. Cano (Eds.), *Novel Food Processing Technology*. Boca Raton (Fl): CRC Press.

Condón, S., Mañas, P., & Cebrian, G. (2010). Manothermosonication for microbial inactivation. In H. Feng & J. Weiss (Eds.), *Ultrasound technologies for food and bioprocessing*. New York: Springer.

Conesa, R., Periago, P. M., Esnoz, A., López, A., & Palop, A. (2003). Prediction of *Bacillus subtilis* spore survival after a combined non-isothermal isothermal heat treatment. *European Food Research and Technology,* 217 (4), 319-324.

Conesa, R., Andreu, S., Fernández, P. S., Esnoz, A., & Palop, A. (2009). Non-isothermal heat resistance determinations with the thermoresistometer Mastia. *Journal of Applied Microbiology,* 107, 506-513.

Cook, A. M., & Gilbert, R. J. (1968). Factors affecting the heat resistance of *Bacillus stearothermophilus* spores. II. The effect of sporulating conditions and nature of the heating medium. *Journal of Food Technology,* 3, 295-302.

Dallyn, H., Falloon, W. C., & Bean, P. G. (1977). Method for the immobilization of bacterial spores in alginate gel. *Laboratory Practice*, 26, 773-775.

Daudin, J. D., & Cerf, O. (1977). Influence des chocs thermiques sur la destruction des spores bacteriennes par la chaleur. *Lebensmittel Wissenschaft und Technologie* 10, 203-207.

David, J. R. D., & Merson, R. L. (1990). Kinetic parameters for inactivation of *Bacillus stearothermophilus* at high temperatures. *Journal of Food Science*, 55 (2), 488-493.

Davies, F. L. (1975). Heat resistance of *Bacillus* species. *Journal of the Society of Dairy Technology*, 28 (2), 69-78.

De Cordt. S., Hendrickx, M., Maesmans, G., & Tobback, P. (1992). Immobilized alfa-amylase from *Bacillus licheniformis*: a potential enzymatic time-temperature integrator for thermal processing. *International Journal of Food Science and Technology*, 27(6), 661-673.

Edwards, J. L., Busta, F. F., & Speck, M. L. (1965). Heat injury of *Bacillus subtilis* spores at ultrahigh temperatures. *Applied Microbiology,* 13, 858-864.

Espina, L., Somolinos, M., Pagán, R., & García-Gonzalo, D. (2010). Effect of citral on the thermal inactivation of *Escherichia coli* O157:H7 in citrate phosphate buffer and apple juice. *Journal of Food Protection,* 73, 2189-2196.

Esty, J. R., & Willians, C. C. (1924). Heat resistance studies. I A new method for the determination of heat resistance of bacterial spores. *Journal of Infectious Diseases,* 34, 516-528.

Fernández, A., Ocio, M. J., Fernández, P. S., Rodrigo, M., & Martínez, A. (1999). Application of nonlinear regression analysis to the estimation of kinetic parameters for two enterotoxic strains of *Bacillus cereus* spores. *Food Microbiology,* 16 (6), 607-613.

Fernández, A., Alvarez-Ordóñez, A., López, M., Bernardo, A. (2009). Effects of organic acids on thermal inactivation of acid and cold stressed *Enterococcus faecium. Food Microbiology,* 26 (5), 497-503.

Fernández, P. S., Ocio, M. J., Sánchez, T., & Martínez, A. (1994). Thermal resistance of *Bacillus stearothermophilus* spores heated in acidified mushroom extract. *Journal of Food Protection,* 57 (1), 37-41.

Gaillard, S., Leguerinel, I., & Mafart, P. (1998). Model for combined effects of temperature, pH and water activity of thermal inactivation of *Bacillus cereus* spores. *Journal of Food Science,* 63, 887-889.

Garza, S., Teixido, J. A., Sanchís, V., Viñas, I., & Condón, S. (1994). Heat-resistance of *Saccharomyces cerevisiae* strains isolated from spoiled peach puree. *International Journal of Food Microbiology,* 23 (2), 209-213.

Goldoni, J. S., Kogima, S., Leonard, S., & Meil, J. R. (1980). Growing spores of P. A. 3679 in formulations of beef heart infusion broth. *Journal of Food Science,* 45, 467-470, 475.

González, I., López, M., Mazas, M., González, J., & Bernardo, A. (1995). The effect of recovery conditions on the apparent heat resistance of *Bacillus cereus* spores. *Journal of Applied Bacteriology,* 78 (5), 548-554.

Hassani, M., Álvarez, I, Raso, J., Condón, S., & Pagán, R. (2005). Predicting heat inactivation of *Listeria monocytogenes* under non-isothermal treatments. *Journal of Food Protection,* 68 (4), 736-743.

Hassani, M., Cebrián, G., Mañas, P., Condón, S., & Pagán, R. (2006). Induced thermotolerance under nonisothermal treatments of heat sensitive and resistant strain of *Staphylococcus aureus* in media of different pH. *Letters in Applied Microbiology,* 43, 619-624.

Hassani, M., Condón, S., & Pagán, R. (2007). Predicting microbial heat inactivation under nonisothermal treatments. *Journal of Food Protection,* 70 (6), 1457-1467.

Hassani, M., Lázaro, R., Pérez, C., Condón, S., & Pagán, R. (2008). Thermostability of oxytetracycline, tetracycline, and doxycycline at ultrahigh temperatures. *Journal of Agricultural and Food Chemistry,* 56 (8), 2676-2680.

Hunter, G. M. (1972). Continuous sterilization of liquid media containing suspended particles. *Food Technology Australia,* 24, 158-165.

Kooiman, W. J., & Geers, J. M. (1975). Simple and accurate technique for determination of heat resistance of bacterial spores. *Journal of Applied Bacteriology*, 38 (2), 185-189.

Leguerinel, I., & Mafart, P. (2001). Modelling the influence of pH and organic acid types on thermal inactivation of *Bacillus cereus* spores. *International Journal of Food Microbiology*, 63, 29-34.

Lekogo, B. M., Coroller, L., Mathot, A. G., Mafart, P., & Leguerinel, I. (2010). Modelling the influence of palmitic, palmitoleic, stearic and oleic acids on apparent heat resistance of spores of *Bacillus cereus* NTCC 11145 and *Clostridium sporogenes* Pasteur 79.3. *International Journal of Food Microbiology*, 141, 242-247.

Levine, S. (1956). Determination of the thermal death rate of bacteria. *Food Research*, 21, 295-301.

López, P., Sánchez, A. C., Vercet, A., Burgos, J. (1997). Thermal resistance of tomato polygalacturonase and pectinmethylesterase at physiological pH. *Zeitschrift fur Lebensmittel Untersuchung und Forschung A,* 204 (2), 146-150.

Loss, C. R., & Hotchkiss, J. H. (2004). Determination of thermal inactivation kinetics of microorganisms with a continuous microflow apparatus. *Journal of Food Protection,* 67 (11), 2560-2564.

Mañas, P., Pagán, R., Raso, J., & Condón, S. (2003). Predicting thermal inactivation in media of different pH of *Salmonella* grown at different temperatures. *International Journal of Food Microbiology,* 87, 45-53.

Menegazzi, G. S., & Ingledew, W. M. (1980). Heat processing of spent brewer's yeast. *Journal of Food Science,* 45, 182-186.

Mikolajcik, E. M., & Rajkowsky, K. T. (1980). Simple technique to determine heat resistance of *Bacillus stearothermophilus* spores in fluid systems. *Journal of Food Protection*, 43 (10), 799-804.

Montville, T. J., & Sapers, G. M. (1981). Thermal resistance of spores from pH elevating strains of *Bacillus licheniformis*. *Journal of Food Science,* 46, 1710-1712.

Murrell, G. W. (1964). Heat resistance of microorganisms. A discussion on heat resistance in relation to bacteriological aspects of heat sterilisation. *Australian Journal of Pharmacology*, 30, S40-S46.

National Canners Association Research Laboratory. (1968). *Laboratory Manual for Food Canners and Processors. Vol 1.* Wesport, Connecticut: The AVI Publishing Company, Inc.

Ocio, M. J., Sánchez, T., Fernández, P. S., Rodrigo, M., & Martínez, A. (1994). Thermal resistance characteristics of PA3679 in the temperature range of 110-121ºC as affected by pH, type of acidulant and substrate. *International Journal of Food Microbiology,* 22, 239-247.

Ocio. M. J., Fernández, P. S., Rodrigo, M., Periago, P., & Martínez, A. (1997). A time temperature integrator for particulated foods: thermal process evaluation. *Zeitschrift fur Lebensmittel-Untersuchung und-Forschung A*, 205, 325-328.

Odlaug, T. C., & Pflug, I. J. (1977). Thermal destruction of *Clostridium botulinum* spores suspended in tomato juice in aluminum thermal death time tubes. *Applied and Environmental Microbiology,* 34, 23-29.

Ordoñez, J. A., Aguilera, M. A., García, M. L., & Sanz, B. (1987). Effect of combined ultrasonic and heat treatment (thermoultrasonication) on the survival of a strain of Staphylococcus aureus. *Journal of Dairy Research,* 54, 61-67.

Pagán, R., Condón, S., & Sala, F. J. (1997). Effect of several factors on the heat-shock-induced thermotolerance of *Listeria monocytogenes*. *Applied and Environmental Microbiology*, 63: 3225.

Pagán, R., Alvarez, I., Mañas, P., & Sala, F.J. (1998) Heat resistance in different heating media of *Listeria monocytogenes* ATCC 15313 grown at different temperatures. *Journal of Food Safety,* 18, 205-219.

Pagán, R., Mañas, P., Raso, J., & Sala, F. J. (1999a). Heat resistance of *Yersinia enterocolitica* grown at different temperatures and heated in different media. *International Journal of Food Microbiology*, 47 (1-2), 59-66.

Pagán, R., Mañas, P., Álvarez, I., & Condón, S. (1999b). Resistance of *Listeria monocytogenes* to ultrasonic waves under pressure at sublethal (Manosonication) and lethal (Manothermosonication) temperatures. *Food Microbiology,* 16, 139-148.

Pagán, R., Mañas, P., Palop, A., & Sala, F.J. (1999c). Resistance of heat-shocked cells of *Listeria monocytogenes* to mano-sonication and mano-termo-sonication. *Letters in Applied Microbiology, 28,* 71-75

Palop, A., Raso, J., Condón, S., & Sala, F. J. (1996a). Heat resistance of *Bacillus subtilis* and *Bacillus coagulans*: effect of sporulation temperature in foods with various acidulants. *Journal of Food Protection*, 59 (5), 487-492.

Palop, A., Raso, J., Pagán, R., Condón, S., & Sala, F. J. (1996b). Influence of pH on heat resistance of *Bacillus licheniformis* in buffer and homogenised foods. *International Journal of Food Microbiology*, 29, 1-10.

Palop, A., Sala, F. J., & Condón, S. (1997). Occurrence of a highly heat-sensitive spore subpopulation of *Bacillus coagulans* STCC 4522 and its conversion to a more heat-stable form. *Applied and Environmental Microbiology*, 63, 2246-2251.

Palop, A., Raso, J., Pagán, R., Condón, S., & Sala, F. J. (1999). Influence of pH on heat resistance of *Bacillus licheniformis* in buffer and homogenized foods. *International Journal of Food Microbiology*, 46, 243-249.

Peleg, M., & Cole, M. B. (1998). Reinterpretation of microbial survival curves. *Critical Reviews in Food Science.* 38, 353-380.

Periago, P. M., Leontidis, S., Fernández, P. S., Rodrigo, C., & Martínez, A. (1998). Note. Kinetic parameters of *Bacillus stearothermophilus* spores under isothermal and non-isothermal heating conditions. *Food Science and Technology International*, 4 (6), 443-447.

Periago, P. M., Abee, T., & Wouters, J.A. (2002a). Analysis of the heat-adaptative response of psychrotrophic *Bacillus weihenstephanensis*. *International Journal of Food Microbiology,* 79, 17-26.

Periago, P. M., van Schaik, W., Abee, T., & Wouters, J. A. (2002b). Identification of proteins involved in the heat stress response of *Bacillus cereus* ATCC 14579. *Applied and Environmental Microbiology,* 68, 3486-3495.

Perkin, A. G., Burton, H., Underwood, H. M., & Davies, F. L. (1977). Thermal death kinetics of Bacillus stearothermophilus spores at ultra high temperatures. Journal *of Food Technology,* 12, 131-148.

Pflug, I. J., & Esselen, W. B. (1953). Development and application of an apparatus for study of thermal resistance of bacterial spores and thiamine at temperatures above 250 degrees F. *Food Technology*, 7 (6), 237-241.

Pflug, I. J., & Odlaug, T. E. (1978). A review of Z and F values used to ensure the safety of low-acid canned food. *Food Technology*, 32 (6), 63-70.

Raso, J., Pagán, R., Condón, S., & Sala, F. J. (1998a). Influence of temperature and pressure on the lethality of ultrasound. *Applied and Environmental Microbiology*, 64 (2), 465-471.

Raso, J., Palop, A., Pagán, R., & Condón, S. (1998b). Inactivation of *Bacillus subtilis* spores by combining ultrasonic waves under pressure and mild heat treatment. *Journal of Applied Microbiology*, 85, 849-854.

Reichart, O. (1979). A new experimental method for the determination on the heat destruction parameters of microorganism. *Acta Alimentaria*, 8 (2), 131-155.

Reichart, O. (1983). Experimental method for the determination of the thermal death parameters of microorganisms in a continuous system . *Acta Alimentaria*, 12(1), 35.

Rodrigo, C., Mateu, A., Alvarruiz, A., Chinesta, F., & Rodrigo, M. (1998). Kinetic parameters for thermal degradation of green asparagus texture by unsteady-state method. *Journal of Food Science*, 63 (1), 126-129.

Rodrigo, M., Martínez, A., Sánchez, T., Peris, M. J., & Safón, J. (1993). Kinetics of *Clostridium sporogenes* PA3679 spore destruction using computer-controlled thermoresistometer. *Journal of Food Science*, 58 (3), 649-652.

Sagarzazu, N., Cebrián, G., Pagán, R., Condón, S., & Mañas, P. (2010). Resistance of *Campylobacter jejuni* to heat and pulsed electric fields. *Innovative Food Science and Emerging Technologies*, 136 (1), 110-118.

Sala, F.J., Ibarz, P., Palop, A., Raso, J., & Condón, S. (1995). Sporulation temperature and heat resistance of *Bacillus subtilis* at different pH's. *Journal of Food Protection*, 58, 239-243.

Sánchez, L., Peiro, J. M., Castillo, H., Pérez, M. D., Ena, J. M., & Calvo, M. (1992). Kinetic parameters for denaturation of bovine-milk lactoferrin. *Journal of Food Science*, 57 (4), 873-879.

Schmidt, C. F. (1957). Thermal resistance of microorganisms. In G. F. Reddish (Ed.), *Antiseptics, Disinfectants, Fungicides and Sterilization* (2nd ed.). Philadelphia: Lea & Febiger.

Silla, M. H., Torres, J., Arranz, A., & Peris, M. J. (1993). Citric acid lowers heat resistance of *Clostridium sporogenes* PA 3679 in HTST white asparagus purée. *International Journal of Food Science*, 28, 603-610.

Stern, J. A., & Proctor, B. E. (1954). A micro-method and apparatus for the multiple determination of rates of destruction of bacteria and bacterial spores subjected to heat. *Food Technology*, 8 (3), 139-143.

Stumbo, C. R., Gross, C. E., & Vinton, C. (1945). Bacteriological studies relating to thermal processing of canned meat. *Food Research*, 10, 260-272.

Stumbo, C. R. (1948). A technique for studying resistance of bacterial spores to temperatures in the higher range. *Food Technology*, 2 (3), 228-240.

Stumbo, C. R. (1973). *Thermobacteriology in food processing*. (2nd ed.). New York: Academic Press.

Townsend, C. T., Esty, J. R., & Baselt, F. C. (1938). Heat resistance in relation to the determination of safe processes for canned foods. *Food Research*, 3, 323-326.

Valdramidis, V. P., Geeraerd, A. H., Bernaerts, K., & Van Impe, J. F. (2006). Microbial dynamics versus mathematical model dynamics: the case of microbial heat resistance induction. *Innovative Food Science and Emerging Technologies*, 7, 80-87.

Valdramidis, V. P., Geeraerd, A. H., & Van Impe, J. F. (2007). Stress-adaptative responses by heat under the microscope of predictive microbiology. *Journal of Applied Microbiology,* 103, 1922-1930.

van Derlinden, E., Lule, I., Bernaerts, K., & Van Impe, J. F. (2010). Quantifying the heterogeneous heat response of *Escherichia coli* under dynamic temperatures. *Journal of Applied Microbiology,* 108, 1123-1135.

van Zuijlen, A., Periago P. M., Amézquita, A., Palop, A., Brul, S., & Fernández, P. S. (2010). Characterization of *Bacillus sporothermodurans* IC4 spores; putative indicator microorganism for optimization of thermal processes in food sterilization. *Food Research International,* 43, 1895-1901.

Vercet, A., López, P., & Burgos, J. (1997). Inactivation of heat-resistant lipase and protease from *Pseudomonas fluorescens* by manothermosonication. *Journal of Dairy Science,* 80, 29-36.

Vercet, A, & López, P. (2002). The effects of manothermosonication on tomato endopolygalacturonase beta-subunit. *European Food Research and Technology,* 214 (4), 310-312.

Westhoff, D. C., & Dougherty, S. L. (1981). Characterization of *Bacillus* species isolated from spoiled ultrahigh temperature processed milk. *Journal of Dairy Science,* 64, 572-580.

Williams, C. C., Merrill, C. H., & Cameron, C. J. (1937). Apparatus for determination of spore-destruction rates. *Food Research,* 2, 369-375.

Yura, T. Kanemori, M., & Morita, M.Y. (2000). The heat-shock response: regulation and function. In G. Storz & R. Hengge-Aronis (Eds.), *Bacterial stress responses*. (pp. 3-18). Washington DC: ASM Press.

In: Progress on Quantitative Approaches in Thermal ... ISBN: 978-1-62100-842-2
Editors: V. P. Valdramidis and J. F. M. Van Impe © 2012 Nova Science Publishers, Inc.

Chapter 3

(OPTIMAL) EXPERIMENTAL DESIGN FOR MICROBIAL INACTIVATION

Eva Van Derlinden [1,*], ***Eva Balsa-Canto*** [2,†] ***and Jan F. Van Impe*** [1,‡]
[1]BioTeC - Chemical and Biochemical Process Technology and Control,
Department of Chemical Engineering,
Katholieke Universiteit Leuven, Leuven, Belgium
[2]Process Engineering Group IIM-CSIC
(Spanish Council for Scientific Research), Vigo, Spain

Abstract

Mathematical model building encloses three basic steps: (*i*) identification of the system (explanatory and response variables), (*ii*) model structure selection, and (*iii*) parameter estimation. Both the selection of the explanatory and response variables and the model structure, as well as the model parameter estimation are data-driven processes.

During *parameter estimation*, model parameter values are modified such that the difference between model simulations and experimental measurements is as small as possible. This is obtained by minimizing a cost function that quantifies the distance between the models output and experimental data. Different cost functions as well as methods to quantify parameter uncertainty are listed in this chapter.

Design of experiments (DOE) is an experimental approach that enables to determine the relation between (environmental) factors, their interactions and statistical properties. In the domain of predictive microbiology, the technique of DOE is mostly used: (*i*) to scan an extended region for which a probabilistic model is built, (*ii*) to collect data in an extended region to build response surface models, and (*iii*) to collect data in a specified region to obtain accurate and reliable parameter estimates of existing models. In this chapter, frequently used design schemes are discussed, i.e., full factorial design, fractional factorial design, Box-Behnken design and central composite design. Examples of implementations are given.

Optimal experiment design for model structure discrimination (OED/MD) selects one or more (dynamic) experiments that improve the discrimination between competitive model structures. A search for the correct model structure can help to identify the

*E-mail address: eva.vanderlinden@cit.kuleuven.be
†E-mail address: ebalsa@iim.csic.es
‡E-mail address: jan.vanimpe@cit.kuleuven.be

underlying mechanisms of the studied system, as many hypotheses often exist. This chapter explains the basic concepts of OED/MD and lists some criteria that can be used for experiment design.

Optimal experiment design for parameter estimation (OED/PE) is a mathematical technique that enables to pick a small set of highly informative, static and/or dynamic experiments, resulting in unique and accurate parameter estimates. OED/PE assumes that the model structure taken is valid. When applying dynamic experiments, this approach also guarantees parameter estimates that are valid under varying, more realistic conditions. This section not only deals with the mathematical background of the OED/PE technique, i.e., the Fisher information matrix and the optimization criteria, it also discusses mathematical solution methods and implementations within the domain of predictive microbiology.

1. Introduction

Mathematical modeling is an essential tool for optimal operation and control of bioprocesses. Following the general model building procedure, the development of a mathematical model encloses three basic steps (Asprey and Macchietto, 2000; Tarantola, 2005). Initially, relevant a priori knowledge is collected and a set of independent variables (e.g., physical quantities) is determined which defines the investigated system (parameterization of the system). Next, one or more mathematical model structures are proposed that are able to predict the systems' behavior (forward modeling). A precondition is that the models are structurally identifiable, i.e., all model parameters can be identified uniquely. In the third step, the predictive quality of the models is improved by adjusting the intrinsic model parameters to the actual measurements (parameter estimation). Subsequently, the validity of the model structure and the parameter estimates is tested against new experimental data. When the descriptive quality of the model is insufficient, an adaptation of the model structure and/or parameter estimates is needed. As such, model structure selection and parameter estimation are combined in an iterative model building cycle. Finally, a model is obtained that agrees with reality as closely as possible.

Both the selection of an appropriate model structure and the idenfication of accurate model parameters are data-driven processes, i.e., the efficiency and accuracy of these procedures are determined by the quality of the experimental data. Experimental data should enclose sufficient information with respect to the studied system. As a result, a well-thought experimental scheme can significantly improve model building and parameter estimation.

Different mathematical techniques/approaches exist that can increase and/or optimize the information contained in a (series of) experiment(s). (*i*) **Design of experiments** (DOE) is a statistically funded approach that enables to determine the relation between (environmental) factors, their interactions and statistical properties. Instead of studying the system at single levels for one environmental factor at a time, more information can be collected in a full factorial experiment design, in which all combinations of the selected discrete levels of the different explanatory variables are considered. The number of experiments can be further reduced by a well-thought selection of these levels that unravel the most information about the studied system. (*ii*) Via **optimal experiment design for model structure discrimination** (OED/MD), one or more (dynamic) experiments can be designed to distinguish between possible model structures. In this approach, experiments are designed such that the difference in the output of the various models is maximized. (*iii*) Presuming model validity, the mathematical technique of **optimal experiment design for parameter estimation** (OED/PE) forms an excellent starting point for the selection of a small set of highly informative, static and/or dynamic experiments, resulting in unique and accurate parameter estimates. When applying dynamic experiments, this approach also guarantees parameter estimates which are valid under varying, more realistic conditions.

DOE and OED are frequently used techniques in bioprocess modeling to (*i*) determine relations with environmental factors, and (*ii*) improve the parameter estimation cycle. This chapter starts with a review on the background of parameter estimation, and techniques to evaluate the parameter estimation uncertainty (Section 2.). The following sections discuss DOE (Section 3.), OED/MD (Section 4.1.) and OED/PE (Section 4.2.) in more detail.

2. Parameter Estimation

During parameter estimation, model parameter values are modified such that the difference between model simulations and experimental measurements is as small as possible. This is obtained by minimizing a cost function that quantifies the distance between the model output and experimental data. Different cost functions exist.

The most commonly used parameter estimation criterion is a quadratic cost function. Taking an experiment with duration t_f and continuous measurements, the identification function reads as follows

$$J(\mathbf{p}) = \int_0^{t_f} (\mathbf{y}(\mathbf{p},t) - \mathbf{y}_{exp}(t))^T \, \mathbf{Q} \, (\mathbf{y}(\mathbf{p},t) - \mathbf{y}_{exp}(t)) \, dt \tag{1}$$

and for n_t discrete measurements

$$J(\mathbf{p}) = \sum_{i=1}^{n_t} (\mathbf{y}(\mathbf{p},t_i) - \mathbf{y}_{exp}(t_i))^T \, \mathbf{Q}_i \, (\mathbf{y}(\mathbf{p},t_i) - \mathbf{y}_{exp}(t_i)) \tag{2}$$

with $\mathbf{y}_{exp}$ the experimental data, $\mathbf{y}$ the model prediction for the observed quantities, $\mathbf{p}$ the model parameters, t the time with t_i a particular sampling time corresponding n_t, the number of samples, and $\mathbf{Q}$ the weighting matrix.

This cost function calculates the cumulative sum of the quadratic differences between the experimental data ($\mathbf{y}_{exp}$) and the model simulations ($\mathbf{y}(\mathbf{p})$). Weighting is introduced via the matrix $\mathbf{Q}$ which is typically reduced to the inverse of the measurement error variance covariance matrix. When measurements are uncorrelated, the quadratic cost function becomes the so-called *weighted sum of squared errors*. Equation 1 and Equation 2 reduce to the *sum of squared errors* (SSE) when $\mathbf{Q}$ is set equal to the unity matrix. In this case, the errors on the experimental data ($\mathbf{y}_{exp}$) have the same order of magnitude and no weighting is necessary. Use of the SSE criterion is allowed for additive, independent, uncorrelated and normally distributed noise with zero mean and a constant variance. This variance can be approximated by the *mean sum of squared errors* (MSE), i.e.,

$$MSE = \frac{SSE}{n_t - n_p} \tag{3}$$

with n_t and n_p the number of experimental data points and model parameters, respectively. As the MSE is a measure for the average difference between experimental data and model simulations, it is also a criterion that can be used to evaluate the quality of fit. The minimization of the SSE is the most commonly and widely used criterion for parameter estimation as it is the most easy to implement, i.e., no a priori knowledge is required.

Other parameter estimation criteria are: (i) least modulus, (ii) maximum likelihood and (iii) bayesian parameter estimation. In the least modulus approach, a weighted sum of absolute errors is minimized. In maximum likelihood, parameter values are determined that maximize a known likelihood distribution. As such, maximum likelihood can only be solved when the measurement error distribution is known. If the errors on the model simulations are normally distributed with a known variance, the maximum likelihood approach reduces to the (weighted) sum of squared errors (Walter and Pronzato, 1997). The Bayesian parameter estimation approach assumes knowledge of the probability distribution of the parameter vector **p** and of the experimental measurements with respect to the parameter vector **p**.

Parameter estimation uncertainty. Parameter estimates are always characterized by a certain degree of uncertainty arising from (i) the variability of the modeled system, (ii) the noise on the experimental measurements and (iii) the estimation error. Various methods exist to quantify parameter uncertainty.

The quality of a parameter estimate, or vice versa, its uncertainty, can be quantified by its variance or standard deviation, which is a measure for the spreading of the parameter distribution. Parameter variances can be extracted from the main diagonal of the variance covariance matrix **P**

$$\mathbf{P} = E[(\mathbf{p}^* - \hat{\mathbf{p}})(\mathbf{p}^* - \hat{\mathbf{p}})^T] \tag{4}$$

with $\mathbf{p}^*$ and $\hat{\mathbf{p}}$ the true and estimated parameter vector. The variance covariance matrix **P** can be approximated by the inverse of the Fisher information matrix (see further).

The variances on the main diagonal of **P** can be used to determine the (1-α) 100% confidence interval for each parameter estimate

$$\left[\hat{p}_i - t_{(1-\frac{\alpha}{2}, n_t - n_p)}\sqrt{(s_i^2)},\ \hat{p}_i + t_{(1-\frac{\alpha}{2}, n_t - n_p)}\sqrt{(s_i^2)}\right] \tag{5}$$

with t the student value for $n_t - n_p$ degrees of freedom, (1-α) the confidence level, n_t the number of experimental data points and n_p the number of estimated parameters.

The remaining off-diagonal elements of **P** are the covariances which can be used to calculate the correlation coefficients.

$$-1 \leq \frac{\mathbf{P}_{ij}^{-1}}{\sqrt{\mathbf{P}_{ii}^{-1}}\sqrt{\mathbf{P}_{jj}^{-1}}} \leq 1 \tag{6}$$

If the correlation coefficient is equal to zero, parameters are uncorrelated, i.e., their estimates are independent. However, the more the value tends towards 1 or -1, the more the parameters are correlated. In the latter case, the effect of varying one parameter can be neutralized by a change in the other parameter.

Joint confidence regions. For simultaneously determined parameter estimates, the confidence intervals as calculated with Equation 5 are only valid for independent, uncorrelated parameter estimates. If these confidence intervals are combined, a hypercube is formed which, most often, does not correspond with the real joint confidence region. Hereto, joint confidence regions for simultaneously estimated parameters are calculated following

$$SSE(\mathbf{p}) \leq SSE(\hat{\mathbf{p}}) \left[1 + \frac{n_p}{n_t - n_p} F_{(n_p, n_t - n_p, 1-\alpha)}\right] \quad (7)$$

with SSE(**p**) the sum of squared errors with respect to the parameter vector **p**, SSE($\hat{\mathbf{p}}$) the least sum of squared errors, i.e., with respect to the best parameter estimates $\hat{\mathbf{p}}$, and $F(1-\alpha, n_p, n_t - n_p, 1-\alpha)$ the associated value from the F-distribution. For a two parameter estimation problem, circular confidence regions point at uncorrelated parameter estimates. Ellipsoids with a positive or negative slope correspond with correlated parameters: if p_1 increases/decreases, the second parameter also increases/decreases. As this method is based on the F-test, it is only valid for linear models or models with a valid linear approximation. Furthermore, the real form of the confidence region can deviate from the symmetrical ellipsoid and the confidence level (1-α) is only an approximation of the real probability level (Bates and Watts, 1988).

Monte Carlo simulations. This approach is based on the knowledge that when experiments are repeated under identical conditions, small differences in the experimental data sets can be observed due to variability of the system and noise on the experimental measurements. As a result, derived parameter vectors are slightly different. Based on the estimated parameters, new fictitious data sets are generated via model simulation with the addition of noise. New parameter estimates are identified from these data sets. To obtain a correct parameter distribution, a large number of fictitious data sets has to be generated making the Monte Carlo approach computationally intensive. Furthermore, correct quantification of the measurement noise is needed.

More information on parameter estimation can be found in Chapter 7 which focuses specifically on the case study of parameter identification under continuously changing temperature conditions.

2.1. Parameter Estimation and Experiment Design

In general, the uncertainty of a model is determined by the uncertainty of the model structure and the uncertainty of model input(s) and model parameter(s), which arises from the natural variability of the biological system and the lack of knowledge of the true underlying mechanisms. Once an acceptable and valid model structure is defined, the remaining objective is the accurate and unique identification of model parameters. Parameter uncertainty arises, e.g., from the intrinsic variability of the explored system and the parameter estimation error. Note that, due to the complexity of microbial behavior, only the intrinsic and extrinsic factors which are most influential are selected to describe the system. The error resulting from this simplification is referred to by Baranyi and Roberts (1995) as the *completeness error*.

Unique and accurate estimation of model parameters based on given experimental data is often not straightforward due to (one or a combination of) the following reasons: (1) a too small amplitude of system output sensitivities with respect to the model parameters, (2) a high model sensitivity to experimental noise, (3) model parameter correlation, (4) measurements with limited accuracy and/or small measurement frequency, (5) a lack of measurements for certain (biologically important) state variables and (6) the fact that model and reality are not perfectly corresponding since the model omits variables in order to control model complexity (e.g., Bernaerts and Van Impe, 2004; Pinto, 1998). A well-thougth experimental design can increase the information content such that unique and accurate parameter estimation is obtained. In the following sections, two approaches are discussed which yield a statistically based set of experiments that can significantly improve the accuracy, reliability and identifiability of models and their parameters, i.e., (1) *design of experiments*, and (2) *optimal experiment design for parameter estimation*.

3. Design of Experiments

When building a model, information must be gathered about the effect of one or more explanatory variables on the considered, independent response variable(s). A random, unstructured data collection approach might yield experimental data with insufficient information to unravel all underlying interactions. However, well-founded experimental schemes, developed following a *design of experiments* approach, generally result in informative data sets that are effective for model building.

The general formulation *design of experiments* includes different-statistically inspired-approaches that result in the structured and controlled collection of information with respect to one or more variables, their variability and interactions. Correct analysis of the experimental results will unravel which variables mainly control the studied system and how these variables interact. Starting from explanatory factors that will be studied at a set of constant levels, the basic idea behind DOE is to hold certain factors constant and to alter the levels of one or more other variables. By selecting certain experimental conditions, the extraction of useful information is guaranteed. It is important to note that DOE relates to (a combination of) static conditions.

The technique of DOE can be of interest during different stages of the experimental process (Eriksson et al., 2008):

- *Familiarization:* first analysis of a system's response, e.g., at the start of a new research.
- *Screening:* analysis of multiple factors to identify which have the greatest effect on the process outcome.
- *Finding optimal regions:* reducing the experimental region to the most interesting ranges.
- *Optimization:* accurate identification of a minimum or a maximum in the system's response.
- *Mechanistic model building.*

When starting the experimental procedure, the most important design questions are: (1) what is the experimental objective, (2) which variables are selected and what are the intervals/levels taken, (3) which responses are measured, (4) what kind of model is considered, and (5) which experimental scheme is selected? Also, the choice of the DOE approach depends on (*i*) the objectives of the model and experiments, (*ii*) the number of factors to be investigated, and (*iii*) the maximum number of experiments allowed.

In the domain of predictive microbiology, the technique of DOE is mostly used for the following purposes. (*i*) To scan an extended region for which a probabilistic model is built, e.g., a model that quantifies the chance of survival and/or growth of pathogens or spoilage microorgansims given certain processing conditions (e.g., Belletti et al., 2007; Gao et al., 2006; Uljas et al., 2001). Taking into account possible interactions between different environmental conditions is of extreme importance for food microbiology, see, e.g., the hurdle concept (Leistner, 2000). (*ii*) The extended data collection is used to build response surface models, i.e., via polynomial regression, to study the effect of the environmental factors on selected variables describing microbial dynamics. With respect to inactivation, these response surface models can describe the effect of environmental conditions on parameters like the first decimal reduction D or the time for four decimal reductions t_{4D} (e.g., Maks et al., 2010). (*iii*) In some cases, DOE is also used to rigorously collect data in a specified region to finally obtain accurate and reliable parameter estimates of existing models (e.g., Jagannath et al., 2005; Miller et al., 2009b).

3.1. Models

After performing all experiments identified with the design of experiments, experimental results are analyzed by means of a mathematical model. The selection of an appropriate model is an important aspect of the experimental procedure. Different types of models can be applied for data processing, depending on the aim of the study.

Emperical models. Polynomial models are often applied to describe a system's response in relation to the selected explanatory variables. In terms of DOE, response surface designs are specifically developed to unravel the effect of variables and their interactions using second-order polynomial regression, e.g.,

$$x = \beta_0 + \beta_1 \cdot f_1 + \beta_2 \cdot f_2 + \ldots + \beta_k \cdot f_k + \beta_{12} \cdot f_1 \cdot f_2 + \beta_{13} \cdot f_1 \cdot f_3 - \ldots \quad (8)$$
$$+\beta_{k-1,k} \cdot f_{k-1} \cdot f_k + \beta_{11} \cdot f_1^2 + \beta_{22} \cdot f_2^2 + \ldots + \beta_{kk} \cdot f_k^2 \quad (9)$$

with x the independent variable, β_i constants and f_1, f_2, ..., f_k the environmental factors. In this case, at least three levels must be taken for each environmental factor. In many cases, the response surface model (RSM) approach is implemented to define a second order polynomial relation that can identify the optimum of a specific response in relation to selected factors, e.g., to optimize a production process.

Response surface models, i.e., polynomials, are usually used in predictive microbiology to quantify the effect of multiple environmental factors on selected variables describing microbial dynamics, e.g., the growth rate or the inactivation rate. These model types are useful to describe relations between static conditions. However, polynomial regression models may not be considered for dynamics evolving in time. As such, time varying conditions (such as processing or storage temperatures) may not be easily handled. Moreover, using RSM models to define optimal processing conditions can lead to wrong results, particularly when polynomial approximations are not valid.

Semi-emperical kinetic models. The above model type yields empirical black box models, i.e., no information with respect to underlying mechanisms is included. Next to this, the collected experimental information can also be used to optimize existing semi-empirical grey box models, i.e., to improve the model parameter values and/or their accuracy. These models have previously been developed to describe relations between microbiologically relevant parameters, e.g., growth or inactivation rate, and one or more environmental factors. Examples are the Arrhenius model or the model of Bigelow (see Chapter 5) which describe the effect of temperature on the maximum specific inactivation rate. It is generally assumed that these model structures are valid for a wide range of micro-organisms. New experimental data can be used to (1) confirm the strain validity, and (2) define strain specific model parameters.

In the following sections, four highly relevant DOE approaches are discussed, i.e., full factorial design, fractional factorial design, central composite design and Box-Behnken design. When possible, examples from the domain of (predictive) microbiology, and specifically with respect to inactivation, are given. All information on DOE and specific design approaches can be found in Anderson and McLean (1974); Eriksson et al. (2008); Selwyn (1996).

3.2. Full Factorial Design

Globally, *full factorial designs* are the experimental schemes most often applied in many research areas. In this approach, experiments are performed at all levels of the selected explanatory variables and also at all the possible combinations of variables. For an experimental study with two factors, each having two levels, this results in four experiments, for three factors with two levels, 8 experiments are to be performed (see Figure 1).

Full factorial design has the advantage that underlying interactions are most likely unraveled. Also, taking all possible combinations yields a balanced experimental scheme in which the individual conditions are considered multiple times. However, the approach is rather conservative, and very labor-intensive and costly, certainly when a high number of variables and/or an extended range of levels is considered. Taking two environmental conditions under study, the number of experiments to be performed is equal to 2^k, with k the number of levels taken.

A myriad of examples of full factorial design to study microbial inactivation kinetics exists. Chawla et al. (1996) modeled the effect of plate count medium composition and storage on the survival of *Listeria monocytogenes* using a 5^3 design with NaCl (5 levels), pH (5 levels) and temperature (5 levels) as controlled variables. Chhabra et al. (2002) built a model that describes the thermal resistance of *Listeria monocytogenes* as a function of growth kinetics, based on a 3×3 factorial design, i.e., three temperature levels, three pH levels, and three levels of milkfat. Calicioglu et al. (2003) used a $2 \times 2 \times 5 \times 5 \times 2$ design, i.e., acid-adapated versus not-adapted (2), replicates (2), predrying treatments (5), drying treatments (5), and agar type (2), to study the effect of acid adaptation on *Salmonella* behavior during storage in treated beef products. Leguerinel (2007) studied and modeled the effect of heat treatment, pH and recovery medium on the survival of *Salmonella typhimurium* based on a complete factorial design. Black and Jaczynski (2008) applied a 3×3 factorial design, i.e., three sample types (beef, chicken and fish) and three water activity levels, to study the D_{10} values of *Escherichia coli* when subjected to an electron beam treatment. Other examples can be found in Jagannath et al. (2005); Krishnamurth et al. (2008); Miller et al. (2009a); Periago et al. (1998).

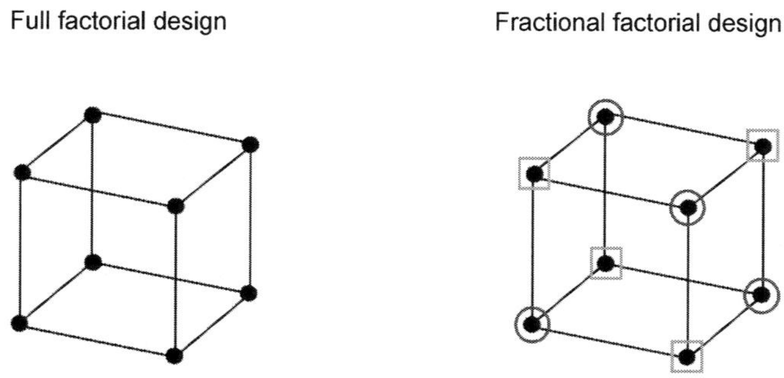

Figure 1. Illustration of (left) a full factorial, and (right) two options (○ and □) for a corresponding fractional factorial design for three factors with two levels (adapted from the DOE guide of the SAS Institute Inc. (2008)).

3.3. Fractional Factorial Design

When the experimental load is too high, a well-selected part of the full factorial design can be retained, i.e., a *fractional factorial design*. This approach is highly relevant when multiple environmental conditions and/or levels are considered. Generally, fractional designs are deduced from the full design, mainly directed by the number of experiments that can be retained. The experimenter, for example, can choose to maintain the full design for a small number of factors and add some additional experiments focussing on the interactions with the remaining environmental factors considered. Usually, this selection is based on a priori knowledge or assumptions about the most important factors and/or expected interactions. An important aspect when reducing a full factorial design is the balance of the design, i.e., for all variables, both low and high levels must be retained, the number of experiments with respect to each factors should be in the same order of magnitude, etc.

Most often, fractional designs are of the power 1/2. In this case, the number of experiments is reduced to the half. An example is given in Figure 1: starting from the eight experiments of the full factorial design in the case of three factors with two levels, 2 equal fractional designs can be deduced enclosing four experimental conditions. When many factors and/or levels are considered, the experimental load can be further reduced to a 1/4 fractional factorial design.

The disadvantage of fractional factorial design is that certain higher order interaction terms (≥ 2) can not be determined accurately. In this case, they are confounded in each other, in other (single) model parameters, or lower level combinations. By reducing the during storage in treated beef products. Leguerinel (2007) studied and modeled the effect of heat treatment, pH and recovery medium on the survival of *Salmonella typhimurium* based on a complete factorial design. Black and Jaczynski (2008) applied a 3×3 factorial design, i.e., three sample types (beef, chicken and fish) and three water activity levels, to study the D_{10} values of *Escherichia coli* when subjected to an electron beam treatment. Other examples can be found in Jagannath et al. (2005); Krishnamurth et al. (2008); Miller et al. (2009a); Periago et al. (1998).

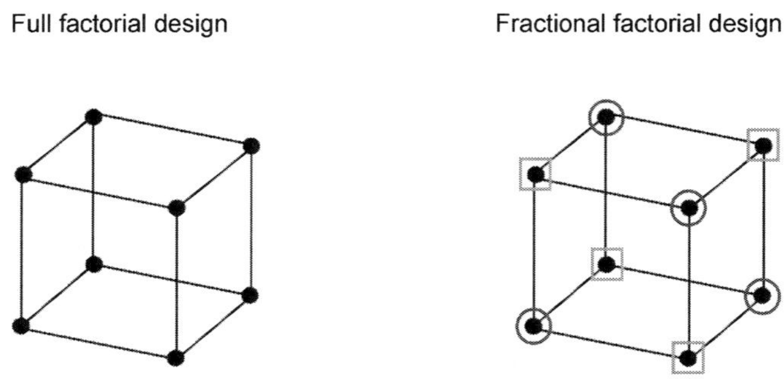

Figure 1. Illustration of (left) a full factorial, and (right) two options (○ and □) for a corresponding fractional factorial design for three factors with two levels (adapted from the DOE guide of the SAS Institute Inc. (2008)).

3.3. Fractional Factorial Design

When the experimental load is too high, a well-selected part of the full factorial design can be retained, i.e., a *fractional factorial design*. This approach is highly relevant when multiple environmental conditions and/or levels are considered. Generally, fractional designs are deduced from the full design, mainly directed by the number of experiments that can be retained. The experimenter, for example, can choose to maintain the full design for a small number of factors and add some additional experiments focussing on the interactions with the remaining environmental factors considered. Usually, this selection is based on a priori knowledge or assumptions about the most important factors and/or expected interactions. An important aspect when reducing a full factorial design is the balance of the design, i.e., for all variables, both low and high levels must be retained, the number of experiments with respect to each factors should be in the same order of magnitude, etc.

Most often, fractional designs are of the power 1/2. In this case, the number of experiments is reduced to the half. An example is given in Figure 1: starting from the eight experiments of the full factorial design in the case of three factors with two levels, 2 equal fractional designs can be deduced enclosing four experimental conditions. When many factors and/or levels are considered, the experimental load can be further reduced to a 1/4 fractional factorial design.

The disadvantage of fractional factorial design is that certain higher order interaction terms (≥ 2) can not be determined accurately. In this case, they are confounded in each other, in other (single) model parameters, or lower level combinations. By reducing the obtained. Together with the center points, these experiments render information with respect to second order model terms.

Starting from a three-level factorial set-up, the design can be represented starting from a cube (see Figure 2). The advantage of composite design (based on fractional factorial design) compared to full factorial designs is the reduction of the experimental load and its higher informative character with respect to the second-order terms in the response surface model that describe the factor interactions. Two disadvantages are (1) estimating effects with unequal variances, and (2) fewer degrees of freedom as, analogous to fractional designs, less information about interactions terms is included.

Implementations of CCD designs in (predictive) microbiology mainly relate to the building of response surface models. Debs-Louka et al. (1999) constructed an RSM model describing the efficiency of compressed CO_2 on the survival of *E. coli*, *S. cerevisiae* and *Enterococcus faecalis* based on a 10 experiments CCD with CO_2 pressure and holding time as the explanatory variables. Gao et al. (2006) identified in a first step, based on a fractional design, that pressure, temperature and holding time are the most influential factors with respect to high hydrostatic pressure (HHP) inactivation of *Bacillus subtilis* spores. In a second step, a CCD was implemented to build an RSM describing the efficiency of HHP and heat treatment on *B. subtilis* spore inactivation. Papathomopoulou and Kotzekidou (2009) studied the effect of temperature and glucose concentration on the antimicrobial characteristics of *Lactobacillus sakei* against *E. coli* and *L. monocytogenes*. They used a centered CCD to build a response surface model to finally identify the optimal glucose concentration given a low temperature to achieve the required pathogen inactivation. Maks et al. (2010) used a CCD approach to build an RSM that can predict the effect of sodium lactate, sodium diacetate, and pediocin on the thermal inactivation of *Listeria monocytogenes* on bologna. Bover-Cid et al. (2010) used a central composite design to build an RSM to describe the high hydrostatic pressure effect on the inactivation of *Listeria monocytogenes* as a function of the pressure, the holding time and the fluid temperature.

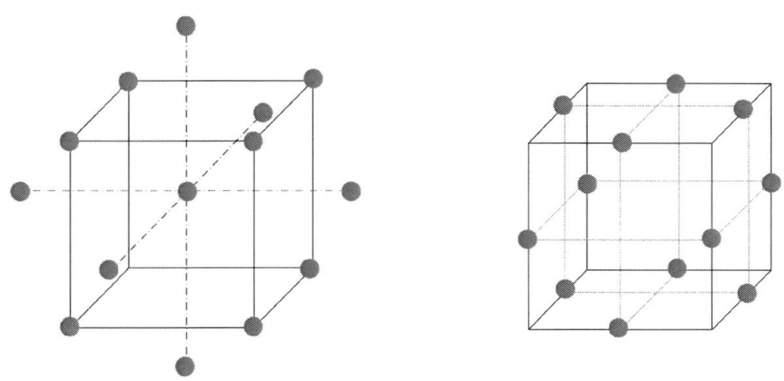

Figure 2. Illustration of (left) a central composite design, and (right) a Box-Behnken design for 3 factors.

3.3.2. Box-Behnken Design

A Box-Behnken design is not based on a full or fractional factorial design but only contains center points and points in the middle of each side of the cube, i.e., no corner points - extremes - are included. A Box-Behnken design is sphere-like, i.e., all the points are situated on a radius, except for the center point. This approach is very useful when corner points, i.e., at the extremes, are practically or financially unfavorable.

Implementations of Box-Behnken in the domain of predictive microbiology are scarce. Buzrul et al. (2008) selected experimental data from an extended data set based on the Box-Behnken design method to compare the performance of an RSM model and a Weibull model to describe the inactivation kinetics of *Listeria monocytogenes* subjected to high hydrostatic pressure. First, Gao et al. (2005) identified that pressure, temperature and pressure holding time are the most influencing factors with respect to the inactivation of *Staphylococcus aureus* by HHP. Next, a Box-Behnken design was implemented to build an RSM describing the efficiency of HHP and heat treatment on the inactivation of *S. aureus*.

4. Optimal Experiment Design

Optimal experimental design aims to select a (series of), most often dynamic, experiment(s) with an optimal information content that can be used for model structure identification and/or model parameter estimation.

Optimal experiment design for structure discrimination (Section 4.1.) selects one or more (dynamic) experiments to yield experimental data that enable to distinguish between possible model structures. In this approach, experiments are designed such that the difference in the output of the various models is maximized.

Optimal experiment design for parameter estimation (Section 4.2.), which starts from a model structure presumed valid, selects a (small) set of highly informative, static and/or dynamic experiments, resulting in unique and accurate parameter estimates. When applying dynamic experiments, this approach also guarantees parameter estimates which are valid under varying, more realistic conditions.

4.1. Optimal Experiment Design for Structure Discrimination

One of the starting points of optimal experiment design for parameter estimation (see Section 4.2.) is the selection of a model structure and the assumption of its validity. In many research topics, different models, based on different assumptions, can describe the studied process equally. In this case, it is useful to perform a first experimental run, aimed at the discrimination of the different model structures. A search for the correct model structure can help to identify the underlying mechanisms of the studied system, as many hypotheses can be possible. Once selected, the appropriate model can be used to predict the dynamics of the system under different circumstances. Generally, OED/MD selects (a set of) experiments for which the difference in the output of the models under study is large. Different optimization criteria exist.

$max(y_1(u,t,\mathbf{p}_1) - y_2(u,t,\mathbf{p}_2))^2$ with y_1 and y_2 predictions by model 1 and model 2, respectively, given the experimental input u. This method selects the set of allowable settings and process inputs in order to maximize the so-called L^2 distance between the outputs of the two selected models. This T-optimality is generally used to discriminate between two or several linear regression with normal errors. Disadvantage of this approach is that the uncertainty on the model output is not considered (Hunter and Reiner, 1965).

$max \sum_{n=1}^{M-1} \sum_{n=m+1}^{M} P_m P_n \left[\frac{(\sigma_m^2 - \sigma_n^2)^2}{(\sigma^2+\sigma_m^2)+(\sigma^2+\sigma_n^2)} + (y_m - y_n)^2 (\frac{1}{\sigma^2+\sigma_n^2} + \frac{1}{\sigma^2+\sigma_m}) \right]$ takes into account the probability of models m and n, namely, P_m and P_n, and the corresponding variances on the model predictions, namely, σ_m^2 and σ_n^2, with σ^2 the experimental variance (Box and Hill, 1967).

$max\ min(J_{I1}, J_{I2})$ with $J_{I1} = \int_0^T e_1^T Q_1 e_1 dt$ and $e_1 = y_1(p_1^1, u^2) - y_2(p_2^2, u^2)$, and $J_{I2} = \int_0^T e_2^T Q_2 e_2 dt$ and $e_2 = y_2(p_2^1, u^2) - y_2(p_1^2, u^2)$. The parameter vectors p_1^1 and p_2^1 are retrieved from a first experiment. The parameter vector p_1^2 is identified for model 1 on the basis of data generated for model 2 ($y_2(p_1^1, u^2)$), assuming that model 2 is correct. The parameter vector p_2^2 is identified for model 2 on the basis of data generated for model 1 ($y_1(p_1^1, u^2)$) assuming the model 1 is correct. Q_1 and Q_2 are weighting matrices (Munack, 1992).

$max\ min(D_{p1}, D_{p2})$ This criterion is based on the same hypothesis as the previous one, i.e., one of the two models is correct. The distance considered in this case is the Mahalanobi distance which is given by $D_{pi} = (p - p_i)^T \mathrm{V}_i^{-1}$ (p-p$_i$) with V_i^{-1} the covariance, p_i the parameters identified from the first experiment and p the current parameter estimates (Munack, 1992).

4.2. Optimal Experiment Design for Parameter Estimation

4.2.1. The Fisher Information Matrix

As explained in Section 2., parameter estimation is the result of the minimization of a quadratic cost function (Equation 1 and Equation 2). This identification functional depends on the parameter vector **p** and can be linearized using a first order Taylor expansion with respect to parameters **p** slightly different from the true parameters **p***(e.g., Munack, 1989; Walter and Pronzato, 1997).

$$\mathbf{y}(\mathbf{p},t) \cong \mathbf{y}(\mathbf{p}^*,t) + \frac{\partial \mathbf{y}(\mathbf{p},t)}{\partial \mathbf{p}}|_{\mathbf{p}=\mathbf{p}^*} \Delta\mathbf{p}^T \tag{10}$$

Assuming that the measurements are characterized by white Gaussian noise, i.e., serially uncorrelated random measurement errors following a Guassion (normal) distribution with zero mean and constant variance, the expected value of the parameter identification cost for any set of parameters **p** in the neighborhood of $\mathbf{p}^*$ can be formulated as follows

$$\begin{aligned} E(J(\mathbf{p})) &= E(J(\mathbf{p}^*)) + \Delta\mathbf{p}^T \\ & \left[\int_0^{t_f} \left(\frac{\partial \mathbf{y}(\mathbf{p},t)}{\partial \mathbf{p}} \right)^T |_{\mathbf{p}=\mathbf{p}^*} \mathbf{Q} \left(\frac{\partial \mathbf{y}(\mathbf{p},t)}{\partial \mathbf{p}} \right) |_{\mathbf{p}=\mathbf{p}^*} dt \right] \Delta\mathbf{p} \end{aligned} \tag{11}$$

To obtain an identification cost valley with a steep slope and a unique minimum, the difference in the identification cost J for the two slightly different parameter vectors $\mathbf{p}$ and $\mathbf{p}^*$ has to be maximal. This can be obtained by maximizing the term between brackets, i.e., the **Fisher information matrix** (**F**). So, given a certain experimental input, the information content of an experiment with duration t_f and continuous output measurements with regard to parameter estimation, can be quantified by the Fisher information matrix.

$$\mathbf{F}(\mathbf{p}) = \int_0^{t_f} \left(\frac{\partial \mathbf{y}(\mathbf{p},t)}{\partial \mathbf{p}}\right)^T |_{\mathbf{p}=\mathbf{p}^*} \mathbf{Q} \left(\frac{\partial \mathbf{y}(\mathbf{p},t)}{\partial \mathbf{p}}\right) |_{\mathbf{p}=\mathbf{p}^*} \, dt \tag{12}$$

As the true values $\mathbf{p}^*$ are unknown, **F** is evaluated at the best guess of $\mathbf{p}$, i.e., the nominal values ($\mathbf{p}^\circ$). In the case of discrete output measurements, the Fisher information matrix reads as follows

$$\mathbf{F}(\mathbf{p}) = \sum_{i=1}^{n_t} \left(\frac{\partial \mathbf{y}(\mathbf{p},t_i)}{\partial \mathbf{p}}\right)^T |_{\mathbf{p}=\mathbf{p}^*} \mathbf{Q} \left(\frac{\partial \mathbf{y}(\mathbf{p},t_i)}{\partial \mathbf{p}}\right) |_{\mathbf{p}=\mathbf{p}^*} \tag{13}$$

with n_t the number of model output measurements.

F combines information on (i) the error on the output measurements (typically **Q** is defined as the inverse of the measurement error variance matrix), and (ii) the sensitivities of the model output ($\mathbf{y}(\mathbf{p},t)$) to small variations in the model parameters $\mathbf{p}$ (expressed in the sensitivity matrix $\frac{\partial \mathbf{y}(\mathbf{p},t)}{\partial \mathbf{p}}$). If a small change in a parameter value results in a significant change in the model output, the model is said to be *sensitive* to this parameter and a very accurate parameter estimate is needed to obtain a reliable model prediction. An analysis of the sensitivity functions can reveal which experimental conditions enclose the most information for the accurate estimation of specific parameters. Given a certain experimental set-up, a sensitivity functions analysis can give an indication which parameters can be identified accurately from the resulting measurements.

The Fisher information matrix has an attractive property. According to the Cramer-Rao inequality, the inverse of **F** approximates the lower boundary of the parameter estimation variance covariance matrix **P** thus quantifying the parameter estimation accuracy related to the proposed experiment given unbiased estimators and uncorrelated white Gaussian noise (see, e.g., (Ljung, 1999).

$$\mathbf{P} \geq \mathbf{F}^{-1} \tag{14}$$

4.2.2. Optimization Criteria

For a given experiment (or set of experiments), **F** returns information on the identification cost surface (see Equation 11) as well as the parameter estimation quality (see Equation 14). Therefore, **F** forms the basis of OED/PE in which some scalar function of **F** is minimized or maximized. The selected scalar function determines the focus of the design. Widely known and used criteria are A-, D-, E- and modified E-optimality (Asprey and Macchietto, 2002; Vanrolleghem et al., 1995; Walter and Pronzato, 1997).

A-criterion: min[trace($\mathbf{F}^{-1}$)]. A-optimal designs minimize the arithmetic mean of the parameter estimation errors. This corresponds with the minimization of the sum of the squared axes of the asymptotic joint confidence region, i.e, minimizing the frame enclosing this confidence region (see Figure 3). To lower the computational burden, the A-criterion is often reformulated as maximizing the trace of **F**.

D-criterion: max[det(F)]. The D-criterion minimizes the geometric mean of the parameter estimation errors. D-optimal design aims at the minimization of the parameter estimation variance covariance, i.e., minimization of the joint confidence region on **p** (see Figure 3) via the maximization of the determinant of **F**.

E-criterion: max[λ_{min}(F)]. E-optimality focuses on the minimization of the largest parameter error (i.e., maximization of the smallest eigenvalue) (see Figure 3), and as such, neglects uncertainty on the remaining parameters. This corresponds to minimizing the longest axis of the joint confidence region.

ME-criterion: min[$\frac{\lambda_{max}(\mathbf{F})}{\lambda_{min}(\mathbf{F})}$]. In modified E-optimal designs, the ratio between the largest and smallest eigenvalue of **F**, i.e., the condition number Λ, is minimized. The optimum of this criterion is known, i.e., the condition number equals one, which corresponds with circular confidence regions for a two-parameter estimation problem. The condition number being one, parameter estimates are unique and uncorrelated but the volumes of the joint confidence regions can still be large, i.e., parameter uncertainty can be large.

Other, less frequently used criteria exist, e.g., the G-criterion and the L-criterion. G-optimal design minimizes the maximum variance on the prediction (Steinberg and Hunter, 1984). The L-criterion is a variant of the A-criterion and aims at the minimization of [trace(W $\mathbf{F}^{-1}$)] with W a positive semidefinite weighting matrix. More complex optimization criteria have been developed with respect to problems encountered in optimization of specific problems (e.g., Baltes et al., 1994; Franceschini and Macchietto, 2008; Vila and Gauchi, 2007).

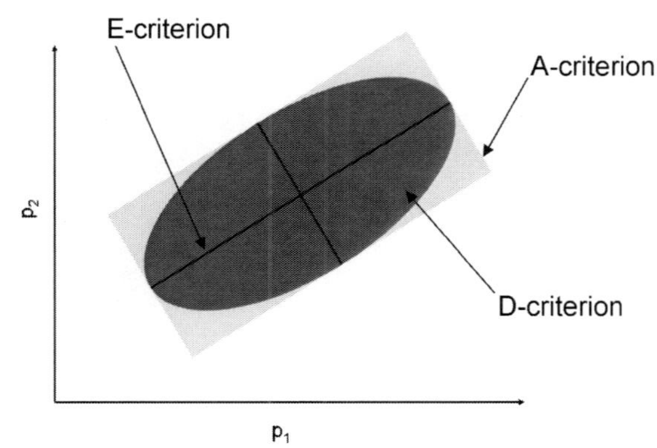

Figure 3. Schematic representation of the optimization criteria, adapted from Asprey and Macchietto (2002)).

Recently, a **maximin approach** has been implemented to obtain more robust designs. In this approach, only a certain degree of a priori knowledge is necessary, i.e., the lower and upper bound on the parameter domain of the nominal vector have to be known. For a given

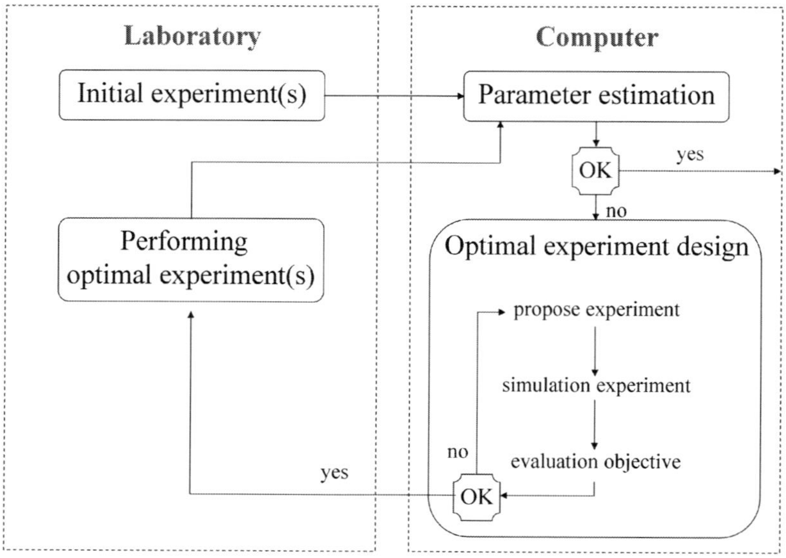

Figure 4. Schematic illustration of the sequential optimal experiment design cycle (adapted from Dochain and Vanrolleghem (2001)).

experimental set-up, the nominal parameter vector is located for which the selected criterion value is the lowest. This corresponds with the least information that can be obtained. Finally, the design is calculated for which the criterion value is maximal with respect to these worst nominal parameters. This way, the loss of accuracy due to the unknown nominal values is minimized (Dette et al., 2005).

Iterative OED/PE. For nonlinear models, the Fisher information matrix is determined by the unknown true parameters $\mathbf{p}^*$ and is therefore evaluated at the nominal values $\mathbf{p}^\circ$. As a consequence, the optimality of the design is only locally valid, i.e., for parameters close to $\mathbf{p}^\circ$. When the nominal values differ significantly from the true parameters, convergence to the latter values is not guaranteed. A sequential OED/PE strategy was proposed to overcome the possible convergence problem. In this approach, optimal experiment design and parameter estimation are alternated to improve the nominal values. This iterative procedure is illustrated in Figure 4. A first series of experiments is used to validate the model structure and to obtain realistic nominal values. Next, the optimization criterion, the parameterization of the optimization problem and the bounds on the degrees of freedom are selected. The resulting optimal experiments are performed and new parameter estimates are derived. These new parameter values are then evaluated with respect to the nominal values and literature. When the new parameter estimates differ significantly from the nominal values or do not correspond with realistic parameter values, a new OED/PE is started after replacing $\mathbf{p}^\circ$ by the new parameter estimates. The OED/PE cycle is repeated until convergence to the true parameters is obtained. Balasko et al. (2006) and Körkel et al. (1999) suggested to include previous experiments in the new designs. In this case, the new optimal experiments are complementary to earlier obtained information and experiment design becomes more robust and efficient.

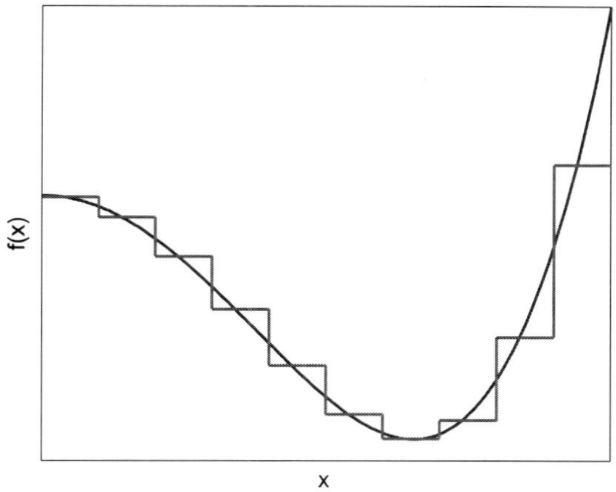

Figure 5. Graphical representation of control vector parameterization, i.e., approximation of the continuous profile by a sequence of constants.

4.2.3. Solving Optimal Experiment Design Problems

Most often dynamic OED/PE optimization problems are solved using a direct method, namely via *control vector parameterization* for which the finite number of degrees of freedom is optimized with respect to the selected optimization criterion. The optimization problem is reformulated as a nonlinear programming problem (NLP) by discretization of the controls (CVP) or both the control and state variables (CP, complete parameterization). The experimental time-horizon is subdivided and in each time-period the control and/or state variable is described by a basic function, e.g., piece-wise linear and piece-wise constant. An illustration is given in Figure 5. These parameterization approaches have two major disadvantages. (1) The resulting NLP is often multimodal, i.e., local optima exist, which hinders the identification of the global optimum. (2) When the NLP encloses a large number of parameters (due to a more complex parameterization), the computational burden seriously increases. Using CP for OED/PE not only needs states' sensitivities but also parametric sensitivities to compute the objective function with the consequent increase in the number of constraints. For the case of complete parameterization, constraints may be imposed so as to guarantee continuity of the solution at the optimum which requires suitable optimization methods. Generally, CVP is more convenient for optimal experiment design since it allows to handle large dynamic systems or large number of parameters without solving excessively large constrained NLPs.

Once the optimization problem is defined, an optimization algorithm must be selected to efficiently perform the experiment optimization. Optimization algorithms are generally grouped into deterministic and stochastic/probabilistic algorithms, based on their mode of operation.

Deterministic algorithms. Most classic optimization algorithms are iterative, gradient-based methods. These algorithms, that use the gradient of the cost function with respect to the degrees of freedom to be optimized, are known to be successful for unimodal objective functions, e.g., convex optimization problems. In this case, the algorithm converges quickly to the global optimum. Starting from the initial estimates, the optimization evolves in a systematic predefined way, i.e., following the gradient, until a minimum is reached. When

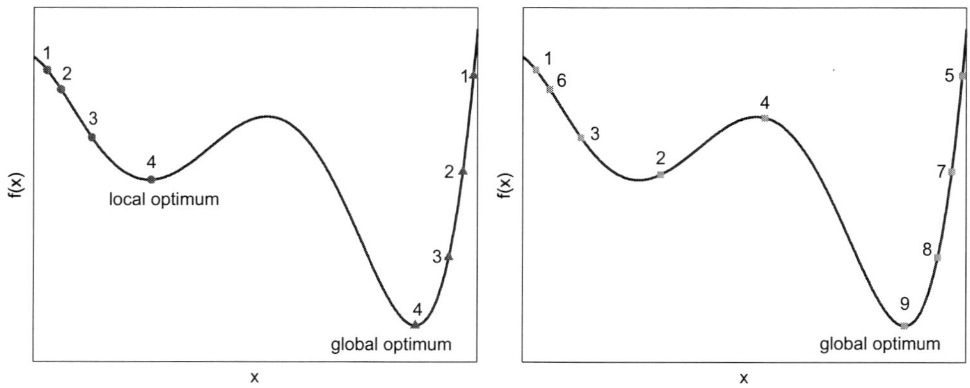

Figure 6. Illustration of the principle of (left) a deterministic and (right) a stochastic optimization algorithm.

the cost function is highly multimodal, good initial values are required, i.e., close to the global optimum. Otherwise, the algorithm will most probably get stuck in a local optimum. This is illustrated in Figure 6 (left). A local optimum or a global optimum (point (4)) is yielded depending on the selected initializations (point (1)). Gradient-based deterministic algorithms are often referred to as local optimization methods.

Next to these local deterministic methods, global deterministic optimization algorithms have been developed. One of the first global procedures is the *multistart* approach. Here, the local optimization is started from a large series of starting values such that a broad region is covered. In practice, however, there is no guideline for the necessary number of random initial estimates and, therefore, no guarantee that the global optimum will be reached. Another drawback of the procedure is that, when many initial estimates are used, the global optimum will be yielded several times, making this method rather inefficient. Recently, more advanced, global, deterministic optimization methods have been developed, e.g., branch & bound and interval methods, that are also gradient and/or Hessian based. These approaches are much more robust than, for instance, the multi-start method. These advanced deterministic methods allow, for some typical optimization problems, to locate the global optimum. However, the price to pay is a rather large computational cost (Floudas, 2000).

Stochastic algorithms. Stochastic methods are global optimization algorithms that can be suggested as efficient alternatives for multimodal optimization problems (Michalewicz and Fogel, 2004). Starting from the initial estimates, a certain degree of randomness is introduced during the generation of the next set of parameter values. Through scanning a wide region of the control vector, stochastic algorithms can trace the neighborhood of the optimum efficiently, even with badly scaled problems or when good starting values are absent [see Figure 6 (right)]. Furthermore, stochastic optimization algorithms are known to be easily implemented and reasonably computationally efficient, although a large number of evaluations may be necessary to locate the global optimum. Amongst the different approaches developed to introduce the randomness in the optimization, some popular classes are: (*i*) evolutionary computation, (*ii*) controlled random search methods, and (*iii*) simulated annealing.

Table 1. Four dynamic temperature profiles used for OED/PE by Versyck et al. (1999)

T-profile 1	$[0, t_f]$	$T(t) = a \cdot t + b$
T-profile 2	$[0, t_f]$	$T(t) = c \cdot t^2 + a \cdot t + b$
T-profile 3	$[0, t_s]$	$T(t) = T_1$
	$[t_s, t_f]$	$T(t) = a \cdot (t - t_s) + b$
T-profile 4	$[0, t_s]$	$T(t) = T_1$
	$[t_s, t_f]$	$T(t) = T_2$

Hybrid algorithms. Hybrid optimization algorithms combine the advantages of both the stochastic and the deterministic algorithms, i.e., the robustness of the stochastic, global methods and the fast convergence of the deterministic, local method (see, e.g., Balsa-Canto et al. (2005)).

4.2.4. Implementations of OED/PE in Predictive Microbiology

The technique of optimal experiment design for parameter estimation was introduced in the domain of predictive microbiology by Versyck et al. (1999). The focus was on the estimation of the parameters of the Bigelow model, a secondary model describing the effect of temperature on the microbial inactivation rate k_{max}

$$k_{max}(T) = \frac{2.303}{D_{ref}} \exp\left[\frac{2.303}{z} T - T_{ref}\right] \tag{15}$$

with D_{ref} [min] the decimal reduction time at T_{ref} [°C] the reference temperature, and z the number of degrees change of temperature required to effect a tenfold change in D_{ref}.

Cell density as a function of time was modeled with the primary inactivation model of Geeraerd et al. (2000)

$$\begin{aligned} \frac{dN(t)}{dt} &= \frac{1}{1+C_c(t)}\left(-k_{max}(T)\right)\left(1-\frac{N_{res}}{N(t)}\right)N(t) \\ \frac{dC_c(t)}{dt} &= -k_{max}\, C_c(t) \end{aligned} \tag{16}$$

with $N(t)$ [CFU/mL] the cell number as a function of time, k_{max} the maximum specific inactivation rate, N_{res} the population density in the tailing phase and C_c a measure for the physiological state of the cells. The maximum specific growth rate as a function of temperature is replaced by (Equation 15) to enable description under dynamic temperature conditions. (More information on the model of Geeraerd et al. and the Bigelow model can be found in Chapter 4 and Chapter 5, respectively.)

Starting from the assumption of continuous and perfect measurements, both static and dynamic experiments are evaluated with respect to their structural identifiability, i.e., can the parameters be identified uniquely and accurately from a data set given a specific temperature profile. A first model analysis showed that both model parameters can not be identified uniquely based on a single isothermal experiment.

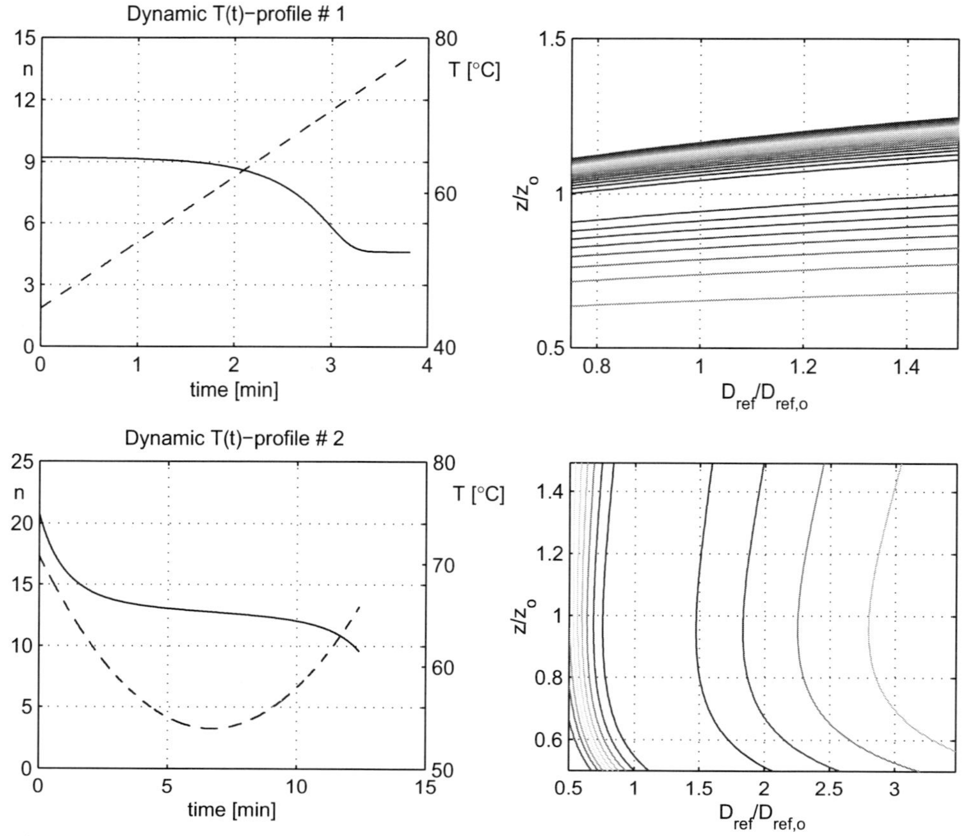

Figure 7. Simulation results for two dynamic experiments. Upper plots: dynamic $T(t)$ profile with a=8.6°C/min and b=45.0°C for an initial population density $N(0)=10^4$ (CFU/mL). Lower plots: dynamic $T(t)$ profile with b=70.8°C, a= -5.0°C/min and c=0.4°C/min^2 for an initial population density $N(0)=10^9$ (CFU/mL). Left plots: full lines represent $n(t)$ and dashed lines represent temperature T [°C]. (Figures reprinted with permission from Versyck et al. (1999)).

In a second step, four types of dynamic temperature profiles are evaluated (see Table 1). For all four cases, the practical identifiability is evaluated based on optimal experiments designed by minimizing the modified E-criterion (optimal value = 1). (i) For the first temperature profile (linear temperature increase or decrease), the obtained value for the ME-criterion is still rather high ($\approx$ 50). The confidence regions are very elongated ellipses which indicates a rather strong correlation between the two parameter estimates (see Figure 7). A fairly good estimate can be obtained for the z parameter, however, a large uncertainty exists for D_{ref}. (ii) The second temperature profile taken has a quadratic structure. Although a more optimal criterion value is obtained (i.e., $\approx$ 20), confidence regions are far from circular (see Figure 7). Again, a strong correlation between the parameter estimates is observed from data collected in this experiment. (iii) In the third temperature profile, a linear phase at T_1 is followed by a linear temperature change. An optimal experiment is obtained with the criterion value equal to one. Confidence regions are almost circular which points at unique and decorrelated estimates (see Figure 8). (iv) The fourth temperature profile is a step experiment with temperature changing from T_1 to T_2 at the time t_s. As

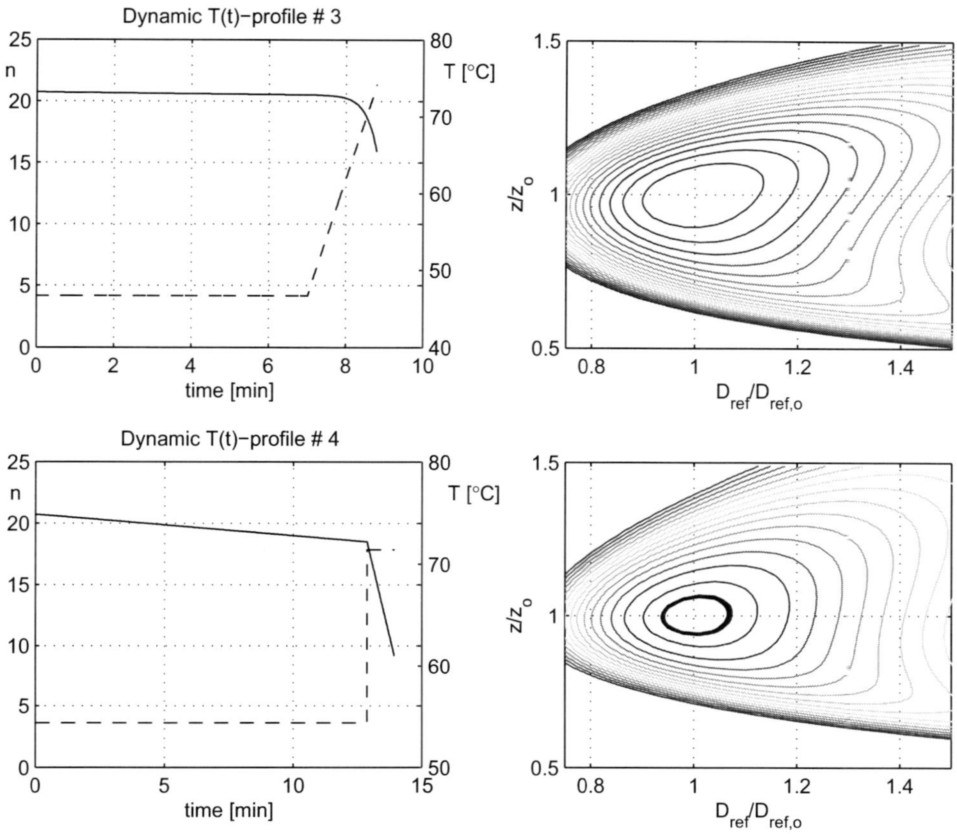

Figure 8. Simulation results for two dynamic experiments. Upper plots: dynamic $T(t)$ profile with T_1=46.7 °C, t_s= 7.010 min and the rate a = 15.3°C/min for an initial population density $N(0)=10^9$ (CFU/mL). Lower plots: dynamic $T(t)$ profile with T_1=54 3°C, t_s=12.87 min and T_2=71.4°C for an initial population density $N(0)=10^9$ (CFU/mL). Left plots: full lines represent $n(t)$ and dashed lines represent temperature T [°C]. (Figures reprinted with permission from Versyck et al. (1999).

for the third profile, a criterion value equal to unity is obtained together with nearly circular confidence regions (see Figure 8).

Both the third and fourth temperature parameterization yield criterion values equal to one which proves that a single dynamic experiment can enable the uncorrelated estimation of both model parameters. As such, this simulation study revealed the undiscovered benefits of the implementation of dynamic OED/PE for the estimation of kinetic model parameters in the domain of predictive microbiology, i.e., a dynamic experiment is more efficient than the traditional, static approach where D and z are derived from a (limited) set of isothermal experiments.

Also based on simulations, Balsa-Canto et al. (2008) further evaluated the application of dynamic optimal experiments for the identification of the two parameters of the Bigelow model (i.e., D_{ref} and z). Hereto, a more general approach was taken, i.e., next to the optimal design of the dynamic temperature profile, also the sampling positions (for a fixed number of samples), experiment duration, initial conditions were optimized and an experimental error was added to the simulated output.

The informative character of a step-profile is evaluated for (1) equidistant sampling of 10, 20, and 50 samples, and (2) optimal positioning of 8 and 10 samples, given three optimization criteria, i.e., the D-, the E- and the modified E-criterion. The simulation results revealed that the modified E-criterion is insensitive for the number of samples. All options yield the same criterion value, but confidence regions have different forms (Figure 9).

D- and E-optimal designs are largely influenced by the selected number of samples. Figure 10 also shows that optimal sample locations significantly improve the parameter estimation procedure, i.e., smaller confidence regions are obtained. Overall, the E-criterion is evaluated as the best optimization objective. Not only are more circular confidence regions obtained, the optimal temperature profile is practically feasible in contrast to the D-optimal profile. The fast temperature decrease makes sampling before the temperature change impossible.

Already in 1997, Cunha et al. published a basic application of OED/PE to determine the parameters of the Bigelow model. Both isothermal, i.e., static, and non-isothermal, i.e., dynamic, experiments were taken. For the isothermal conditions, only the parameter D was considered. The optimal sampling point, next to sampling at $t = 0$, was found when $(N(t)/N(0))$ is equal to 0.368. For non-isothermal conditions, a linearly changing temperature profile was optimized. The best solution was found to be a linearly increasing temperature starting at the lower temperature boundary and increasing at the highest rate allowed.

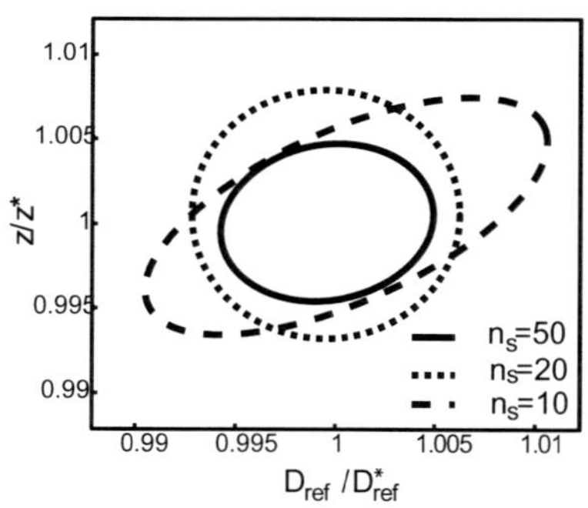

Figure 9. Comparison of the confidence regions for different numbers of equidistant sampling using the Modified E-criterion (with n_s the number of sampling times, D_{ref} the decimal reduction time, D^*_{ref} the optimal value of the decimal reduction time. (Figure reprinted with permission from Balsa-Canto et al. (2008).)

Other examples of OED/PE relevant for the domain of microbiology and predictive microbiology, and inactivation specifically, can be found in Cunha et al. (1998) and Cunha and Oliveira (2000). Cunha et al. (1998) applied OED/PE to determine the parameters of the Weibull probability density function, which can describe microbial inactivation as a function of time:

$$\log_{10} S(t) = -\left(\frac{-t^{\beta}}{\alpha}\right) \tag{17}$$

with $S(t)$ equal to $N(t)/N(0)$, α the time for the first decimal reduction, and β a shape parameter. One or a series of two isothermal experiments were designed with a number of sampling points equal to the number of model parameters, i.e., two. For a design including one experiment, the simulation study showed that these two sampling times correspond to the time points were the fractional concentration $S(t)$ is equal to 0.70322 and 0.19245. For a set of experiments, the two are to be performed at a temperature close to the boundary of the studied region with sampling at $S(t)$ equal to 0.70322 and 0.19245 for the first experiment, and $S(t)$ equal to $\frac{1}{e}$ for the second. Additional samples should be taken (1) at the beginning of the experiment to determine the initial concentration, and (2) at significantly further time period to identify the residual cell concentration.

In 2000, Cunha and Oliveira followed an analogous approach to determine the parameter of the Arrhenius equation which describes the effect of temperature on the microbial inactivation rate (Cunha and Oliveira, 2000). Taking a linearly changing temperature profile, the aspects considered are the initial and final temperature, the heating rate, as well as sampling times. Generally, the optimal experiments start at the lower temperature boundary. However, for each initial temperature, the optimal heating rate differs. Moreover, in practice, this heating rate is often too high such that experiments must be performed at sub-optimal heating rates. Optimal sampling positions are similar to the Bigelow model, i.e., $S(t)$ equal to 0.70322. and 0.19245.

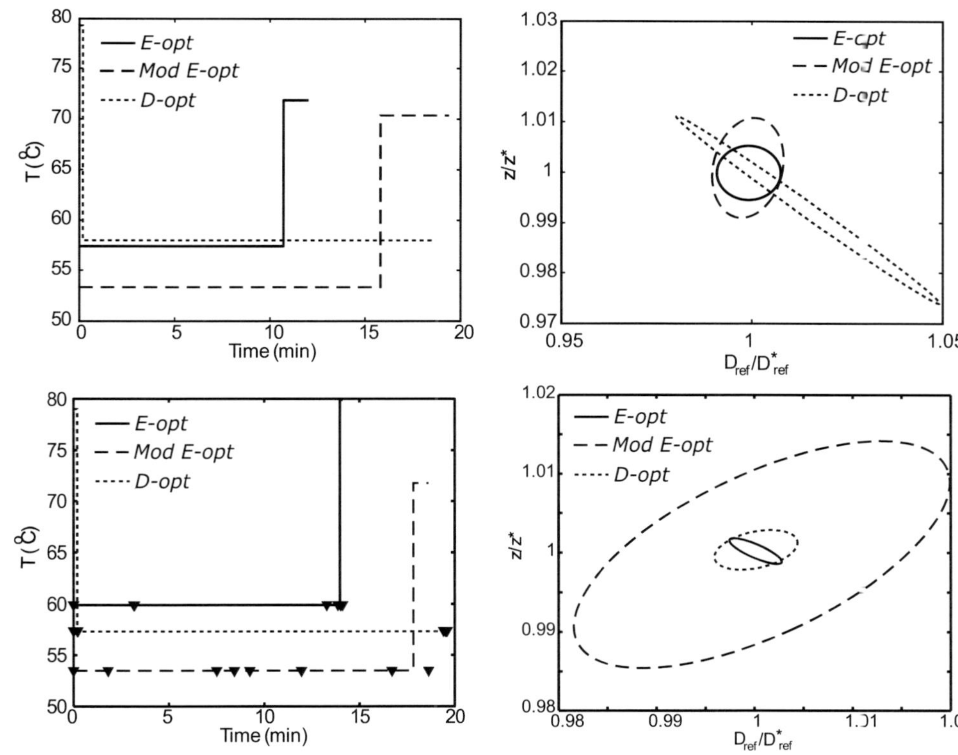

Figure 10. Comparison of the optimal temperature profiles and the confidence regions for the three optimization criteria considered, i.e., D-optimality, E-optimality and Modified E-optimality. (Upper plots) Equidistant positioning of $n_s = 20$ samples. (Lower plots) Optimal positioning of $n_s = 8$ samples. (D_{ref} the decimal reduction time and D^*_{ref} the optimal value of the decimal reduction time). (Figure reprinted with permission from Balsa-Canto et al. (2008).)

Other implementations of OED/PE for predictive modeling of microbial dynamics are situated within the region of **growth dynamics**. Bernaerts et al. adopted the OED/PE technique for the estimation of the two parameters of the square root model (SQRT) (Ratkowsky et al., 1982), modeling the temperature effect on μ_{max} in the suboptimal temperature region (Bernaerts et al., 2002). A dynamic experiment which consists of a sequence of constrained steps was designed for *E. coli* K12 MG1655 with respect to the ME-criterion. Parameters identified from the latter experiment were reliable and accurate with small uncertainties. Analogously, Grijspeerdt and De Reu (2005) designed D-optimal two-step temperature profiles for the estimation of the square root model parameters for *Bacillus cereus* and *E. cloacae* grown in liquid whole egg product. Balsa-Canto et al. (2008) designed a temperature profile enclosing multiple steps with a large number of equidistant samples aiming at the identification of the SQRT parameters. The duration of the experiment was optimized together with the input profile. Using one optimal experiment made it impossible to accurately estimate the model parameters due to the high correlation. Therefore, two dynamic experiments were designed simultaneously. The combination of the two optimal experiments significantly improved the parameter estimation.

Later on, Bernaerts et al. (2004) implemented OED/PE for the identification of the Cardinal Temperature Model with Inflection (CTMI) from a limited set of static experiments. This model describes the effect of temperature on μ_{max} from the minimum to the maximum temperature for growth and encloses four parameters. Taking it one step further, Van Derlinden et al. (2008, 2010) designed optimal dynamic experiments to optimize the estimation of the four parameters of the CTMI from one or a limited series of dynamic bioreactor experiments. Optimal dynamic temperature profiles were designed focusing on the minimization of the parameter estimation (co)variance (D-optimal design). Experiments revealed that the maximum growth temperature could only be estimated accurately from optimal experiments that enclose temperatures at or very close to the T_{max}. This observation underlines the importance of selecting the upper temperature constraint for OED/PE as close as possible to the true T_{max}. Generally, compared to estimates from non-optimized dynamic experiments, more reliable CTMI parameter values were obtained from the optimal experiments.

4.3. Improved Parameter Estimation via Sample Scheme Optimization

Poschet et al. (2004) used the Monte Carlo simulation technique to study the effect of data distribution, and as such the sampling scheme, on the quality of the primary Baranyi and Roberts model (1994) description. Equidistant grouping is compared to grouped data collection. For equidistant grouping, the frequency has a very small effect on the parameter estimates. Only the uncertainty is determined by the number of samples, i.e., the higher the number of samples, the smaller the uncertainty. A higher parameter estimation reliability and accuracy can be obtained by grouping samples: more frequent sampling during the lag and the transition from the exponential to stationary phase reduces the estimation error.

Similar conclusions were drawn by Grijspeerdt and Vanrolleghem (1999). Using the technique of OED/PE, based on the D-criterion, they optimized the sampling schedule for the static model of Baranyi and Roberts (1994). When optimizing the sampling schedule of an experiment, different degrees of freedom can be taken: (*i*) the number of samples, and/or (*ii*) the timing of the samples. In this case, the optimization problem reduces to finding the optimal number and position of the samples with respect to the best parameter reliability and accuracy. So, in this case, the scalar function of the Fisher information matrix is optimized with respect to the sampling scheme. Optimal sampling points were situated in four regions, i.e., in the lag phase, at the begin and end of the exponential growth phase and in the stationary phase. These four regions more or less correspond to the minima and maxima of the sensitivity functions of the Baranyi model as described by Poschet et al. (2004).

These observations can be directly translated to the inactivation model of Geeraerd et al. (2000) which has the same basic structure as the Baranyi and Roberts model. Parameter estimation uncertainty for the Geeraerd model can be minimized by majorly sampling the initial shoulder phase and the transition to the loglinear inactivation phase. When a tailing phase is expected, also the transition from inactivation to tailing is an important sampling point.

In 2005, Poschet also applied the Monte Carlo simulation technique to study optimal sampling schemes for the identification of first order kinetics and the related D- and z-value as shown in Bigelow type models. To accurately determine the parameter D from static, isothermal experiments, an optimal 50-50 sampling scheme was obtained: samples should be distributed at the start and the end of the static experiment. The next step is to determine the series of static experiments which are expected fo finally result in accurate and efficient estimation of the parameters of the secondary Bigelow model. A first major conclusion is that the selection of the T_{ref} value highly influences the further estimation error of D_{ref}. The most optimal situation is to position T_{ref} in the centre of the studied temperature region. Contrarily, the choice of T_{ref} has no influence on the accuracy of z. Generally, the most accurate D_{ref} and z values are obtained when the isothermal experiments are grouped at the boundaries of the studied temperature region.

Analogous, sampling analysis can be performed for other primary and secondary inactivation models. For each model, sensitivity functions can be defined which will already give a good indication of the most informative region in an experiment. Once the sensitivity functions are constructed, an optimal sampling schedule can be calculated based on the Fisher information matrix and the selected scalar function.

Conclusion

After the selection of the explanatory and response variables and the definition of a model structure, the next step in the model building process is the accurate identification of the model parameters, i.e., parameter estimation. Often, accurate model building and, more specifically, parameter identification is hindered due to, e.g., parameter correlation and inaccurate measurements. Only sufficiently informative data can lead to reliable and accurate parameter estimates and, as such, accurate model predictions. Mathematical techniques exist that can increase the informative character of (a set of) experiments which finally improves the efficiency of the model building cycle.

In *design of experiments* (DOE) schemes, specific or all combinations of the selected discrete levels of the different explanatory variables are considered, instead of studying the system at single levels for one environmental factor at a time. Combining experimental levels increases the overall information content of a single experiment and, therefore, also of the total set of experiments. DOE schemes can be applied for every step of the model building cycle, i.e., (1) selection of variables and characterization of their interactions, (2) selection of the model structure, and (3) estimating the model parameters.

When multiple model structures, e.g., based on different underlying hypotheses, can be proposed to describe the dynamic system under study, a second technique, i.e., *optimal experiment design for model structure discrimination* (OED/MD), can be applied to differentiate between the different options. Generally, OED/MD selects one or more (dynamic) experiments which try to maximize the difference in the model outputs given a selected input. Starting from the model structures given, different criteria quantifying this difference can be applied.

Next to DOE, a second method can be applied to optimize the parameter estimation accuracy. *Optimal experiment design for parameter estimation* (OED/PE) has been proven an excellent tool to improve the informative character of experiments and consequently improve the parameter estimation. In this technique, the experimental information content is quantified via the Fisher information matrix and experiments are optimized through the minimization or maximization of a scalar function of the Fisher information matrix. In contrast to the DOE technique, the OED/PE method selects experiments starting from the model assumed valid, i.e., the model structure is embedded in the Fisher information matrix. This way, the experimental conditions are chosen in relation to the sensitivities of the model under study.

Whereas DOE techniques are regularly applied in the domain of (predictive) microbiology to obtain informative data sets for the construction of polynomial models or the identification of model parameters, OED/MD and OED/PE are rarely implemented. Although different models can be published in literature to describe a specific microbial phenomenon, aspects of model discrimination are never applied. Since the end of the last century, OED/PE has found its way to predictive microbiology. Although the results are very promising and clearly illustrate the advantages of a well-structured experimental design, actual implementations are limited.

References

U. M. Alvarez, A. M. Loske, E. Castaño-Tostado, and F. E. Prieto. Inactivation of *Escherichia coli* O157:H7, *Salmonella* typhimurium and *Listeria monocytogenes* by underwater shock waves. *Innovative Food Science and Emerging Technologies*, **5**:459–463, 2004.

V. L. Anderson and R. A. McLean. *Design of experiments. A realistic approach*. Marcel Dekker, Inc., New York, 1974.

S. P. Asprey and S. Macchietto. Statistical tools for optimal dynamic model building. *Computers and Chemical Engineering*, **24**:1261–1267, 2000.

S. P. Asprey and S. Macchietto. Designing robust optimal dynamic experiments. *Journal of Process Control*, **12**:545–556, 2002.

B. Balasko, J. Madar, and J. Abonyi. Additive sequential evolutionary design of experiments. *Artificial Intelligence and Soft Computing - ICAISC 2006, Proceedings*, **4029**: 324–333, 2006.

E. Balsa-Canto, V. S. Vassiliadis, and J. R. Banga. Dynamic optimization of single and multi-stage systems using a hybrid stochastic-deterministic method. *Industrial and Engineering Chemical Research*, **44**:1514–1523, 2005.

E. Balsa-Canto, A. A. Alonso, and J. R. Banga. Computing optimal dynamic experiments for model calibration in predictive microbiology. *Journal of Food Process Engineering*, **31**:186–206, 2008.

M. Baltes, R. Schneider, C. Sturm, and M. Reuss. Optimal experimental design for parameter estimation in unstructured growth models. *Biotechnology Progress*, **10**:480–488, 1994.

J. Baranyi and T. A. Roberts. A dynamic approach to predicting bacterial growth in food. *International Journal of Food Microbiology*, **23**:277-294, 1994.

J. Baranyi and T. A. Roberts. Mathematics of predictive food microbiology. *International Journal of Food Microbiology*, **26**:199–218, 1995.

D. M. Bates and D. G. Watts. *Nonlinear Regression Analysis and Its Applications*. Wiley Series in Probability and Statistics, New York, 1988.

N. Belletti, S. S. Kamdem, F. Patrignan, R. Lanciotti, A. Covelli, and F. Gardini. Antimicrobial activity of aroma compounds against Saccharomyces cerevisiae and improvement of microbiological stability of soft drinks as assessed by logistic regression *Applied and Environmental Microbiology*, **73**:5580–5586, 2007.

K. Bernaerts and J. F. Van Impe. Data-driven approaches to the modelling of bioprocesses. *Transactions of the Institute of Measurement and Control*, **26**:349–372, 2004.

K. Bernaerts, R. D. Servaes, S. Kooyman, K. J. Versyck, and J. F. Van Impe. Optimal temperature input design for estimation of the Square Root model parameters: parameter accuracy and model validity restrictions. *International Journal of Food Microbiology*, **73**:145–157, 2002.

K. Bernaerts, E. Dens, K. Vereecken, A. H. Geeraerd, A. R. Standaert, F. Devlieghere, J. Debevere, and J. F. Van Impe. Concepts and tools for predictive modeling of microbial dynamics. *Journal of Food Protection*, **67**:2041–2052, 2004.

J. L. Black and J. Jaczynski. Effect of water activity on the inactivation kinetics of *Escherichia coli* O157:H7 by electron beam in ground beef, chicken breast meat, and trout fillets. *International Journal of Food Science and Technology*, **43**:579–586, 2008.

S. Bover-Cid, N. Belletti, M. Garriga, and T. Aymerich. Model for *Listeria monocytogenes* inactivation on dry-cured ham by high hydrostatic pressure processing. *Food Microbiology*, 28:804-809, 2010.

G. Box and W. Hill. Discrimination among mechanistic models. *Technometrics*, **9**:57–71, 1967.

S. Buzrul, M. Cevik, and H. Alpas. Comparison of response surface methodology and the proposed weibull model for inactivation of *Listeria innocua* by high hydrostatic pressure. *Journal of Food Safety*, **28**:142–156, 2008.

M. Calicioglu, J. N. Sofos, J. Samelis, P. A. Kendall, and G. C. Smith. Effect of acid adaptation on inactivation of *Salmonella* during drying and storage of beef jerky treated with marinades. *International Journal of Food Microbiology*, **89**:51–65, 2003.

C. S. Chawla, H. Chen, and C. W. Donnelly. Mathematically modeling the repair of heat-injured *Listeria monocytogenes* as affected by temperature, pH and salt concentration. *International Journal of Food Microbiology*, **30**:231–242, 1996.

A. T. Chhabra, W. H. Carter, R. H. Linton, and M. A. Cousin. A predictive model that evaluates the effect of growth conditions on the thermal resistance of *Listeria monocytogenes*. *International Journal of Food Microbiology*, **78**:235–243, 2002.

L. M. Cunha and F. A. R. Oliveira. Optimal experimental design for estimating the kinetic parameters of processes described by the first-order Arrhenius model under linearly increasing temperature profiles. *Journal of Food Engineering*, **46**:53–60, 2000.

L. M. Cunha, F. A. R. Oliveira, T. R. S. Brandão, and J. C. Oliveira. Optimal experimental design for estimating the kinetic parameters of the Bigelow model. *Journal of Food Engineering*, **33**:111-128, 1997.

L. M. Cunha, F. A. R. Oliveira, and J. C. Oliveira. Optimal experimental design for estimating the kinetic parameters of processes described by the Weibull probability distribution function. *Journal of Food Engineering*, **37**:175–191, 1998.

E. Debs-Louka, N. Louka, G. Abraham, V. Chabot, and K. Allaf. Effect of compressed carbon dioxide on microbial cell viability. *Applied and Environmental Microbiology*, **65**: 626–631, 1999.

H. Dette, V. B. Melas, A. Pepelyshev, and N. Strigul. Robust and efficient design of experiments for the Monod model. *Journal of Theoretical Biology*, **234**:537–550, 2005.

D. Dochain and P. Vanrolleghem. *Dynamical Modelling and Estimation in Wastewater Treatment Processes*. IWA Publishing, London, 2001.

L. Eriksson, E. Johansson, N. Kettaneh-Wold, C. Wikstrom, and S. Wold. *Design of experiments: principles and applications*. Umetrics Academy, Boca Raton, Florida, USA, 2008.

C. A. Floudas. *Deterministic global optimization: Theory, methods and applications*. Kluwer Academics, The Netherlands, 2000.

G. Franceschini and S. Macchietto. Novel anticorrelation criteria for model-based experiment design: Theory and formulations. *AICHE Journal*, **54**:1009–1024, 2008.

Y.-L. Gao, Y.-X. Wang, and H.-H. Jiang. Effect of high pressure and mild heat on *Staphylococcus aureus* in milk using response surface methodology. *Process Biochemistry*, **40**: 1849–1854, 2005.

Y.-L. Gao, X.-R. Ju, and H.-H. Jiang. Studies on inactivation of *Bacillus subtilis* spores by high hydrostatic pressure and heat using design of experiments. *Journal of Food Engineering*, **77**:672–679, 2006.

A. H. Geeraerd, C. H. Herremans, and J. F. Van Impe. Structural model requirements to describe microbial inactivation during a mild heat treatment. *International Journal of Food Microbiology*, **59**:185–209, 2000.

K. Grijspeerdt and K. De Reu. Practical application of dynamic temperature profiles to estimate the parameters of the square root model. *International Journal of Food Microbiology*, **101**:83–92, 2005.

K. Grijspeerdt and P. Vanrolleghem. Estimating the parameters of the Baranyi model for bacterial growth. *Food Microbiology*, **16**:593–605, 1999.

W. Hunter and A. Reiner. Designs for discriminating between two rival models. *Technometrics*, **7**:307–323, 1965.

A. Jagannath, T. Tsuchido, and J.-M. Membre. Comparison of the thermal inactivation of Bacillus subtilis spores in foods using the modified Weibull and Bigelow equations. *Food Microbiology*, **22**:233–239, 2005.

V. K. Juneja and B. S. Eblen. Predictive thermal inactivation model for *Listeria monocytogenes* with temperature, pH, NaCl and sodium pyrophosphate as controlling factors. *Journal of Food Protection*, **62**:986–993, 1999.

S. Körkel, I. Bauer, H. G. Bock, and J. P. Schlöder. A sequential approach for nonlinear optimum experimental design in DAE systems. In F. Keil, W. Mackens, H. Voss, and J. Werther, editors, *Scientific Computing in Chemical Engineering II*, Volume 2. Springer Verlag, Berlin, Heidelberg, 1999.

K. Krishnamurth, S. Jun, J. Irudayaraj, and A. Demirci. Efficacy of infrared heat treatment for inactivation of *Staphylococcus aureus* in milk. *Journal of Food Process Engineering*, **31**:798–816, 2008.
nel2007

I. Leguerinel, I. Spegagne, O. Couvert, L. Corroler, and P. Mafart. Quantifying the effects of heating temperature, and combined effects of heating medium pH and recovery medium pH on the heat resistance of *Salmonella typhimurium*. *International Journal of Food Microbiology*, **116**:88–95, 2007.

L. Leistner. Basic aspects of food preservation by hurdle technology (review). *International Journal of Food Microbiology*, **55**:181–186, 2000.

L. Ljung. *System identification: theory for the user*. Prentice Hall, Inc., Upper Saddle River, N.J., 1999.

N. Maks, L. Zhu, V. K. Juneja, and S. Ravishankar. Sodium lactate, sodium diacetate and pediocin: effects and interactions on the thermal inactivation of *Listeria monocytogenes* on bologna. *Food Microbiology*, **27**:64–69, 2010.

SAS Institute Inc. *JMP 8 Design of experiments guide*. Cary, New York, 2008.

Z. Michalewicz and D. B. Fogel. *How to solve it: Modern heuristics*. Springer, 2004.

F. A. Miller, M. M. Gil, T. R. S. Brandao, P. Teixeira, and C. L. M. Silva. Sigmoidal thermal inactivation kinetics of *Listeria innocua* in broth: influence of strain and growth phase. *Food Control*, **20**:1151–1157, 2009a.

F. A. Miller, B. Ramos, M. M. Gil, T. R. S. Brandao, P. Teixeira, and C. L. M. Silva. Influence of pH, type of acid and recovery media on the thermal inactivation of *Listeria innocua*. *International Journal of Food Microbiology*, **133**:121–128, 2009b.

A. Munack. Optimal feeding strategy for identification of Monod-type models by fed-batch experiments. In N. M. Fish, R. I. Fox, and N. F. Thornhill, editors, *Computer Applications in Fermentation Technology, Modelling and Control of Biotechnological Processes*, pages 195–204. Elsevier, Amsterdam, 1989.

A. Munack. Some improvements in the identification of bioprocesses. In M. Karim and G. Stephanopoulos, editors, *Modelling and Control of Biotechnological Processes*, pages 89–94. Pergamon Press, Oxford, 1992.

K. Papathomopoulou and P. Kotzekidou. Inactivation of verocytotoxigenic *Escherichia coli* and *Listeria monocytogenes* co-cultured with *Lactobacillus sakei* in simulated meat fermentation medium. *Journal of Food Safety*, **29**:331–347, 2009.

P. M. Periago, P. S. Fernandez, M. C. Salmeron, and A. Martinez. Predictive model to describe the combined effect of pH and NaCl on apparent heat resistance of *Bacillus stearothermophilus*. *International Journal of Food Microbiology*, **44**:21–30, 1998.

J. C. Pinto. On the costs of parameter uncertainties. Effects of parameter uncertainties during optimization and design of experiments. *Chemical Engineering Science*, **53**:2029–2040, 1998.

F. Poschet, K. Bernaerts, A. H. Geeraerd, N. Scheerlinck, B. M. Nicolaï, and J. F. Van Impe. Sensitivity analysis of microbial growth parameter distributions with respect to data quality and quantity by using Monte Carlo analysis. *Mathematics and Computers in Simulation*, **65**:231–243, 2004.

F. Poschet, A. H. Geeraerd, A. M. Van Loey, M. E. Hendrickx, and J. F. Van Impe. Assessing the optimal experiment setup for first order kinetic studies by Monte Carlo analysis. *Food Control*, **16**:873-882, 2005.

D. A. Ratkowsky, J. Olley, T. A. McMeekin, and A. Ball. Relationship between temperature and growth rate of bacterial cultures. *Journal of Bacteriology*, **149**:1–5, 1982.

M. R. Selwyn. *Principles of experimental design for the life sciences*. CRC Press, Boca Raton, 1996.

D. M. Steinberg and W. G. Hunter. Experimental design: review and comment. *Technometrics*, **26**:71–97, 1984.

A. Tarantola. *Inverse problem theory and methods for model parameter estimation*. Siam, 2005.

H. G. Uljas, D. W. Schaffner, S. Duffy, Z. L., and S. C. Ingham. Modeling of combined processing steps for reducing *Escherichia coli* O157:H7 populations in apple cider. *Applied and Environmental Microbiology*, **67**:133–141, 2001.

E. Van Derlinden, K. Bernaerts, and J. F. Van Impe. Accurate estimation of cardinal temperatures of Escherichia coli from optimal dynamic experiments. *International Journal of Food Microbiology*, **128**:89-100, 2008.

E. Van Derlinden, K. Bernaerts, and J. F. Van Impe. Simultaneous versus sequential optimal experiment design for the identification of multi-parameter microbial growth kinetics as a function of temperature. *Journal of Theoretical Biology*, **264**:347–355, 2010.

P. A. Vanrolleghem, M. Van Daele, and D. Dochain. Practical identifiability of a biokinetic model of activated sludge respiration. *Water Research*, **29**:2561–2570, 1995.

K. J. Versyck, K. Bernaerts, A. H. Geeraerd, and J. F. Van Impe. Introducing optimal experimental design in predictive modeling: A motivating example. *International Journal of Food Microbiology*, **51**:39–51, 1999.

J.-P. Vila and J.-P. Gauchi. Optimal designs based on exact confidence regions for estimation of a nonlinear regression model. *Journal of Statistical Planning and Inference*, **137**: 2935–2953, 2007.

E. Walter and L. Pronzato. *Identification of Parametric Models from Experimental Data*. Springer, Masson, 1997.

R. C. Whiting, S. Sackitey, S. Calderone, K. Merely, and J. G. Philips. Model for the survival of *Staphylococcus aureus* in nongrowth environment. *International Journal of Food Microbiology*, **31**:231–243, 1996.

L. L. Zaika and J. G. Philips. Model for the combined effects of temperature, pH and sodium chloride concentration on survival of *Shigella flexneri*. *International Journal of Food Microbiology*, **101**:179–187, 2005.

In: Progress on Quantitative Approaches in Thermal ... ISBN: 978-1-62100-842-2
Editors: V. P. Valdramidis and J. F. M. Van Impe © 2012 Nova Science Publishers, Inc.

Chapter 4

MODELLING MICROBIAL INACTIVATION KINETICS: PRIMARY MODELS

A. M. Cappuyns [1]***, V. P. Valdramidis*** [2,*] ***and J. F. M. Van Impe*** [1,†]
[1]Katholieke Universiteit Leuven - Department of Chemical Engineering,
BioTeC - Chemical and Biochemical Process Technology and Control,
Leuven (Heverlee), Belgium
[2]Biosystems Engineering UCD School of Agriculture,
Food Science and Veterinary Medicine University,
College Dublin Belfield, Dublin, Ireland

1. Introduction

The understanding of the survival and inactivation of pathogens and spoilage organisms during food processing or food treatment is critical with regard to food safety and quality. Heating remains the principal method of microbial inactivation and modeling the response of microorganisms to a temperature-time trajectory, can be of great aid in the design of efficient heat treatments. In this chapter, the focus is on primary models for thermal inactivation.

Primary models describe the evolution of a microbial population, i.e., growth, inactivation or survival, as a function of time and can be described in a very general way by the following set of differential equations (Bernaerts et al., 2004; Van Impe et al., 2005)

$$\frac{dN_i(t)}{dt} = \mu\left(N_i(t), < P(t) >, < S(t) >, < phys(t) >\right) \cdot N(t) \quad (1)$$

with N [cfu/mL] the cell density at time instant t and μ $[h^{-1}]$ the specific (growth/inactivation) rate. This specific rate μ is function of the interactions within the population ($< N(t) >$), physicochemical environmental conditions ($< env(t) >$), such as process conditions (e.g., temperature) and atmospheric conditions, microbial metabolite concentrations ($< P(t) >$), the physiological state(s) of the cells ($< phys >$), substrate

*E-mail address: vvadram@gmail.com
†E-mail address: Jan.VanImpe@cit.kuleuven.be

concentration(s) ($< S(t) >$), and components governing interspecies/intraspecies interactions. Expression (1) gives the evolution of microbial dynamics in function of time, but if a food product is not homogeneous, the microbial evolution is dependent on space as well (Dens and Van Impe, 2001). μ is positive in the case of microbial growth and negative in the case of microbial inactivation. In this chapter, focus is on primary models for microbial inactivation, i.e., more precisely inactivation caused by temperature.

Primary models are often used in combination with secondary models, i.e., models that describe the influence of environmental conditions on primary model parameters. As such, the evolution of a microbial population is described in a two step or in one step procedure (Valdramidis et al., 2005; van Zuijlen et al., 2010). More in formation on secondary models for microbial resistance during thermal treatments can be found in chapter 5.

2. Loglinear Inactivation

Chick (1908) investigated the effect of temperature in the context of disinfection and concluded that the *reaction velocity of disinfection* increases with rise of temperature in a manner similar to that of a chemical reaction. As such, the effect of temperature can be described in analogy with a chemical reaction as a first order decay in the microbial population:

$$\frac{dN}{dt} = -k \cdot N \tag{2}$$

with N the microbial population and k the specific inactivation rate. This equation can be seen as a simplification of the general description (equation 1). The effect of temperature on the specific inactivation rate can be taken into account by the formula of Arrhenius and as such the effect of a certain time-temperature profile on a microbial population can be calculated.

Rudra et al. (2010) evaluated different models, i.e., including both loglinear and and non loglinear models, to describe thermal inactivation of *Bacillus stearothermophilus* spores in coriander leaf puree. They concluded that a loglinear inactivation model combined with the formula of Arrhenius is the most appropriate to describe inactivation of *B. stearothermophilus* spores at UHT temperatures (105°C-145°C).

Bigelow and Esty (1920) and Bigelow (1921) published important results on the effect of temperature on bacterial spores and how to quantify and describe the inactivation of microorganisms during a sterilisation treatment. Bigelow and Esty (1920) defined *thermal death point in relation to time* as the time necessary to completely destroy a definite concentration of spores in a medium with a known hydrogen-ion concentration. Bigelow (1921) used the term *thermal death time* to express the same idea. If the time needed to destroy a given number of microorganisms at a given temperature is determined, the whole inactivation curve can be reconstructed.

By plotting the inactivation curves on semilog paper, Bigelow (1921) pointed out the logarithmic nature of survival curves for the temperatures investigated in that article. In the thermal death time approach, the decimal reduction time D_T, i.e., the time needed for a 10-fold (i.e., decimal of 1-log) reduction in the number of microorganisms at a given temperature, is calculated as follows

$$D_T = \frac{t}{\log_{10} N_0 - \log_{10} N_t} \tag{3}$$

with N_t the number of microorganisms at time t and N_0 the number of microorganisms at time 0. Remark that this concept is equivalent to the description published by Chick (1908).

As mentioned above, the D_T-value is temperature dependent. The z-value, i.e., the increase in temperature required to change a D_T value by a factor of ten, allows to calculate the D_T-value at different temperatures based on the information of D_T at a reference temperature. By plotting D_T values against temperature, a loglinear relationship is usually obtained indicating a z value which is independent of temperature. The z value can be seen as a measure of the relative resistance of the microorganisms to the temperature treatment.

In the food industry, especially in the canning industry (e.g., Stumbo et al., 1975) the concept of Thermal Death Time is widely used with the F-value as a basis for designing heat sterilisation procedures. The F-value is a measure for the accumulated lethality and is expressed as the time needed at a specific temperature to reduce a microbial population by 90% in a given food product under specified conditions.

$$F = \int_0^t 10^{(T(t)-T(ref))/z} dt \tag{4}$$

with T(t) the product temperature at a time t and T(ref) is a reference temperature. In the canning industry, the F-value is usually set at 12D-values, i.e., to obtain at 12 log reduction of the microbial population. More information on the lethality concept and how this has been further developed to in cases of non-linear inactivation kinetics refer to Chapter 10.

3. Deviations Form Loglinearity

3.1. A Shift in Thermal Food Treatment

Linear kinetics (the D/z-concept) have been been used world wide especially in the canning industry (e.g., Stumbo et al., 1975) with satisfying results, but these models do not satisfy modern food treatment. The last decades a shift has occurred in the processing of food products, i.e., heat treatments are less intensive than the ones used in the canning industry. Focus of the heat treatment is double: *(i)* pathogenic organisms and spoilers should be minimized and *(ii)* the texture and quality of the food products should be maintained. An optimal trade off has to be obtained between textural and sensorial quality retention on one hand, and reduction of pathogenic and spoilage vegetative cells on the other hand. Nevertheless, due to its broad applicability, the D/z-concept remains an appropriate method to obtain a first impression on the performance of an inactivation process (van Asselt and Zwietering, 2006). This is especially useful for the food industry where elaborated knowledge and the necessary tools for complicated models are not available.

While most survival curves of microorganisms subjected to intense heat treatments, e.g., in a sterlisation process, with some exceptions (refer to Mafart et al. (2010)) are loglinear, milder heat treatments can result in survival curves displaying more evidently a non loglinear shape. Some commonly observed shapes of survival curves are displayed in Figure 1 (Geeraerd et al., 2005).

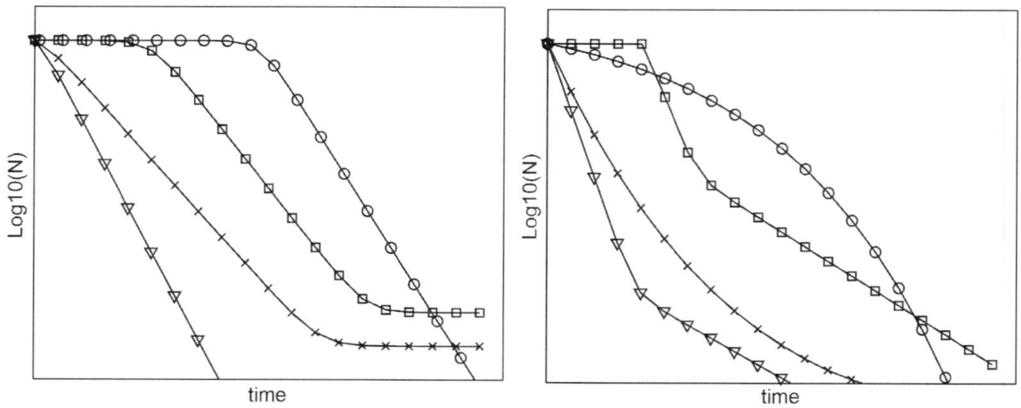

Figure 1. Commonly observed types of inactivation curves. Left plot: linear (∇), linear with tailing (×), sigmoidal-like (□), linear with a preceding shoulder (o). Right plot: biphasic (∇), concave (×), biphasic with a shoulder (□), and convex (o) (reprinted with permissions by Geeraerd et al. (2005)).

Predictive models cannot provide exact predictions of what will happen in food microbiology because, by nature, uncertainty and variability play an important role in both foods and microbiology (Nauta, 2006). Halder et al. (2007) investigated whether the added complexities of Weibull models compared to first order models are a worthwhile investment once the effect of uncertainties in various microbiological and process parameters is included. They concluded that due to the large effect of variability in the input parameters on the final log reduction, the effort to move towards more complex kinetic models needs to be weighed against inclusion of variability.

3.2. Tails and Shoulders

The most observed deviation from loglinear survival curves are a smooth initiation of the inactivation, i.e., a shoulder, and the persistence of a small fraction of the population which seems to have a considerably lower death rate, i.e., a tail. To be able to find an accurate model for non loglinear curves, it is useful to look at possible explanations of the occurrence of tails and shoulders.

Shoulder Effect

Possible explanations for the occurence of a shoulder in an inactication curve were summarized by Mossel et al. (1995).

- A shoulder can be due to microorganisms being present in groups or clumps, each clump being represented, upon culturing in a solid medium, by one colony. The lenght of the shoulder reflects the time before all but one organism in such a clump have been killed.
- Alternatively, the shoulder represents a period during which the cells are able to

resynthesize a vital component and death ensues only when the rate of destruction exceeds the rate of synthesis.

- The composition of the medium, e.g., pH and the presence of proteins and fats, can have an important effect on the heat resistance and as such influence the length of a shoulder. The resistance to heat is the greatest for pH levels near neutrality. Proteins could reduce the loss of solutes, stabilize the membrane, or have a buffering effect on low pH values. Fat particles can increase heat resistance by an indirect reduction of water activity due to the increasing solubility of water in fat at raising temperatures.

- Moats et al. (1971) formulated a fourth possible explanation for the occurrence of shoulders in survival curves. The authors start from the assumption that a larger number of the same critical molecules, or a large number of different critical molecules, needs to be inactivated. In such a case the type of damage is cumulative rather than instantly lethal.

Tailing Effect

Cerf (1977) investigated the occurrence of tails in survival curves, summarized possible explanations and concluded with following classification.

The vitalistic theory The basic assumption of the vitalistic theory is that individuals in a population are not identical. Therefore, resistance to heat (or other lethal agents) can be possessed to different degrees by different individuals in a microbial population. It is also assumed that the difference in degree of resistance is permanent within a pure culture. According to this theory, survival curves should be sigmoidal or concave upwards.

The mechanistic theory According to the mechanistic theory, (i) there is a good similarity in resistance between the different individuals of the microbial population (the resistance of a population is homogeneous) , (ii) the process of destruction is an orderly time process presenting a close analogy to chemical reactions, (iii) therefore at any given time only a portion of the individuals is in position to participate in the inactivation reaction, and (iv) as a consequence the inactivation may be regarded as a reaction of first order within a reaction of higher order.

Within the mechanistic theory, Cerf (1977) defines three interpretations for tails in survival curves.

- Tailing is a normal feature, bound to the mechanism of inactivation, i.e , some of the microorganisms are inaccessible for heat.

- Tailing is a normal feature, bound to the mechanism of resistance, i.e., some of the microorganisms have adapted to heat during the treatment.

- Tailing is an artefact independent from the mechanism of inactivation. Tails as an artefact can be due to, e.g., genetically heterogeneity of the population, heterogeneity of the treatment, clumping of microorganisms, errors in the enumeration method.

The possible effect of clumping causing a lag in heat transfer is debated in the literature. Grant et al. (1999) concluded that a longer holding time is more likely to achieve the complete inactivation of *Mycobacterium paraturbeculosis* in milk than a higher pasteurisation temperature and attributed this observation to the delay of the heat penetration due to the creation of clump of cells or due to some other mechanisms (e.g., production of heat shock proteins within the clump). On the contrary, Hasting et al. (2001) has shown that a thermal process delivered during pasteurisation of milk (for the same microorganism) is relatively unaffected by particle diameter and thus clumping of cells is unlikely to reduce the effectiveness of the normal pasteurisation process. It is also worth mentioning that in the case of spores several studies consider a direct relation between formation of clumps and inactivation rate. For example Furukawa et al. (2005) suggested that heat treatment increases hydrophobicity of spore surfaces contributing to the formation of spore clumps resulting in an increase of the (spores) heat resistance. Therefore, it is of high interest to quantify tailing and account for it in industrial processes.

4. Suggestions for Structural Properties of Thermal Inactivation Models

As mentioned before, the classical first order inactivation models ar not adequate for describing non loglinear survival curves. Geeraerd et al. (2000) formulated the following model design properties, in order to describe accurately inactivation curves under static and dynamic conditions during a mild heat treatment.

Dynamic Character of the Model

The majority of thermal processes are dynamic in their nature. This means that there is a time period during which temperature increases till the desired level is reached (e.g., Valdramidis et al. (2011); Miller et al. (2011)). It is therefore desired that the developed primary models are dynamic, i.e, under the form of one or more differential equations, or it should be possible to derive a dynamic version by differentiating the static model with respect to time t. This model requirement is of great importance for heat treatments with fluctuating environmental conditions (Valdramidis et al., 2008). For dynamic models, the influence of a temperature increase or decrease within a food product can easily be described by integrating the model equation over the time course, taking into account the time dependency of the environmental conditions. In general, a differential equation has a set of constant parameters, e.g., representing the maximum specific inactivation rate. The initial condition $N(0)$ and the initial time t_0 is desirable not to appear explicitly in the differential equation, because the differential equation is supposed to be valid for a whole range of $[t_0, N(0)]$ couples. Additionally, a time dependency on the model structure may have implications on the prediction of the momentary microbial population. This means that the predictive capability of the model will be influenced by the chosen time increments. A study that illustrates this phenomenon has been reported by Janssen et al. (2008).

The explicit solution of a dynamic model for a specific $[t_0, N(0)]$ couple, results in the static version of the model and is only valid during non-varying environmental conditions. When trying to identify a static model on dynamical experimental data, this can only be done by resetting certain values, e.g., t_0 and $N(0)$, at every environmental change, a procedure which becomes rather complex or even impossible for continuously changing environmental conditions.

Description of Absolute Populations

A primary model is desirable to describe absolute populations N, and not the relative population $\frac{N}{N(0)}$. A first reason is that the implicit occurrence of $N(0)$ will give problems when modeling subsequent stages of microbial growth/inactivation. A second reason is that from a statistical point of view, describing the microbial population relatively to the initial population holds the assumption that the first experimental data point, related to this initial population, is very accurate. This could be circumvented if the experimental design is built in such a way that separated $N(0)$s are collected for each kinetic point (see e.g., Valdramidis et al. (2006)).

Initial Condition

For the static equation describing $N = f(t)$, the value of N at t_0, denoted as $N(t_0)$, is required to be equal to the initial condition $N(0)$, otherwise the initial condition will not be completely separated from the model parameters and will confound their estimation (France and Thornley, 1984).

Shoulder and Tail

As mentioned in the previous sections, a lot of microbial survival curves have been reported and described with shapes deviating from loglinear. Non loglinear inactivation curves often display a smooth initiation or shoulder during which the microbial concentration does not decrease yet. Another observed phenomenon is the tailing effect, i.e., a fraction of the population which survives the inactivation treatment. The model, representing $\log_{10}(N)$ or $\ln(N)$ as a function of time, should be able to simulate if necessary a smooth initiation (shoulder) and/or saturation (tail) of the inactivation curve. The length of the shoulder, the inactivation rate, and the level of tailing should not be dependent upon each other. This feature implies that the model should encompass a (more or less) loglinear inactivation by the selection of specific parameter values.

In the following sections some well known models for non loglinear microbial inactivation are reviewed and analysed with respect to the model properties as mentioned in Section 4. and with their ability to describe different types of survival curves.

5. Non Loglinear Modelling Structures

5.1. Biphasic Models

Cerf (1977) proposed a two-fraction model, which can be formulated as follows (see also (Xiong et al., 1999)).

$$\frac{dN_1(t)}{dt} = -k_{max1} \cdot N_1(t) \tag{5}$$

$$\frac{dN_2(t)}{dt} = -k_{max2} \cdot N_2(t) \tag{6}$$

or for a static conditions:

$$N(t) = N(0) \cdot (f \cdot \exp(-k_{max1} \cdot t) + (1-f) \cdot \exp(-k_{max2} \cdot t)) \tag{7}$$

with $N(t) = N_1(t) + N_2(t)$, f the fraction of the initial population in a major subpopulation ($f \cdot N(0) = N_1(0)$), $(1-f)$ the fraction of the initial population in a subpopulation ($(1-f) \cdot N(0) = N_2(0)$) (which is more heat resistant than the previous one), and k_1 and k_2 [1/min] the inactivation rates for the two fractions, respectively.

The model describes biphasic curves, which were generally considered to represent a mix of two species or strains having different heat resistances. This approach follows the mechanistic theory as it is considered that the resistance of individuals is homogeneous but two genetically different populations exist in the initial population. The dynamic version of the model makes it applicable under time-varying environments. Biphasic thermal inactivation kinetics were also studied without the use of a fraction factor (Humpheson et al., 1998). This makes the transition to the second (more resistant) population less smooth than in the original Cerf model (Equation (6)).

5.2. The Modified Gompertz Model

The modified Gompertz model (Gibson et al., 1987) has been proved to be suitable to decribe growth curves (Zwietering et al., 1990) and has also been tested to describe non-linear inactivation curves (Bhaduri et al., 1991; Linton et al., 1995). The modified Gompertz equation describes a sigmoidal function with two horizontal asymptotes and is written as follows for the case of inactivation (Mossel et al., 1995)

$$\log_{10}(CFU(t)) = A - C \exp[-\exp(-B[t-M])] \tag{8}$$

with A the asymptotic log count as t decreases indefinitely, C the asymptotic amount of death that occurs as t increases indefinitely, M the time instant at which the absolute death rate is maximal and B the relative death rate at time M.

Equation 8 can be rewritten to contain parameters with a biological meaning (Zwietering et al., 1990):

$$\log N = \log N_0 - \log\left(\frac{N_0}{N_{res}}\right) \cdot \exp\left(-\exp\left(\frac{k \cdot \exp(1)}{\log\left(\frac{N_0}{N_{res}}\right)} \cdot (L-t) + 1\right)\right) \tag{9}$$

with L the length of the shoulder and k the maximum inactivation rate.

The results of Bhaduri et al. (1991) and Linton et al. (1995) indicate the usefulness of this model for describing sigmoidal inactivation curves. Van Impe et al. (1992) derived a dynamic version of the Gompertz model to describe growth, a dynamic description can be derived in analogy for inactivation curves. This approach was used by Gil et al. (2006) to describe the inactivation behaviour under time-varying temperature conditions at the surface of a food product. Miller et al. (2011) used a Gompertz inspired model to describe the effect of a linearly increasing temperature on the inactivation of *Listeria innocua* while Huang (2009) compared three mathematical models, i.e., linear, Weibull-type, and modified Gompertz, to describe the inactivation of *L. monocytogenes* in ground beef under both isothermal and dynamic conditions. Under isothermal conditions, both the Weibull and modified Gompertz model provided good results. For these specific case studies and under dynamic conditions (linear temperature increase), however, the modified Gompertz model was more accurate than the linear and Weibull-type models.

Two major modelling problems are inherent to this modified Gompertz type model description: (i) $N(t = 0)$ is not equal to $N(0)$, and (ii) the differential equation does explicitly depend on $N(0)$, which is to be avoided as explained in Section 4. A critical review with respect to the use of the (modified) Gompertz model and its drawbacks and advantages for describing thermal inactivation kinetics can be found in Gil et al. (2011).

5.3. Model of Geeraerd et al. (2000)

Geeraerd et al. (2000) developed a mathematical model to describe microbial inactivation during a mild heat treatment. This model was also tested for several microorganisms.

$$\frac{dN}{dt} = k_{max} \cdot \left(\frac{1}{1+C_c}\right) \cdot \left(1 - \frac{N_{res}}{N}\right) \cdot N \quad (10)$$

$$\frac{dC_c}{dt} = -k_{max} \cdot C_c \quad (11)$$

with N the microbial cell density, N_res the residual population, C_c a parameter related to the physiological state of the cells and k_{max} the maximum specific inactivation rate. This model has four degrees of freedom: two parameters (k and N) and two initial states ($N(0)$ and $C(0)$).

The first factor at the right-hand side of Equation (10), i.e., $kmax$, models the loglinear part of the inactivation curve and reflects the classical first-order inactivation kinetics. The second factor, i.e., $(\frac{1}{1+C_c})$ describes the shoulder effect and is based on the hypothesis of the presence of a whole pool of protective components around or in each cell (Mossel et al., 1995). Gradually, this pool is destroyed. In case of a shoulder, $(\frac{1}{1+C_c})$ takes on a small (positive) value. Towards the end of the shoulder region $(\frac{1}{1+C_c})$ becomes (approximately) equal to one, due to the component C_c undergoing heat inactivation following a first-order relationship (see Equation (11)). Finally, the last factor of Equation (10), i.e., $(1 - \frac{N_{res}}{N})$, implies the existence of a more resistant subpopulation N_{res} [mechanistic conception according to Cerf (1977)]. The model encompasses loglinear inactivation by the selection (after identification on experimental data) of a very low value for $C(0)$ and N_{res}, implying the absence of a shoulder and a tail, respectively.

For static conditions, the model can be written as

$$N(t) = (N(0) - N_{res}) \cdot \exp(-k_{max} \cdot t) \cdot \left(\frac{1}{1 + C_c(0)}\right) \exp(-k_{max} \cdot t) + N_{res} \quad (12)$$

This model can be interpreted or, if necessary, adapted from a mechanistic point of view if, e.g., the critical component(s) and/or the mechanisms of their inactivation would be identified, or, if the tailing phenomenon would be elucidated.

Among others Velliou et al. (2011) used the Geeraerd model to quantify the effect of acid treatment on the heat tolerance of *Escherichia coli* K12 at 54°C and 58°C (refer also to Chapter 8). Cells were exposed to acid treatment and the effect on heat tolerance was quantified by the change the inactivation parameters, i.e., shoulder length and inactivation rate.

5.4. Weibull Type Models

Survival curves can be seen as an expression of the underlying statistical distribution of inactivation times t_c (van Boekel, 2002). Different models have been proposed to study microbial inactivation in a probabilistic way, and it turns out that the Weibull distribution is a simple and appropriate model for the analysis of survival curves (van Boekel, 2002; Peleg and Cole, 1998). The Weibull probability density function is

$$f(t) = \frac{\beta}{\alpha}\left(\frac{t}{\alpha}\right)^{\beta-1} \exp\left(-\left(\frac{t}{\alpha}\right)^{\beta}\right) \tag{13}$$

with $t, \alpha, \beta > 0$, and the cummulative distribution is given by

$$F(t) = \exp\left(-\left(\frac{t}{\alpha}\right)^{\beta}\right) \tag{14}$$

with α a scale parameter and β the so-called shape parameter.

In terms of survival curves, the cumulative function is

$$S(t) = \exp\left(-\left(\frac{t}{\alpha}\right)^{\beta}\right) \tag{15}$$

and

$$\log(S(t)) = -\frac{1}{2.303} \cdot \left(\frac{t}{\alpha}^{\beta}\right) \tag{16}$$

with $S(t) = \frac{N}{N_0}$.

The Weibull model corresponds to a concave upwards survival curve for $\beta < 1$ and to a concave downward curve for $\beta > 1$. For $\beta = 1$, the Weibull distribution reduces to an exponential distribution.

van Boekel (2002) published a study concerning the use of the Weibull model to describe thermal inactivation. The authors conclude that the Weibull model is a more flexible model than the classical thermal death time model using D and z values. The Weibull model is able to describe some frequently observed nonlinearities. Even though it is an empirical model, the value of the shape parameter can to some extent be linked to physiological effects, such as adaptation ($\beta < 1$) or accumulated damage ($\beta > 1$) and as such the model can signal those effects. Another effect that would be signaled using the Weibull model is that microorganisms may change thermoresistance depending on conditions. For instance, their thermal behaviour may be quite different when they have been adapted to certain stress conditions in foods or when they have grown in laboratory conditions.

van Boekel (2002) also checked the temperature dependency of the Weibull parameters α and β. The temperature dependence of the scale parameter α can be modelled as in the classical thermal dead time model, i.e., $\log(\alpha)$ is linearly related to temperature. As for the shape parameter β, there was no temperature dependency in a systematic way for the cases under study. The authors, however, note that possible parameter-temperature relationship may have gone unnoticed because of insufficient or too inaccurate data.

When comparing the Weibull model described above to the model requirements of Geeraerd et al. (2000), some limitations become visible. This specific structure of the Weibull model is not able to describe tailing effects and it is not given in a dynamic format. Moreover, the model (see Equations above) describes relative populations instead of absolute populations.

A Modified Weibull Model

Albert and Mafart (2005) published a reparametrization and an extension of the Weibull model to describe inactivation curves that include tailing effects. The modified model reads as follows

$$N\left(\frac{t}{\delta}\right) = (N_0 - N_{res}) \cdot 10^{\left(-\left(\frac{1}{\delta}\right)^p\right)} + N_{res} \tag{17}$$

The authors modified the Weibull model to meet the model requirements as listed in Geeraerd et al. (2000). The dynamic version of the model reads as follows

$$\frac{dN}{dt} = -\frac{\ln(10)}{\delta^p} \cdot p \cdot t^{p-1} \cdot \left(1 - \frac{N_{res}}{N}\right) \cdot N \tag{18}$$

with $\delta > 0$ and $p > 0$.

The modification presented by Albert and Mafart (2005) enable the use of the Weibull model under time varying condition. However, the modified Weibull model has the independent variable time t appearing in the right-hand side of the differential equation (equation 18). As a consequence, the momentary inactivation rate depends on the absolute time instance t (Janssen et al., 2008) (In that work the dynamic versions of Peleg and Penchina (2000); Peleg (2005) are assessed).

Gonzales et al. (2009) used the modified Weibull model to describe the survival of *Campylobacter jejuni* in minced chicken meat. The GInaFiT software (see Section 8.) was used to compare the Weibull model with the loglinear model. The Weibull model resulted in the best fit for the survival curves in this case study. Deboosere et al. (2010) took the modified Weibull model as published by Albert and Mafart (2005) as basis to describe the thermal inactivation of Hepatitis A virus in red berries.

5.5. Other Multiphasic Models

One of the first reported multiphasic models is that of Whiting (1993) which was derived from the logistic based model of Kamau et al. (1990). The model includes two populations, i.e., a major population and a subpopulation, one of them with larger thermal resistance.

This leads to a survivor curve with two distinct regions of linearly decreasing populations, the second with a less negative slope decreasing populations can be observed (which is, in this formulation, designated as tailing). Geeraerd et al. (2000) derived the dynamic form of this model.

$$\frac{dN}{dt} = -b_1 \cdot \left(1 - \frac{N(0)(1-F_1)}{N}\right) \cdot \left[1 - \frac{N - N(0)(1-F_1)}{N(0)F_1(1+\exp(-b_1 t_l))}\right] \cdot N \tag{19}$$

with b_1 the maximum specific death rate of the major population, b_2 the maximum specific death rate of the subpopulation, $F1$ the fraction of the initial population in the major population, F_2 = (1-F_1) the fraction of the initial population in the subpopulation, and t_l the shoulder or lag period. This model encompasses classical linear inactivation kinetics by choosing $t_l = -\infty$ (although this parameter value is physically unrealistic) and $F_1 = 1$.

Geeraerd et al. (2000) assessed some of the characteristics of the Whiting (1993) model and found the following limitations.

- A model without a shoulder can only be correctly attained by selecting $t_l = -\infty$, which is not a realistic biological value.
- This model does not meet all model requirements, since $N(0)$ and F_1, which is related to $N(0)$, are present in the right-hand side of Equation (19). As a result, the applicability of the model is limited to static conditions.
- A minor remark is the observation that t_l does not visually coincide with a time period where the bacterial population remains at the inoculum level

Geeraerd et al. (2005) proposed a model structure similar to the Whiting model but eliminating any previously observed constrains. In this model a shoulder followed by two different populations with respect to their resistance (biphasic survival curves with a shoulder). The two distinct survivor curve represents a decreasing thermal resistance of two microbial sub-populations. This model combines the interesting features of the Geeraerd et al. (2000) model and the Cerf (1977) model and reads as follows:

$$\frac{dN_1(t)}{dt} = -k_{max1} \cdot \frac{1}{1+C_c(t)} \cdot N_1(t) \tag{20}$$

$$\frac{dN_2(t)}{dt} = -k_{max2} \cdot N_2(t) \tag{21}$$

$$\frac{dC_c(t)}{dt} = -k_{max1} \cdot C_c(t) \tag{22}$$

As in the Cerf model, the total population N [cfu/mL] equals the sum of the two subpopulations N_1 and N_2, and a fraction f is defined as $N_1(0)/N_2(0)$. The model is based on the same mechanistic theory as the one presented for the Cerf (1977) model, i.e., a mix of two species or strains having different heat resistances are considered. The less heat resistant population exhibits a shoulder feature due to the protective environment established in the population. By the time that the most heat resistant population is dominant within the microbial environment the effect of the shoulder is eliminated.

In static conditions with the parameter S_l the model reads as follows

$$\begin{aligned}\log_{10}(N(t)) &= \log_{10}(N(0)) \\ &+\log_{10}\left(f \cdot e^{-k_{max1} \cdot t} \cdot \frac{e^{k_{max1} \cdot S_l}}{1+(e^{k_{max1} \cdot S_l}-1) \cdot e^{-k_{max1} \cdot t}} \right.\\ &\left. +(1-f) \cdot e^{-k_{max2} \cdot t} \cdot \left(\frac{e^{k_{max1} \cdot S_l}}{1+(e^{k_{max1} \cdot S_l}-1) \cdot e^{-k_{max1} \cdot t}} \right)^{\frac{k_{max1}}{k_{max2}}} \right)\end{aligned} \tag{23}$$

Herein, the parameters are $N(0)$, f, k_{max1}, k_{max2}, and S_l. All parameter values have a clear significance and the different segments of the survivor curve (shoulder, biphasic pattern) are easily recognisable. If $S_l = 0$ (after identification on experimental data), the model reduces in a natural way to the Cerf model.

6. Mixed Effect Inactivation Models

In a very general definition, mixed effect models are statistical models that take both fixed effects and random effects into account. In case of fixed effects, the values of explanatory variables are fixed and each level is of interest. For random effects, on the other hand, the values arise from random causes and the levels represent only a sample from the population. In the case of random effects, it is the variability between the different levels which is of interest, rather than the specific level. In general mixed models can be formulated as follows

$$y = X \cdot \beta + Z \cdot u + e \tag{24}$$

with y the observations, e.g., cell concentration, X the explanatory variables with fixed effects β, Z the explanatory variable with random effects u and e an error term.

Juneja and Marks (2005) used non-linear mixed effect models to describe the heat resistance kinetics among various isolates of *E. coli*. They assumed that the studied serotypes represent a random sample from a population of *E. coli* strains or serotypes. The underlying distribution was assumed to be normal. Probability intervals of the expected lethality for random selected serotypes were computed from the derived estimates. In another article by the same authors, mixed models were used to calculate thermal inactivation of *Salmonella* in poultry of different fat levels (Juneja et al., 2001). In this research, the observations for a given temperature and for a given fat level were considered as random variables. Jordan et al. (2011) evaluated different expressions to describe the inactivation kinetics for a four-strain composite of *Salmonella* Enteritidis and Oranienburg in commercial liquid egg yolk and used mixed effects nonlinear regression to fit the survival curves. The samples at a specific time were considered a random variable following a Poisson distribution.

7. Overview of Inactivation Models

The inactivation models discussed in this chapter are summarised in Table 1. A first major subdivision is based on the suitability of the models for use in dynamic conditions. Next, columns are used to indicate the possibility (or absence of the possibility) to describe specific survivor type curve, as introduced in Figure 1.

Table 1. Model features (proposed for vegetative microorganisms) with respect to the different types of inactivation curves (see Figure 1) and their design characteristics. +: for models describing the specific type of curves, -: not able to describe the type of curves. Table content is partly originating from Xiong et al. (1999); Geeraerd et al. (2000); Devlieghere et al. (2004)

	Loglinear	Loglinear + tail	Sigmoidal -like	Loglinear +shoulder	Biphasic	Concave (multiphasic)	Biphasic +shoulder	Convex (multiphasic)
Models suitable for use in both constant and in changing environmental conditions								
Loglinear	+	-	-	-	-	-	-	-
Cerf	+	+	-	-	+	-	-	-
Geeraerd	+	+	+	+	-	-	-	-
Extended Geeraerd	+	+	+	+	+	-	+	-
Models only to be used under constant environmental conditions or by approximating a time-varying environment as successive static conditions (that is, the model is inherently static, or inconsistent at time zero, or inconsistent at changing conditions)								
Modified Gompertz	-	-	+	-	-	-	-	-
Whiting	+[a]	+	+	+	+	-	+	-
Weibull	+	-	-	-	-	+	-	+
Modified Weibull	+	+	-	-	-	+[b]	-	+[b]

[a]The model only approximates loglinear behaviour
[b]The model can also describe a subsequent tailing effect

8. Modelling Softwares

Geeraerd et al. (2005) created a freeware add-in for Microsoft Excel (GInaFiT) (http://cit.kuleuven.be/biotec/downloads.php) which can describe ten different types of microbial survival curves: (i) classical loglinear curves, (ii) curves displaying a shoulder before a loglinear decrease, (iii) curves displaying a tail after a loglinear decrease, (iv) survival curves displaying both shoulder and tailing behavior, (v) concave curves, (vi) convex curves, (vii) convex/concave curves followed by tailing, (viii) biphasic inactivation kinetics, (ix) biphasic inactivation kinetics preceded by a shoulder, and (x) curves with a double concave/convex shape. Next to the derivation of parameter estimates and standard errors, some goodness-of-fit indicators are also calculated. The GInaFiT tool can help end-users to communicate the performance of food preservation processes using the concept of the number of log cycles of reduction, rather than the classical D-value. Baty and Delignette-Muller developed a similar tool (survivalmodels) for describing bacteria survival curves of the previously described shapes in an R environment (package 'nlstools').

Other softwares include the tool for simulation studies by the use of Weibull model at static and dynamic conditions (http://www-unix.oit.umass.edu/aew2000/GrowthAndSurvival/GrowthAndSurvival.html) (Peleg et al., 2008). The Pahtogen Modeling Program (PMP) and the AMI Process Lethality Determination Spreadsheet are also widespread simulation tools, including simulations for thermal inactivation. PMP (pmp.arserrc.gov) offers a user interface for simulation of some predictive models described in literature. These predictions are specific to certain bacterial strains and specific environments (e.g., culture media, food, etc.) that were used to generate the models. Additionally, the AMI Process Lethality Determination Spreadsheet (www.amif.org) is an excel sheet which allows food processors to evaluate the efficiency of a specific cooking process, by calculating $D/z/F$-values.

ComBase (http://www.combase.cc) and Sym'Previous (www.symprevius.net) are two initiatives which offer a database of microbial responses to different food environments. Besides the database, simulation and/or modelling tools are provided. The ComBase modelling toolbox offers a collection of software tools based on the ComBase database, e.g., the ComBase Predictor and DMFit. The ComBase predictor is a simulation tool which includes thermal death models for foodborne pathogenic and spoilage microorganisms. DMfit allows to fit a growth/survival model to experimental data. Both tools are based on the model of Baranyi and Roberts (1994). Also Sym'Previous contains several modules besides a database, e.g., the thermal destruction simulator for some important pathogenic species. This module calculates the reduction rate at the end of a specified treatment and the probability of a microorganism to survive. In another study Halder et al. (2010) analyzed data from published literature to construct a database with associations of foods and target microorganisms. The aim of this type of database was to obtain group specific kinetic data which should allow to extend simulations to a broader range of products and processes. Halder et al. (2011) presented an integrated software which combines physics-based models of the process, e.g., heat transfer trough the food product, with the kinetics of microbiological and chemical changes during processing.

Conclusion

Accurate modelling of thermal inactivation of pathogens and spoilage organisms in food can be a very valuable tool in assessing food safety and quality. Survival curves are traditionally modelled by loglinear kinetics and this delivers satisfying results for inactivation at high temperatures, e.g., during a sterilization process. For modern food treatments, however, deviations from loglinearity are observed and a different model structured is required. In this chapter, an overview and evaluation is given of widely used primary model structures for thermal inactivation, i.e., two fraction model of Cerf (1977), the modified Gompertz model, the model of Geeraerd et al. (2000), Weibull type models, multiphasic models and mixed effects models.

The loglinear model, i.e., the D/z-concept remains of great value to obtain a first impression on the performance of an inactivation process, e.g., in cases were elaborated knowledge and the necessary tools for complicated models are not available. When modelling microbial dynamics in food, uncertainty and variability of both food and microbiology play an important role and the effort to move towards more complex kinetic models needs to be weighed against inclusion of variability.

The past decade some user-friendly software tools have become available which make (inactivation) models accessible to a broader public, e.g., food producers. Basic knowledge of predictive microbiology, however, is required since the added value of the software will dependent on the correct interpretation of the results.

References

I Albert and P Mafart. A modified Weibull model for bacterial inactivation. *International Journal of Food Microbiology*, **100**(1-3):197–211, 2005.

J Baranyi and TA Roberts. A dynamic approach to predicting bacterial-growth in food. *International Journal of Food Microbiology*, **23**(3-4), 1994.

K Bernaerts, E Dens, K Vereecken, Geeraerd AH, A Standaert, F Devlieghere, J Debevere, and JF Van Impe. Concepts and tools for predictive modeling of microbial dynamics. *Journal of Food Protection*, **67**(9):2041–2052, 2004.

S Bhaduri, PW Smith, SA Palumbo, CO Turner-Jones, JL Smith, BS Marmer, RL Buchanan, LL Zaika, and AC Williams. Thermal destruction of *Listeria monocytogenes* in liver sausage slurry. *Food Microbiology*, **8**:75–78, 1991.

WD Bigelow. The logarithmic nature of thermal death time curves. *The Journal of Infectious Diseases*, **29**(5):528–536, 1921.

WD Bigelow and JR Esty. The thermal death point in relation to typical thermophylic organisms. *Journal of Infectious Diseases*, **27**:602–617, 1920.

O Cerf. A review. tailing of survival curves of bacterial spores. *Journal of Applied Bacteriology*, **42**:1–19, 1977.

H Chick. An investigation of the laws of disinfection. *Journal of Hygiene*, **8**:92–158, 1908.

N Deboosere, A Pinon, A Delobel, S Temmam, B Morin, G Merle, S Blaise-Boisseau, S Perelle, and M Vialette. A predictive microbiology approach for thermal inactivation of hepatitis a virus in acidified berries. *Food microbiology*, **27**:962–967, 2010.

E Dens and J Van Impe. On the need of another type of predictive models in strucured foods. *International Journal of Food Microbiology*, **64**:247–260, 2001.

F Devlieghere, K Francois, KM Vereecken, AH Geeraerd, JF Van Impe, and J Debevere. Effect of chemicals on the microbial evolution in foods. *Journal of Food Protection*, **67**: 1977–1990, 2004.

JF France and JHM Thornley. Growth functions. In *Mathematical models in agriculture*, book chapter 5, pages 75–94. 1984.

S Furukawa, N Narisawa, T Watanabe, T Kawarai, K Myozen, S Okazaki, H Ogihara, and M Yamasaki. Formation of the spore clumps during heat treatment increases the heat resistance of bacterial spores. *International Journal of Food Microbiology*, **102**(1):107–111, 2005.

AH Geeraerd, CH Herremans, H Van Damme, and J Van Impe. Structural model requirements to describe microbial inactivation. *International Journal of Food Microbiology*, **59**:185–209, 2000.

AH Geeraerd, V Valdramidis, and JF Van Impe. GInaFiT, a freeware tool to assess non-log-linear microbial survivor curves. *International Journal of Food Microbiology*, **102** (1):95–105, 2005.

AM Gibson, N Bratchell, and TA Roberts. The effect of sodium chloride and temperature on the rate and extend of growth of *Clostridium botulinum* type a in pasteurized pork slurry. *Journal of Applied Bacteriology*, **62**:479–490, 1987.

MM Gil, FA Miller, TRS Brandao, and CLM Silva. On the use of the gompertz model to predict microbial thermal inactivation under isothermal and non-isothermal conditions. *Food Engineering Review*, **3**:17–25, 2011.

MM Gil, TRS Brandao, and CLM Silva. A modified Gompertz model to predict microbial inactivation under time-varying temperature conditions. *Journal of Food Engineering*, **76**(1):89–94, 2006.

M Gonzales, PN Skandamis, and M-L Hanninen. A modified weibull model for describing the survival of *Campylobacter jejuni* in minced chicken meat. *International Journal of Food Microbiology*, **136**:52–58, 2009.

IR Grant, HJ Ball, and MT Rowe. Effect of higher pasteurization temperatures, and longer holding times at 72 degrees C, on the inactivation of *Mycobacterium paratuberculosis* in milk. *Letters in Applied Microbiology*, **28**(6):461–465, 1999.

A Halder, AK Datta, and SSR Geedipalli. Uncertainty in thermal process calculations due to variability in first-order and weibull kinetic parameters. *Journal of Food Science*, **72** (4):E155–E167, 2007.

A Halder, DG Black, PM Davidson, and A Datta. Development of Associations and Kinetic Models for Microbiological Data to Be Used in Comprehensive Food Safety Prediction Software. *Journal of Food Science*, **75**(6):R107–R120, AUG 2010. doi: 10.1111/j.1750-3841.2010.01687.x.

A Halder, A Dhall, AK Datta, DG Black, PM Davidson, J Li, and S Zivanovic. A user-friendly general-purpose predictive software package for food safety. *Journal of Food Engineering*, **104**(2):173–185, MAY 2011. doi: 10.1016/j.jfoodeng.2010.11.021.

APM Hasting, CDW Blackburn, and JS Crowther. *Mycobacterium paratuberculosis* and the commercial pasteurization of milk. *Food and Bioproducts Processing*, **79**(C2):83–88, 2001.

L Huang. Thermal inactivation of listeria monocytogenes in ground beef under isothermal and dynamic temperature conditions. *Journal of Food Engineering*, **90**:380–387, 2009.

L Humpheson, MR Adams, WA Anderson, and MB Cole. Biphasic thermal inactivation kinetics in *Salmonella enteritidis* PT4. *Applied and Environmental Microbiology*, **64**(2): 459–464, 1998.

M Janssen, A Verhulst, V Valdramidis, F Devlieghere, JF Van Impe, and AH Geeraerd. Inactivation model equations and their associated parameter values obtained under static acid stress conditions cannot be used directly for predicting inactivation under dynamic conditions. *International Journal of Food Microbiology*, **128**:136–145, 2008.

JS Jordan, JB Gurtler, HM Marks, DR Jones, and WK Shaw. A mathematical model of inactivation kinetics for a four-strain composite of salmonella enteritidis and oranienburg in commercial liquid egg yolk. *Food Microbiology*, **28**:67–75, 2011.

VK Juneja and HM Marks. Heat resistance kinetics variation among various isolates of *Escherichia coli*. *Innovative Food Science & Emerging Technologies*, **6**(2):155–161, 2005.

VK Juneja, BS Eblen, and HM Marks. Modeling non-linear survival curves to calculate thermal inactivation of Salmonella in poultry of different fat levels. *International Journal of Food Microbiology*, **70**(1-2):37–51, OCT 22 2001. ISSN 0168-1605.

DN Kamau, S Doores, and KM Pruitt. Enhanced thermal destruction of *Listeria monocytogenes* and *Staphylococcus aureus* by the lactoperoxidase system. *Applied and Environmental Microbiology*, **56**(9):2711–2716, 1990.

RH Linton, WH Carter, MD Pierson, and CR Hackney. Use of a modified gompertz equation to model nonlinear survival curves for *Listeria monocytogenes* scott a. *Journal of Food Protection*, **58**(9):946–954, 1995.

P. Mafart, I. Leguerinel, O. Couvert, and L. Coroller. Quantification of spore resistance for assessment and optimization of heating processes: A never-ending story. *Food Microbiology*, **27**(5):568 – 572, 2010. ISSN 0740-0020. doi: DOI: 10.1016/j.fm.2010.03.002.

FA Miller, BF Ramos, MM Gil, TRS Brandao, P Teixeira, and CLM Silva. Heat inactivation of *Listeria innocua* in broth and food products under non-isothermal conditions. *Food Control*, **22**:20–26, 2011.

Fatima A. Miller, Barbara F. Ramos, Maria M. Gil, Teresa R. S. Brandao, Paula Teixeira, and Cristina L. M. Silva. Heat inactivation of Listeria innocua in broth and food products under non-isothermal conditions. *FOOD CONTROL*, **22**(1):20–26, JAN 2011. ISSN 0956-7135. doi: 10.1016/j.foodcont.2010.06.004.

WA Moats, R Dabbah, and VM Edwards. Interpretation of nonlogarithmic survivor curves of heated bacteria. *journal of Food Science*, **36**:523–526, 1971.

DAA Mossel, JEL Corry, CB Struijk, and RM Baird. *Essentials of the microbiology of foods: A textbook for advanced studies*. John Wiley & Sons, New York, 1995. 0 471 93036 9.

MJ Nauta. *Modelling Microorganisms in Food*, chapter Uncertainty and variability in predictive models of microorganisms in food. CRC Press, 2006.

M Peleg and MB Cole. Reinterpretation of microbial survival curves. *Critical Reviews in Food Science and Nutrition*, **38**(5):353–380, 1998.

Micha Peleg and Claude M. Penchina. Modeling microbial survival during exposure to a lethal agent with varying intensity. *Critical Reviews in Food Science and Nutrition*, **40** (2):159–172, 2000. doi: 10.1080/10408690091189301.

Micha Peleg, Mark D. Normand, and Maria G. Corradini. Interactive software for estimating the efficacy of non-isothermal heat preservation processes. *International Journal of Food Microbiology*, **126**(1-2):250 – 257, 2008. ISSN 0168-1605. doi: DOI: 10.1016/j.ijfoodmicro.2008.05.004.

Normand M.D. Corradini M.G. Peleg, M. Generating microbial survival curves during thermal processing in real time. *Journal of Applied Microbiology*, **98**(2):406–417, 2005. cited By (since 1996) 33.

SG Rudra, S Chhibber, and US Shivhare. Thermal death kinetics of b. stearothermophilus in coriander leaf puree. *Biosystems Engineering*, **106**:544–550, 2010.

CR Stumbo, KS Purohit, and TV Ramakrishnan. Thermal process lethality guide for low-acid food in metal containers. *journal of Food Science*, **40**:1316–1323, 1975.

V.P. Valdramidis, N. Belaubre, R. Zuniga, A.M. Foster, M. Havet, A.H. Geeraerd, M.J. Swain, K. Bernaerts, J.F. Van Impe, and A. Kondjoyan. Development of predictive modelling approaches for surface temperature and associated microbiological inactivation during hot dry air decontamination. *International Journal of Food Microbiology*, **100**

(1-3):261 – 274, 2005. ISSN 0168-1605. doi: DOI: 10.1016/j.ijfoodmicro.2004.10.025. The Fourth International Conference on Predictive Modelling in Foods.

V.P. Valdramidis, A.H. Geeraerd, J.E. Gaze, A. Kondjoyan, A.R. Boyd, H.L. Shaw, and J.F. Van Impe. Quantitative description of listeria monocytogenes inactivation kinetics with temperature and water activity as the influencing factors; model prediction and methodological validation on dynamic data. *Journal of Food Engineering*, **76**(1):79 – 88, 2006. ISSN 0260-8774. doi: DOI: 10.1016/j.jfoodeng.2005.05.025. Bugdeath.

V.P. Valdramidis, A.H. Geeraerd, K. Bernaerts, and J.F.M. Van Impe. Identification of non-linear microbial inactivation kinetics under dynamic conditions. *International Journal of Food Microbiology*, **128**(1):146 – 152, 2008. ISSN 0168-1605. doi: DOI: 10.1016/j.ijfoodmicro.2008.06.036. 5th International Conference on Predictive Modelling in Foods.

V.P. Valdramidis, A.H. Geeraerd, B.K. Tiwari, P.J. Cullen, A. Kondjoyan, and J.F. Van Impe. Estimating the efficacy of mild heating processes taking into account microbial non-linearities: A case study on the thermisation of a food simulant. *Food Control*, **22** (1):137 – 142, 2011. ISSN 0956-7135. doi: DOI: 10.1016/j.foodcont.2010.05.007.

ED van Asselt and MH Zwietering. A systematic approach to determine global thermal inactivation parameters for various food pathogens. *International Journal of Food Microbiology*, **107**:73 – 82, 2006.

MAJS van Boekel. On the use of the Weibull model to describe thermal inactivation of microbial vegetative cells. *International Journal of Food Microbiology*, **74**(1-2):139–159, 2002.

JF Van Impe, BM Nicolai, T Martens, J Debaerdemaeker, and J Vandewalle. Dynamic mathematical-model to predict microbial-growth and inactivation during food-processing. *Applied and Environmental Microbiology*, **58**(9):2901–2909, 1992.

JF Van Impe, F Poschet, AH Geeraerd, and KM Vereecken. Towards a novel class of predictive microbial growth models. *International Journal of Food Microbiology*, **100** (1-3):97–105, 2005.

Andre van Zuijlen, Paula M. Periago, Alejandro Amezquita, Alfredo Palop, Stanley Brul, and Pablo S. Fernandez. Characterization of bacillus sporothermodurans ic4 spores; putative indicator microorganism for optimisation of thermal processes in food sterilisation. *Food Research International*, **43**(7):1895 – 1901, 2010. ISSN 0963-9969. doi: DOI: 10.1016/j.foodres.2009.11.011. Climate Change and Food Science.

E. G. Velliou, E. Van Derlinden, A. M. Cappuyns, E. Nikolaidou, A. H. Geeraerd, F. Devlieghere, and J. F. Van Impe. Towards the quantification of the effect of acid treatment on the heat tolerance of Escherichia coli K12 at lethal temperatures. *Food Microbiology*, **28**(4, Sp. Iss. SI):702–711, JUN 2011. ISSN 0740-0020. doi: 10.1016/j.fm.2010.06.007. 6th International Conference of Predictive Modeling in Foods (6ICPMF), Washington, DC, SEP 08-12, 2009.

RC Whiting. Modeling bacterial survival in unfavorable environments. *Journal of Industrial Microbiology*, **12**:240–246, 1993.

R Xiong, G Xie, AE Edmondson, and MA Sheard. A mathematical model for bacterial inactivation. *International Journal of Food Microbiology*, **46**:45–55, 1999.

MH Zwietering, I Jongenburger, FM Rombouts, and K van 't Riet. Modeling of the bacterial growth curve. *Applied and Environmental Microbiology*, **56**(6):1875–1881, 1990.

In: Progress on Quantitative Approaches in Thermal … ISBN: 978-1-62100-842-2
Editors: V. P. Valdramidis and J. F. M. Van Impe © 2012 Nova Science Publishers, Inc.

Chapter 5

Modelling Microbial Resistance during Thermal Treatment: Secondary Models

Pierre Mafart, Ivan Leguerinel, Louis Coroller and Olivier Couvert

Université Européenne de Bretagne, France
Université de Brest, Laboratoire Universitaire de Biodiversité et Ecologie Microbienne, Quimper, France

Abstract

During an early period, the only existing secondary models were the Arrhenius equation and the Bigelow relationship (*z*-concept) which ignored other environmental factors than the heating temperature. The first model combining heating temperature and pH of the medium appeared as late as 1978. Since this date, some polynomial or modular multi-factorial models aiming to describe the effect of the main environmental factors upon the thermal resistance of microorganisms were proposed. First generation models consider the effects of some prominent environmental factors such as pH, water activity of the heating medium, the presence of antimicrobial compounds etc., regardless of the recovery conditions after the heat treatment. Second generation models take the conditions of recovery into account for describing their effects on the so called apparent heat resistance.

While the polynomial models have the main advantage of taking the possible interactions between factors into account, the modular approach yields more robust and parsimonious models including parameters easily interpretable. Furthermore, this type of models allows flexible combinations between mono-factorial sub-models.

If it is relatively easy to account for the major part of the variance by considering a few prevalent factors, the account of the remaining part of variance would require drastic complications of models. For this reason, it is worth simplifying and standardizing heat processes calculations. Despite the imperfection of new multi-factorial models, it is clear that they are likely to bring substantial improvements with respect to the early simple thermal models.

Introduction

While a number of mechanistic or empirical primary models were developed for describing the different shapes of survival curves, there is a limited number of secondary models describing the resistance parameters of primary models as functions of environmental factors. Moreover, regarding the secondary models, researchers were satisfied with purely descriptive empirical models. The famous Arrhenius equation which was considered as the first secondary model is in reality a simple adjustment function, even though it is often wrongly regarded as an equation with a theoretical basis.

For a long period, the only considered factor was the heating temperature. Although, the effect of some environmental characteristics of the heating medium such as the pH, water activity and salts content upon the heat resistance of bacteria has been recognised for a number of decades, very few attempts were made to develop multi-factorial models in order to quantify these effects. The dramatic influence of temperature tended to screen the effect of other factors, so that improvements of predictive modelling have been very scarce since the early mono-factorial models of the early 20th century. However, drastic effects of non thermal environmental factors on decimal reduction times can be observed: for example, a *D*-value of 0.3 minute was found for a strain of *Saccharomyces cerevisiae* heated at 60°C in grape juice, while a *D*-value of 0.0015 minute was found for the same yeast species heated at the same temperature in dry wine of pH 3.5 and 12% ethanol (Courtoisier, 1984). Except for observations by Jordan and Jacobs (1948) who noticed a linear relationship between pH of the heating medium and the logarithm of the decimal reduction time of *Escherichia coli*, no further modelling of the influence of non thermal factors had been developed before 1978, when Davey et al. proposed the first bi-factorial model aiming to quantify the combined effects of temperature and pH on the destruction rate of *Clostridium botulinum.*

After a recall of the classical mono-factorial equations describing the effect of the heating temperature on the heat resistance of cells, this chapter presents the first multi-factorial generation models which ignore recovery conditions and lastly, the second generation models which take into account the conditions of recovery.

1. Classical Thermal Equations

1.1. The Arrhenius Equation (1889)

It can be said that, historically, the first secondary model applied to predictive food microbiology was the Swedish chemist Arrhenius equation which was recovered by microbiologists for linking the specific rate of destruction *k* of bacterial spores and the heat treatment temperature:

$$k = Be^{-\frac{E_a}{RT}} \tag{1}$$

Where *T* is the absolute temperature, *R*, the perfect gas 'constant (8.314 J.mole^{-1}°K^{-1}), E_a, the activation energy and *B*, an empirical pre-exponential constant. For example, in the case of *Clostridium botulinum*, the following parameter values were found (Simpson and Williams, 1974):

$$E_a = 310.1 \text{ kJ.mole}^{-1}$$

$$B = 2.10^{40} \text{ s}^{-1}$$

As it can be seen, the parameter *B* has no real meaning and its order of magnitude incites to eliminate it. By denoting k^* the value of the specific inactivation rate at a reference temperature T^*, equation (1) can be easily modified into the following form:

$$k = k^* \exp\left(-\frac{E_a}{R}\left(\frac{1}{T} - \frac{1}{T^*}\right)\right) \tag{2}$$

In the cited example, the pre-exponential constant *B* is then replaced by the following parameter which has a clear meaning: $k^* = 0.158\ s^{-1}$ for a reference temperature of 121.1°C.

1.2. The Chick Equation (1910)

Ten years before Bigelow, Harriet Chick found an alternative model by observing a linear relationship between the logarithm of the specific inactivation rate and temperature. Her equation can be written as follows:

$$Lnk = a + (Lnc)T \tag{3}$$

Where *a* is an empirical constant and *c* is, what the author called *temperature coefficient* which corresponds to the multiplication factor of the inactivation rate yielded by an increase of the heating temperature of 1°C. For example, she found for *Bacillus thyphosus* a temperature coefficient of 1.62°C^{-1}, (which is equivalent to a z-value of 4.8°C). Comparing her equation with that of Arrhenius, she could not find any difference of goodness of fit.

1.3. The Bigelow Equation (1921)

Based on the same linear relationship to that of Chick, the Bigelow equation could have been written as follows:

$$\log k = \log k^* + \frac{T - T^*}{z} \tag{4}$$

In reality, Bigelow seems to have ignored the specific inactivation rate concept and characterised the heat resistance of bacterial spores by his well-known *TDT* concept (*Thermal Death Time*) which corresponded to the extinction time of a population at a given temperature, so that the original Bigelow equation has then to be written as follows:

$$\log TDT = \log TDT^* - \frac{T - T^*}{z} \tag{5}$$

Where *TDT** is the *TDT* value at the reference temperature T^*, and *z*, the increase of temperature yielding a ten-fold *TDT* reduction.

Nowadays, the Bigelow relationship is generalised to the link of two temperature/time (*T*, *t*) combinations yielding the same destruction ratio:

$$\log t_2 = \log t_1 - \frac{T_2 - T_1}{z} \tag{6}$$

Lastly, since a decimal reduction corresponds to a particular value of any destruction ratio, equation (6) can be obviously applied to the D-value as follows:

$$\log D = \log D^* - \frac{T - T^*}{z} \tag{7}$$

The association of the D and z parameters is so deep-rooted in the minds, that the z concept is often and wrongly associated to log-linear survival kinetics. In reality, equation (6) is valid whatever the shape of survival curves and the Bigelow relationship can be applied to any primary resistance parameter of any non-log-linear model, provided it has the dimensions of a time. For example, this relationship was successfully applied to fit the effect of temperature on the scale parameter of the Weibull frequency distribution modified by Mafart (2002) (Couvert et *al.,* 2005).

2. First Generation Multi-Factorial Models

While the previously cited equations consider only the heating temperature, a first generation of multi-factorial models includes some other important environmental factors such as pH, water activity of the heating medium, the presence of antimicrobial compounds etc., regardless of the recovery conditions after the heat treatment. These models can be built according to two complementary approaches: the classical polynomial approach which was very popular in the end of the previous century and the modular approach which combines several elementary mono-factorial sub-models. It is difficult to deal separately with these two categories of models because there is no clear borderline between both. The main advantage of polynomial models is their ability to consider easily the possible interactions between factors. Moreover, their linearity allows an easy estimation of confidence intervals of parameters. Unfortunately, such models lack of parsimony and are often over-parameterized because their authors forget any statistical screening of the factors having a significant effect. On the one hand, this lack of parsimony is correlated with a lack of robustness. On the other hand, the included parameters have no interpretable physical or chemical meaning.

Some complete quadratic models were developed to describe the heat resistance of spores (Junega et al.,1995; Splittstoesser et al., 1995). Fernandez et al., (1996) studied combined effects of temperature and pH on the heat resistance of *Bacillus stearothermophilus* and *Clostridium sporogenes*. They proposed two alternative models: a complete quadratic model:

$$\log k = a_0 + a_1 T + a_2 pH + a_3 TpH + a_4 T^2 + a_5 pH^2 \tag{8}$$

And a simple linear equation:

$$\log k = C_0 + C_1 T + C_2 pH \tag{9}$$

Results presented by the authors clearly showed that the linear equation could successfully describe the effects of both considered factors and that the complete quadratic model improved very slightly the accuracy of modeling (for example, a R^2 of 0.997 instead of 0.992 for *Bacillus stearothermophilus* heated in a medium acidified with citric acid, or 0.908

instead of 0.900 for *Clostridium sporogenes* at the same conditions). If the pH term is left out of equation (9) it then corresponds to a shape of the Bigelow model, the accuracy of which is well known, and therefore the square temperature term of equation (8) is unnecessary.

Davey and *al.* (1978) added to the logarithmic shape of the Arrhenius equation a pH and a squared pH term to describe the combined effects of temperature and pH on the destruction rate of *Clostridium botulinum* from data of the literature (Xenones and Hutchings, 1965). They obtained the following reduced polynomial equation:

$$Lnk = C_0 + \frac{C_1}{T} + C_2 pH + C_3 pH^2 \qquad (10)$$

where T is the absolute temperature. With respect to a complete quadratic model, the Davey model presents the advantage of being more parsimonious (4 parameters instead of 6) for a satisfactory goodness of fit. However, like in the case of any polynomial model, coefficients remain without any biological meaning. As an example, authors found for a set of data regarding *Clostridium botulinum*:

$$C_0 = 105.23$$

$$C_1 = -\ 3.7041.10^{-4}\ °K$$

$$C_2 = -\ 2.3967$$

$$C3 = 0.1695$$

Following a similar approach, Mafart and Leguérinel (1998) extended the Bigelow model by addition of a pH term and fitted three sets of data of the literature related to *Clostridium botulinum* (Xenones and Hutchings, 1965), *Clostridium sporogenes* (Cameron et al., 1980) and *Bacillus stearothermophilus* (Lopez et *al.,* 1996):

$$\log D = \log D^* - \frac{T - T^*}{z_T} - \left(\frac{pH - pH^*}{z_{pH}} \right)^2 \qquad (11)$$

where T^* is the reference temperature, pH^* is the reference pH = 7 which is close to the pH of maximum heat resistance of spores, z_T is the conventional z-value, z_{pH} is the distance from pH^* which leads to a tenfold reduction of the decimal reduction time. Lastly, D^* is the D-value at T^* and pH^*. For example, fitting the previous cited set of data leads to the following estimates:

$$D^* = 0.139 \text{ min}$$

$$z_T = 9.29°C$$

$$z_{pH} = 3.61$$

As illustrated by the figure 1, the goodness of fit of the Davey and the Mafart model are quite similar.

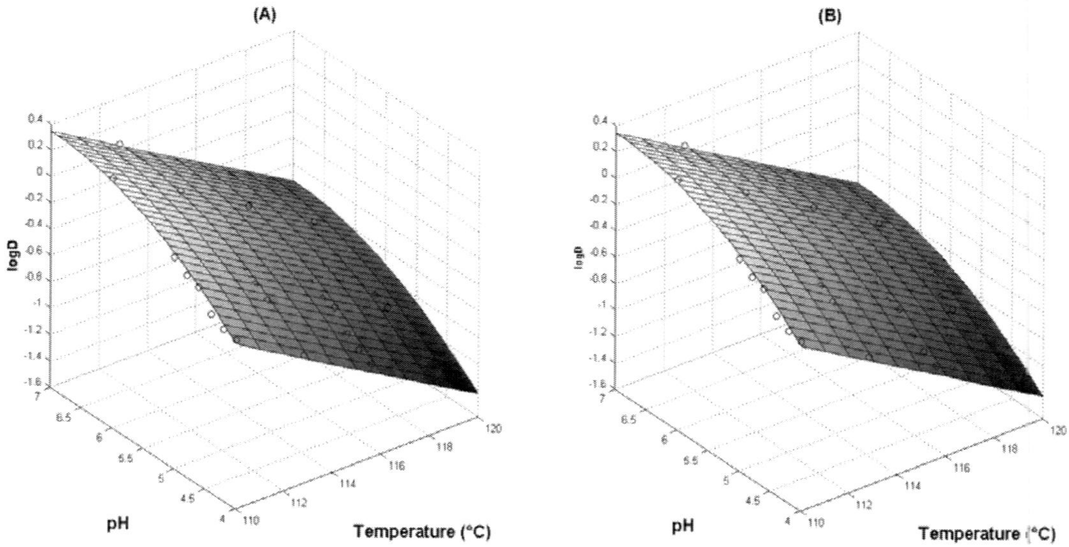

Figure 1. Fitting of the heat resistance of Clostridium botulinum 62A spores from the Davey model (A) and from the Mafart model (B). (Data from Xenones and Hutchings, 1965).

It can be difficult to find the ideal compromise between simplicity and accuracy of models. It can be shown that the four parameter model of Davey is over-parameterized while the three parameter model of Mafart is under-parameterized.

According to the Davey model, the pH of maximal heat resistance is such as:

$$\frac{\partial(Lnk)}{\partial pH} = C_2 + 2C_3 pH = 0 \tag{12}$$

The pH of maximum heat resistance is then:

$$pH_{opt} = -\frac{C_2}{2C_3} \tag{13}$$

Considering that pH_{opt} value is generally close to 7, it becomes clear that parameters C_2 and C_3 are strongly correlated. If pH_{opt} is fixed, either parameter can be then eliminated. Unsurprisingly, a pH_{opt} value of 7.07 can be calculated from equation (13) and the cited values of C_2 and C_3.

While the Mafart model assumes a quadratic relationship between the logarithm of *D*-value, and pH of the heating medium, several authors observed a simple linear relationship between both factors (Jordan and Jacob, 1948; Mazas et al., 1998; Pirone et *al.*, 1987). That means that the model is under-parameterized and would be more general under the following shape:

$$\log D = \log D^* - \frac{T - T^*}{z_T} - \left(\frac{pH - pH^*}{z_{pH}}\right)^n \tag{13}$$

Where n is a further parameter comprised between 1 and 2. However this new equation would present the drawback of being no longer linear, so it is worthwhile to fix the n value to 1 or 2 depending on the type of the data.

The common limitation of the Davey and the Mafart models is that they ignore the possible interactions between factors, while the existence of such interactions is well known. Investigating the heat resistance of *Bacillus cereus*, Gaillard et al. (1998) attempted to circumvent this drawback by modifying equation (11) as follows:

$$\log D = \log D^* + C_1(T - T^*) + C_2(pH - pH^*)^2 + C_3(T - T^*)(pH - pH^*)^2 \quad (14)$$

This modification produced a relatively poor improvement of the goodness of fit of the initial model ($R^2 = 0.985$ instead of 0.977). Such an improvement was judged insufficient for justifying the addition of one more parameter and the loss of the biological interpretability of all parameters, except D^*.

From the neperian version of the Weibull model, Fernandez et al. (2002) modeled the scale parameter α as follows:

$$Ln\frac{1}{\alpha} = Ln\frac{1}{\alpha^*} + \omega(pH - pH^*) - \frac{E_a}{R}\left(\frac{1}{T} - \frac{1}{T^*}\right) \quad (15)$$

At a given pH, the type of involved organic acid has a clear effect on heat resistance in addition to the H^+ ion. For nine organic acids, Leguérinel and Mafart (2001) found, for *Bacillus cereus*, a linear relationship between the calculated z_{pH} value and their lower acid pK_a.

$$z_{pH} = -3.37\,pK_a + 21.23 \quad (16)$$

The first model taking into account the effect of water activity of the medium on the rate of destruction of bacteria is the mechanistic Reichard model (1994) which fitted data regarding *Eschericia coli*. Unfortunately, as this equation is not linear and it especially requires the knowledge of hydrogen and hydroxyl ions concentrations, it cannot be easily applied in the area of predictive food microbiology.

Cerf et al., (1996) fitted data of Reichard according to an extension of the Davey model:

$$\ln k = C_0 + \frac{C_1}{T} + C_2 pH + C_3 pH^2 + C_4 aw^2 \quad (17)$$

Similarly, the Mafart model was extended from data related to *Bacillus cereus* by addition of a water activity term (Gaillard et al., 1998):

$$\log D = \log D^* - \frac{T - T^*}{z_T} - \left(\frac{pH - pH^*}{z_{pH}}\right)^2 - \frac{aw - 1}{z_{aw}} \quad (18)$$

where z_{aw} corresponds to the decrease of water activity value of the treatment medium which would cause a tenfold reduction of the decimal reduction time. The z_{aw} is dependent, not only on the concerned type of bacteria, but also, on the depressor which cause a decrease of water activity. For example, for *Bacillus cereus*, Coroller et al. (2001) found a z_{aw} value of 0.28 with glycerol and glucose and 0.13 with sucrose.

Microbial resistance depends not only of environmental factors, but also on the physiological state of cells and their thermal history. In particular, the sporulation temperature of bacteria has a dramatic impact on the thermal resistance of formed spores. Leguérinel et al. (2006) pointed out, for different *Bacillus* species, a linear Bigelow type relationship between the logarithm of the decimal reduction time and the temperature of sporulation. The corresponding sensibility *z* parameter value is variable but frequently close to 25°C.

3. Second Multi-Factorial Generation Model

Conventional estimates of *D*-values are made by recovering heated surviving cells at optimal incubation conditions. However, many investigators observed that culture medium and incubation temperature both influence the ratio of cell recovery and estimated heat resistance values. At non optimal temperature or in acid or salted recovery media, the estimated *D*-value, sometimes denotes *D'* and called the "apparent" *D* value, is always lower than the estimated values at optimal conditions of recovery. Consequently, as foods may be rich in acid, in salts or stored at unfavourable temperature, the injured spore population has to be considered in calculations. From a number of data from the literature, Mafart and Leguérinel (1997) noticed that, whatever the heat treatment conditions, the ratio *D'/D* remained remarkably constant. From this observation, they derived the following relationship:

$$\log\alpha = -\left(\frac{D}{D'} - 1\right)\frac{F}{D^*} - \varepsilon \qquad (19)$$

Where α is the cell recovery ratio, D^* is the decimal reduction time at the reference temperature, *F* is the sterilisation value (time to heat exposure at the reference temperature which leads to the same destruction ratio than the actual treatment). Lastly, ε is the logarithm of the unheated cell recovery ratio.

Because each environmental factor acts differently during the heating phase and during the incubation following the heat exposure, in order to avoid any confusion, we denoted *X* any factor related to the heating step and *X'* any factor related to the recovery phase. The notation *D* is reserved to the decimal reduction time measured at optimal recovery conditions, while *D'* (the so called apparent decimal reduction time) corresponds to the same parameter measured at an environmental condition *X'* different from X'_{opt}. All models developed in our laboratory present the common following form:

$$\log D' = \log D - \left(\frac{X' - X'_{opt}}{z'_X}\right)^2 \qquad (20)$$

It can be seen that three parameters have to be estimated from a non-linear regression: the optimum value of the considered factor X'_{opt}, the sensitivity value $z_{X'}$ (distance of *X'* from X'_{opt} which leads to a tenfold reduction of the apparent decimal reduction time), and *D* which corresponds to the maximum *D'* value.

This structure presents the advantage of simplicity, parsimony and biological interpretability of parameters, but it imposes a symmetric pattern to the fitted curves. Despite this constraint, the obtained goodness of fit are satisfying.

Equation (20) was successfully applied to temperature of the recovery medium. As an example, it was found the following estimates for *Bacillus cereus*: $D_{95°C} = 2.85$ min; $T'_{opt} = 23.6°C$; $z'_T = 33.7°C$ (Leguérinel et al., 2006).

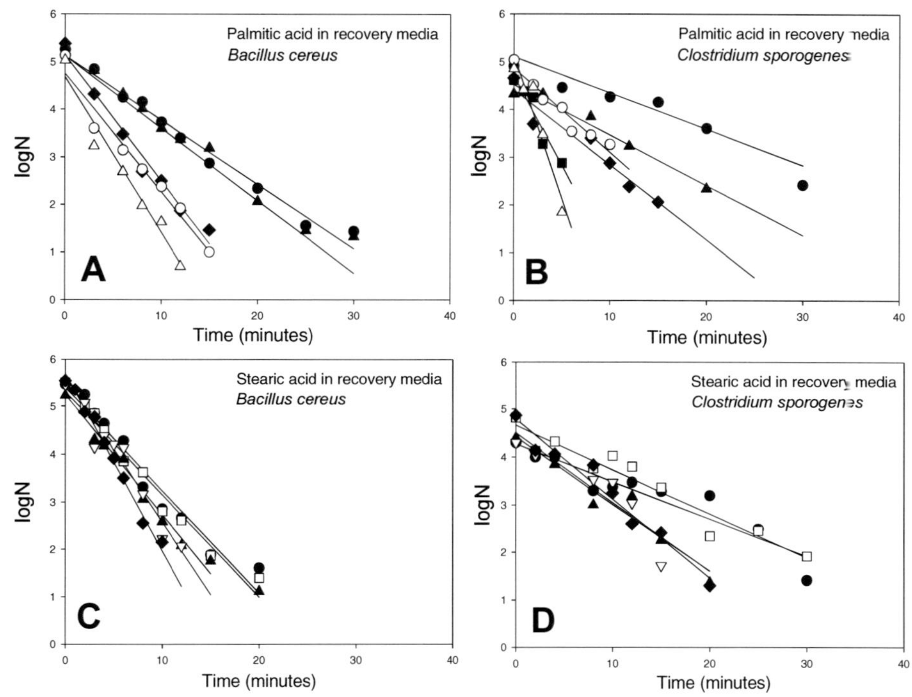

Figure 2. Log N (colony forming units, cfu) vs. time (min) for Bacillus cereus NTCC 11145 at 100°C (A & B) and Clostridium sporogenes Pasteur 79.3 at 90°C (B & D) with, in recovery media, different concentrations of palmitic acid (A & B) or stearic acid (C & D) at different concentration: 0µM ●, 200µM □, 400µM ▲, 600µM ▽, 800µM ◆, 1000µM ○, 1500µM ■, 2000µM△ (Figures reprinted with permission from Mvou Lekogo et al. (2010)).

The same equation is suitable for describing the effect of the recovery medium pH on the apparent heat resistance. Regarding the previously cited strain of *Bacillus cereus*, Couvert et al. (1999) obtained the following estimates: $D_{95°C} = 2.33$ min; $pH'_{opt} = 6.78$; $z'_{pH} = 1.81$.

The same model was also successfully applied for the same strain taking into account the water activity of the recovery medium (Coroller et al., 2001). The authors found an optimal water activity close to 0.98-0.99, while the sensitivity parameter z' was dependent on the nature of the depressor: around 0.1 for dextrose and glycerol, closer to 0.07 for sucrose.

As shown in the figure 2, the presence of free fatty acids in the recovery medium reduces clearly the resistance of bacteria.

This effect was quantified by Mvou Lekogo et al. (2010) according to a Bigelow-like relationship:

$$\log D' = \log D - \frac{[\text{Free Fatty Acid}]}{z'_{FFA}} \tag{21}$$

Table 1 shows some examples of *z'* estimates (mM).

Table 1. z' (mM) values related to Bacillus cereus NTCC 11145 and Closridium sporogenes Pasteur 79.3 for palmitic, palmitoleic, stearic and oleic acids

	Palmitic acid	Palmitoleic acid	Stearic acid	Oleic acid
B. cereus NTCC11145	5.01	0.26	3.19	0.89
C. sporogenes Pasteur 79.3	2.28	0.23	2.36	0.36

4. Modular Combination of Elementary Models and Applications

In the field of predictive microbiology is frequently assumed, a multiplicative effect of each inhibitory factor upon the specific growth rate of bacteria (see the "gamma concept", Zwietering 1992). The same assumption was retained concerning the bacterial heat resistance. As the logarithmic transformation of a product yields a sum, the combination of several mono-factorial elementary models is presented in the structure of equation (19). More generally, a combination of first generation models (related to the heating medium) can be written as follows:

$$\log D = \log D^* - \sum \left(\frac{X_i - X_i^*}{z_{X_i}} \right)^{n_i} \tag{22}$$

where the exponent *n* may be equal to 1 or 2. X^*_i is not a free parameter, but corresponds to a fixed reference value (for example, $T^* = 121.1°C$; $pH^* = 7$; $a^*_w = 1$).

Similarly, a combination of second generation models (related to the recovery medium) can be written as follows:

$$\log D' = \log D - \sum \left(\frac{X'_i - X'_{iopt}}{z'_{X_i}} \right)^2 \tag{23}$$

X'_{iopt} is an estimated parameter which corresponds to environmental factor value related to the maximal apparent bacteria heat resistance. In the particular case of an added inhibiting compound, equation (23) has to be replaced by a Bigelow-type equation.

Lastly, the combined effect of factors *X* related to the heating phase and *X'* related to the recovery conditions yields:

$$\log D' = \log D_{(X^*, X'_{opt})} - \sum \left(\frac{X_i - X_i^*}{z_{X_i}} \right)^{n_i} - \sum \left(\frac{X'_i - X'_{iopt}}{z'_{X_i}} \right)^2 \tag{24}$$

where $D_{(X^*,\ X'_{opt})}$ corresponds to the D value at reference heating condition and optimal recovery condition. This hold-all parameter depends on a number of “minor” factors which are ignored by the secondary modelling system: food matrix, presence of inhibitory or protective compounds in the medium etc. Oppositely, the sensitivity parameters z_{Xi} and z'_{Xi} are, at first approximation, assumed to be independent of food matrix and other “minor” factors. Indeed, such an approximation has the drawback of neglecting the possible interactions between factors. Further works aiming to improve equation (24) by adding interactions terms would be useful.

The practical applications of secondary models by food industries need the availability of parameters values in relevant databases. Two major difficulties then occur:

- The high number of parameters which may be included when several environmental factors are considered. Response to this problem is the simplification of the models when possible.
- The considerable variability of parameters values between strains. This problem can be partially circumvented by a standardisation of calculations.

4.1. Simplification of Models

Two situations can allow the simplification of a model:

- The correlation matrix shows a strong autocorrelation between parameters: the model is then overparameterized and can be simplified,
- Optimal values of some factors remain in a narrow range, with a low variability. In this case the free parameter can be fixed and transformed into a standard or reference value.

For example, Couvert et al. (1999) applied equation (24) for modelling the apparent heat resistance of *Bacillus cereus* spores as functions of the heating medium pH and the recovery medium pH:

$$\log D' = \log D^* - \frac{\left(pH - pH_{opt}\right)^2}{z_{pH}^2} - \frac{\left(pH' - pH'_{opt}\right)^2}{z'^2_{pH}} \qquad (25)$$

The obtained parameter estimates are given in Table 2.

Table 2. Rate coefficients related to equation (25)

Number of data	23
R^2	0.983
Mean square error	0.0253
pH_{opt}	7.15
pH'_{opt}	6.78
$D^*_{95°C}$	2.33 min
z_{pH}	2.81
z'_{pH}	1.81

Not surprisingly, it can be seen that pH_{opt} and pH'_{opt} values are similar and close to 7. Moreover, the correlation matrix showed a strong auto correlation between pH_{opt} and z_{pH} (r = 0.975) and between pH'_{opt} and z'_{pH} (r = 0.956). That means that this model can be simplified by fixing $pH_{opt} = pH'_{opt} = pH^* = 7$. A more robust and parsimonious model with only three parameters instead of five is then obtained:

$$\log D' = \log D^* - \frac{(pH - pH^*)^2}{z_{pH}} - \frac{(pH' - pH^*)^2}{z'_{pH}} \tag{26}$$

Estimations of the simplified model's parameters are shown in Table (3)

Table 3. Rate coefficients related to equation (26)

Number of data	23
R^2	0.945
Mean square error	0.0444
$D^*_{95°C}$	2.46 min
z_{pH}	2.57
z'_{pH}	2.03

A further simplification can be made for pasteurized or sterilized foods which take into account the heating medium and the recovery medium. In this case, $pH = pH'$ and equation (26) can be reduced to:

$$\log D' = \log D^* - \left(\frac{1}{z^2_{pH}} + \frac{1}{z'^2_{pH}}\right)(pH - pH^*)^2 \tag{27}$$

or

$$\log D' = \log D^* - \frac{1}{Z^2_{pH}}(pH - pH^*)^2 \tag{28}$$

with

$$\frac{1}{Z^2_{pH}} = \frac{1}{z^2_{pH}} + \frac{1}{z'^2_{pH}} \tag{29}$$

The overall parameter Z_{pH} takes into account both the effect of the heating conditions and the pH of the recovery medium. According to estimated values of Table 3, $Z_{pH} = 1.59$.

4.2. Standardization of Heat Processes Calculations

While, for a given strain, the optimal factor levels (X_{opt}, X'_{opt}) and the sensitivity parameters (z_X, z'_X) remain relatively constant, it is well known that the main resistance factor (for example D^*) is highly variable. Casolari (1984) reviewed early papers related to the heat resistance of *Clostridium botulinum* 62A and noted $D_{121.1°C}$ values ranging from 0.13 to 0.80 min. Conventional calculations used in thermal processing in the food industry generally input a $D_{121.1°C}$ value of 0.2 min and a z-value of 10°C. That means that, more or less, the standard parameter values of an "ideal strain" are retained. Furthermore, by introducing its F-value concept, Ball (1923) eliminated any $D_{121.1°C}$ value since the calculation of any sterilization value requires, regarding the target strain, only the z-value which is standardized at 10°C.

The classical lethal factor which is implicated in the heat processes calculations is as follows:

$$L(T) = 10^{\frac{T-T^*}{z}} \tag{30}$$

Mafart (2000) proposed to generalize this last equation into the following multi-factorial expression which is the product of a number of partial lethal factors, each one being related to one environmental factor:

$$L(T, pH, a_w ...) = \lambda(T)\lambda(pH)\lambda(a_w)... \tag{31}$$

with for example:

$$\lambda(T) = 10^{\frac{T-T^*}{z_T}} \tag{32}$$

$$\lambda(pH) = 10^{\left(\frac{pH-pH^*}{z_{pH}}\right)^2} \tag{33}$$

$$\lambda(a_w) = 10^{\frac{a_w-1}{z_{a_w}}} \tag{34}$$

A further extension of the lethality concept to environmental factors linked to recovery conditions would need the replacement of free parameters X'_{opt} by fixed relevant reference X'^* values related to typical pathogenic and spoilage strains which remain to be determinate from further research.

Conclusion

Model developers have to be unpretentious and to know that, in any case, their models are imperfect. It is relatively easy to account for the major part of the variance by considering a few prevalent factors, but the explanation of the remaining part of variance would require such complications of models as investigators could be discouraged from any attempt of modeling. Facing this problem, investigators must choose between foregoing modeling or adopting a *simplifying* or a *standardizing* approach.

However, it is comforting to remember that even early mono-factorial models applied in the field of canned food industries remain still very useful and are still largely applied. It is clear that, even simplified, multi-factorial models are likely to bring substantial improvements with respect to the first generation models.

References

Arrhenius, S. (1889). Über die Reaktiongeschwindigkeit bei der Inversion von Rohmzucker durch Saüren. *Z. Phys. Chem.*, 4,226-248.

Ball, C. (1923). Thermal process time for canned food. *Nat. Res. Council,* DC. Bull. 7-1.

Bigelow, W. The logarithmic nature of thermal death curves. *J. Infec. Des.*, 1921, 29,528-536.

Cameron, M., Leonard, S. & Barret, E. (1980). Effect of moderately acidic pH on heat resistance of *Clostridium sporogenes* spores in phosphate buffer and buffered puree. *Appl. Environ. Microbiol.* 39, 943-949.

Casorali, A. (1994). About basic parameters of food sterilization technology. *Food Microbiol.*, 11, 75-84.

Cerf, O., Davey, K. & Sadoudi, A. (1996). Thermal inactivation of bacteria: a new predictive model for the combined effects of temperature, pH and water activity. *Food Res. Inst.*, 299, 219-226.

Chick, H. (1910). The process of disinfection by chemical agencies and hot water. *J. Hyg.*, 10, 237-286.

Coroller, L., Leguérinel, I. & Mafart, P. (2001). Effect of water activity on heating and recovery media on apparent heat resistance of *Bacillus cereus* spores. *Appl. Environ. Microbiol.*, 67, 317-322.

Courtoisier, A. (1984). Action destructive de la chaleur sur les micro-organismes. Calcul pratique d'un traitement thermique et application au vin. *Ind. Alim. Agr.*, 101, 103-114.

Couvert, O., Gaillard, S., Savy, N., Mafart, P. & Leguérinel, I. (2005). Survival curves of heatred bacterial spores: effect of environmental factors on Weibull parameters. *Int. J. Food Microbiol.*, 101,73-81.

Couvert, O., Leguérinel, I. & Mafart, P. (1999). Modelling the overall effect of pH on the apparent heat resistance of *Bacillus cereus* spores. *Int. J. Food Microbiol.*, 49, 57-62.

Davey, K., Lin, S. & Wood, D. (1978). The effect of pH on continuous high temperature/short time sterilization of liquids. *Am. Inst.Chem. Eng. J.*, 24, 537-540.

Fernandez, A., Collado, J., Cunha, L., Ocio, M. & Martinez, A. (2002). Empirical model based on a Weibull distribution to describe the joint effect of pH and temperature on the thermal resistance of *Bacillus cereus* in vegetable substrate. *Int. J. Food Microbiol.*,77, 147-153.

Fernandez, P., Ocio, M., Rodrigo & M. Martinez, A. (1996). Mathematical model for the combined effects of temperature and pH on the thermal resistance of *Bacillus stearothermophilus* and *Clostridium sporogenes* spores. *Int. J. Food Microbiol.*, 32, 225-233.

Gaillard, S., Leguérinel, I. & Mafart, P. (1998). Modelling combined effects of temperature and pH on the heat resistance of spores of *Bacillus cereus. Food microbiol.*, 15, 625-630.

Gaillard, S., Leguérinel, I. & Mafart, P. (1998). Model for combined effects of temperature, pH and water activity on thermal inactivation of *Bacillus cereus* spores. *J. Food Sci.*, 63, 887-889.

Jordan, R. & Jacobs, S. (1948). Studies on the dynamic of disinfection. XIV. The variation of the concentration exponent for hydrogen and hydroxyl ions with the mortality level using standard cultures of *Bact. Coli* at 51°C. *J Hyg. Cambridge.*, 46, 289-295.

Juneja, V., Marmer, B., Phillips, J. & Miller, A. (1995). Influence of intrinsic properties of foods on thermal inactivation of spores of *Clostridium botulinum*: development of a predictive model. *J. Food Safety.*, 15, 349-364.

Leguérinel, I., Couvert, O. & Mafart, P. (2007). Modelling the influence of the sporulation temperature upon the bacterial spore heat resistance, application to heating process calculation. *Int. J. Food Microbiol.*, 114, 100-104.

Leguérinel, I., Couvert, O. & Mafart, P. (2006). Modelling the influence of incubation temperature upon the estimated heat resistance of heated Bacillus spores. *Lett. Appl. Microbiol.*, 43, 17-21.

Leguérinel, I. & Mafart, P. (2001). Modelling the influence of pH and organic acid types on thermal inactivation of *Bacillus cereus* spores *Int. J. Food Microbiol.* ., 63, 29-34.

Lopez, M., Gonzales, I., Condon, S., Martin, J. & Bernado, A. (1996). Effect of pH heating medium on the thermal resistance of *Bacillus stearothermophilus* spores *Int. J. Food Microbiol.* , 28, 405-410.

Mafart, P. (2000). Taking injuries surviving bacteria into account for optimizing heat treatments. *Int. J. Food Microbiol.*, 55,175-179.

Mafart, P., Couvert, O., Gaillard, S. & Leguérinel, I. (2002). On calculating sterility in thermal preservation methods : application of the Weibull frequency distribution model. *Int. J. Food Microbiol.*, 72, 107-113.

Mafart, P. & Leguérinel, (1997). I. Modelling the heat stress and the recovery of bacterial spores. *Int.J. Food Microbiol.*, 37,131-135.

Mafart, P. & Leguérinel, (1998). I. Modelling combined effects of temperature and pH on heat resistance of spores by a linear-Bigelow equation. *J. Food Sci.*, 63, 6-8.

Mazas, M., Lopez, M., Gonzalez, I., Bernado, A. & Martin, R. (1998). Effects of the heating medium pH on the heat resistance of *Bacillus cereus* spores. *J. Food safety.*, 18, 25-36.

Mvou Lekogo, B., Coroller, L., Mathot, A. G., Mafart, P. & Leguérinel, I. (2010). Modelling the influence of palmitic, palmeolitic and oleic acids on apparent heat resistance of *Bacillus cereus* NTCC 11145 and *Clostridium sporogenes* Pasteur 79.3. *Int. J. Food. Microbiol.*, 141, 242-247.

Pirone, G., Mannino, S. & Campanini, M. (1987). Termoresistenza di *Clostridi butirrici* in funzione del pH. *Industria Conserve.*, 62, 135-137.

Reichart, O. (1994). Modelling the destruction of *Escherichia coli* on the base of reaction kinetics. *Int. J. Food Microbiol.*, 23, 449-465.

Simpson, S. & Williams, M. (1974). An analysis of high temperature-short time sterilization during laminar flow. *J. Food Sci.*, 39, 1047-1052.

Splittstoesser, D., Mac Lellan, M. & Churey, J. (1995). Heat resistance of *Escherichia coli* O157:H7 in apple juice. *J. Food Protect.*, 14, 217-229.

Xenones, H. & Hutching, I. (1965). Thermal resistance of *Clostridium botulinum* (62A) spores as affected by fundamental food constituents. *Food Technol.*, 19, 1003-1005.

Zwietering, M., Wijtzes, T., De Wit, J. & Van't Riet, K. (1992). A decision support system for prediction of the microbial spoilage in foods. *J. Food Protect.*, 55, 973-979.

In: Progress on Quantitative Approaches in Thermal … ISBN: 978-1-62100-842-2
Editors: V. P. Valdramidis and J. F. M. Van Impe © 2012 Nova Science Publishers, Inc.

Chapter 6

THE USE OF TRANSFER MODELLING ON THERMAL FOOD PROCESSING MICROBIAL INACTIVATION

***Tomás Norton*[1] *and Brijesh Tiwari*[2]**
[1]Engineering Department, Harper-Adams University College, Newport, Shropshire, UK
[2]Manchester Metropolition University, UK

Abstract

To realise food safety assurance, a systematic preventive approach that addresses physical, chemical, and biological hazards during food manufacturing, distribution and storage is essential. During the processing of foods, thermo-fluid dynamics govern many processes, and are involved in almost every process. For this reason, transfer modelling has played an active part in the design of thermal processes for a number of decades now. In recent years simulations have reached high levels of sophistication. In the current chapter the importance of transfer modelling in food safety assurance is reviewed, including conventional and emerging food processing technologies, with particular emphasis on maintaining a high level of accuracy via circumspect choices made during model development.

1. Introduction

Food safety hazards can occur at any stage during the food production process, so control procedures for food safety are necessary throughout the supply chain. Due to the probability of contamination at any stage in the food chain, measures to ensure food safety have to be implemented on a global scale, necessitating a global approach (Havelaar et al., 2010). Therefore, ensuring food safety is a joint responsibility for all the participating members, meaning that the people must be aware of the potential locations where hazards may occur, and not restrict focus to a single company or process. For example, companies who purchase raw materials, ingredients, and food packaging materials for manufacture of consumer goods, must ensure that these materials are safe and fit for use. On the other hand, consumers must be vigilant of food safety and quality when making purchases, paying greatest attention to handling, storage, preparation, and the final use of the foods (Alli, 2004). While the effectiveness of achieving this aim is often improved through developments in traditional animal and plant sciences, it must also be achieved through the application of chemistry and engineering

Preservation techniques must be applied by a processor to render a food product commercially sterile, as well as to modify sensory characteristics (Richardson, 2001). During the processing of foods, thermo-fluid dynamics govern many processes, and are involved in almost every process. Due to the complexity of the physical mechanisms involved in thermal processing, quantifying the phenomena must involve sophisticated design and analysis tools. Physical experimentation may allow for direct measurement, however, a comprehensive analysis would not only necessitate expensive equipment and require considerable time and expertise, but could also be intrusive and therefore affect the quality of the results (Verboven et al., 2004). Fortunately, ubiquitous processes such as blanching, cooking, drying, sterilisation and pasteurisation, which rely on thermal exchange to raise a product thermal-centre to a pre-specified temperature, are highly amenable to transfer modelling, and have unarguably contributed to its exponential take-up witnessed over recent years. One large breakthrough in transfer modelling of food processes has been seen in the wide-scale application of computational fluid dynamics (CFD), which has led to an improvement in both food quality and safety alongside reducing energy consumption used in industrial processes, and has reduced the amount of empiricism associated with a design process (Norton and Sun, 2007).

In the most recent review aimed specifically at numerical applications in thermal processing, Verboven et al. (2004) highlighted the existing limitations and challenges that face the current users of such modelling. In light of the rapid developments in both thermal processing and numerical modelling, especially via CFD technology, that has taken place over recent years, it is necessary to provide a critical and comprehensive account of the latest advances, made through successful applications in research. Firstly, the need for transfer models in the design of food processes will be presented. The main equations governing the physical mechanisms encountered in thermal processing will then be discussed, followed by an extensive overview of the conventional processes, which have been modelled using mechanistic transfer models. Then a focus of the CFD modelling studies of emerging thermal processing technologies will then be highlighted.

2. The Need for Transfer Modelling in Food Processing

Owing to their complex thermo-physical and in some cases polymorphic properties, heat transfer to and from foods can stimulate complex chemical and physical alteration. Nowadays, with both food preservation and safety being equally important objectives of processing, it is necessary to promote quality characteristics of food whilst eradicating the threat of spoilage. For this to happen efficiently not only must the appropriate temperature and duration of heating be known but other influential phenomena must also be understood. Mathematical models permit the representation of a physical process, from the analysis of measured data or physical properties over a range of experimental conditions, and allow process parameters to then be ascribed as constant or variable, lumped or distributed. Such a representation is necessary in order for model predictions to be generalized in a manner amenable to equipment design and optimization. Both the theoretical and descriptive aspects of food processing have long been regarded as highly important. Heldman and Singh (1981) have described food engineering as being composed of two distinct parts, (a) the descriptive part which deals with detailed descriptions of physical processes and equipment and (b) the theoretical part, which comprises a mathematical description of the processing equipment and the changes that may occur during processing and storage, etc. As these two parts supplement and support each other, mathematical modelling of unit processes has always been an integral part of industrial design.

During food processing, heat is usually transferred to or from the food, and occurs due to one or more of the three mechanisms: *conduction, convection,* and *radiation,* and depending on the unit process one mechanism is usually dominant (Wang and Sun, 2003). As the thermal exchange phenomena are either distributed in space or in space and time, their fundamental representation is by means of partial differential equations (PDEs). However, the full solution of these equations requires complex time-consuming iterative techniques, which have only been attainable within the last number of years. Thus, their simplification has proved, in the past, to be necessary. For example, temperature evolution through a three-dimensional (3D) solid is accurately represented by a parabolic PDE. By assuming that linear conditions prevail, this equation can be integrated to describe 1D steady-state heat transfer between the hot and the cold surface of a solid, and provided that the proportionality constant (which is an intensive property of the conducting material) is adequately represented, a reasonably accurate solution can be obtained. Also in the past, when spatially dependent and transient solutions were sought to optimize the thermal processing of solids, the 1D Fourier equation was used in combination with finite difference methods (Hendrickx et al., 1993). Simplifying conduction to a 1D problem has proved to be a suitable means of describing conduction heat transfer where conditions allow, as heat flows at right angles to the temperature gradient. However, more involved methods are required when the temperature depends on more than one space variable and when heat is transferred through irregular geometry; as generally encountered in the food industry (Tewkesbury et al., 2000).

In contrast to conduction, convection permits heat transfer by mass motion of a fluid system when it is heated and caused to move, carrying energy away from the source of heat, thereby allowing cooler fluid to replace it. Convection above a hot surface occurs because hot fluid expands, becomes less dense, and rises. Depending on the flow velocity, convection can be free (i.e., natural) or forced. Newton's law of cooling describes the energy flow rate due to convective heat transfer at the interface between a solid object and the bulk fluid environment, and is a function of a proportionality constant known as the heat transfer coefficient. Unlike the proportionality constant for conduction, the heat transfer coefficient is an empirically determined parameter with units of W $m^{-2}K^{-1}$, whose value depends on many relevant conditions, which are all process dependent. For convenience, this coefficient can be determined by optimization methods, which iterate through solutions until the heat transfer model for the solid arrives at a good fit to experimental data (Varga and Oliveira, 2000). An average or global heat transfer coefficient has been used in many simulations (Wang and Sun, 2003). However, as discussed by Nicolaï et al. (2001) and Verboven et al. (2004), allowing a global heat transfer coefficient to be representative of the convection process cannot serve as a basis of process improvement when it is inherently a local phenomenon dependent on surface geometry, flow characteristics, and other physical processes such as mass transfer, etc. Nicolaï and de Baerdemaeker (1996) showed that for a small global heat transfer coefficient, small changes in its local value may cause large temperature differences across a food product, thereby suggesting that the global value may find better use as an indicator of product sensitivity in such circumstances. Moreover, in the determination of the global heat transfer coefficient, correlations of the type $Nu = cRe^{m}Pr^{n}$ are widely used in practice. Zheng et al. (2007) noted that with c, m, and n being constants, it is difficult to describe, to good level of accuracy, the change of Nusselt number with Reynolds and Prandtl numbers over a wide range of these parameters. Therefore, such correlations require a comprehensive experimental approach in their validation, as well as information regarding the specific range and conditions from which the correlations were developed. Unfortunately, this is not always the case.

Radiation heat transfer is entirely different to conduction and convection heat transfer as a medium is not required. Radiant exchange occurs at a rate proportional to the fourth power of the absolute temperature difference between two surfaces. Inclusion of radiant heat transfer in calculations is often required when modelling thermal exchange owing to high-temperature surfaces, e.g., in baking ovens.

Microwave or radio frequency radiation in heating or drying operations may need modeling; however, in contrast to thermal radiation, these mechanisms permit the penetration of electromagnetic waves into a food product. Similar to the cases of heat transfer by conduction or convection, and provided conditions allow, the radiant heat transfer can be linearized so that it can be calculated as a function of the temperature difference, the area of transfer, and a proportionality constant of the surfaces (Wang and Sun, 2005). However, radiation can present many difficulties that cannot be modeled by simplified means. For example, bread baked in an oven is subject to radiant energy, which can be blocked by vapor boundary layers that develop during the baking process (Zhou and Therdthai, 2007). Furthermore, color change such as browning influences baking by increasing the absorption of infrared rays (Zhou and Therdthai, 2007). As a consequence, radiation causes localized temperature differentials of an exposed surface, particularly the darkened area (Matz, 1989). In other operations, electromagnetic and radio frequency radiation is used to volumetrically deposit energy into a product, the rate of which is a function of the dielectric properties of the food. To calculate the electric field inside a food, exponential decay has traditionally been used in a lot of oven investigations (Singh and Heldman, 2001). Unfortunately, for many situations, this is a qualitative assumption and can be wrong depending on size of the food and its dielectric properties, a more correct method would be to solve the Maxwell electromagnetic equations for the oven by numerical means, and then obtain the volumetric rate of heating (Verboven et al., 2007).

The transport equations for conduction, convection, and radiation can also be solved analytically for a small number of boundary, and initial conditions (Nicolaï et al., 2001). An analytical procedure has even been developed to solve a conjugate thermal problem associated with radiation, convection, and conduction (Nakayama et al., 2004). However, the localized contributions cannot be determined analytically and numerical analysis is required to establish the distribution of the variables, i.e., temperature, fluid velocity, and mass fraction, during thermal processing. With this in mind, commercial mechanistic transfer modelling codes have been developed around numerical algorithms that solve the nonlinear PDEs governing all fluid flow, heat transfer, and many other physical phenomena. In other words, numerical techniques can be used to build a distributed parameter model that is spatially and temporally representative of the physical system, thereby permitting the achievement of a solution with a high level of physical realism. The accuracy of numerical simulations is a function of many parameters including the level of empiricism involved, i.e., via additional physical models; the assumptions involved; the simplification of both geometry and boundary conditions made in order to reduce processing time; and whether process modifying physical mechanisms is included, i.e., chemical kinetics. The greater the amount of approximations, the less accurate the CFD solution will become (Verboven et al., 2004). However, if used correctly, mechanistic modelling approaches will provide understanding on the physics of a system in detail, and can do so through nonintrusive flow, thermal, and concentration field predictions. Furthermore, after many years of development, commercial modelling codes can now solve advanced problems of combined convection, radiation, and conduction heat transfer flows in porous media, multiphase flows, and problems involving chemical kinetics.

3. Numerical Representation of the Food Processes

3.1. Numerical Techniques

Engineers have a choice of many different numerical techniques to discretise the transport equations. The most important of these include finite difference, finite elements and finite volume. The finite difference technique is the oldest one used, and many examples of its application in the food industry exist. However, due to difficulties in coping with irregular geometry, finite difference is not commercially implemented. Furthermore, the current trend when using commercial finite element (FE) or CFD packages is to use unstructured meshing technology capable of handling the complex three-dimensional geometries encountered in industry. Therefore, while appropriate for simplified cases, the prospects of finite difference being used in complex industrial applications are limited.

Finite element methods have historically been used in structural analysis where the equilibrium of the solution must be satisfied at the node of each element. Nicolaï et al. (2001) provided a short introduction to the use of the finite elements method in conduction heat transfer modelling, and it will not be discussed here. Suffice to say that as a result of the weighting functions used by this method, obtaining a three-dimensional of fluid flow solution with a large number of cells is impractical at present, which conduction heat transfer certainly is suitable. Therefore, finite elements methods have enjoyed use in the modelling of electromagnetic heating in microwave ovens (Verboven et al., 2007; Geedipalli et al., 2007); vacuum microwave drying (Ressing et al., 2007); radio frequency heating of food (Marra et al., 2006); conduction and mass transport during drying (Aversa et al., 2007).

With finite volume techniques the integral transport equations governing the physical process are expressed in conservation form (divergence of fluxes) and the volume integrals are then converted to surface integrals using Gauss's divergence theorem. This is a direct extension of the control volume analysis that many engineers use in thermodynamics and heat transfer applications etc., so it can be easily interpreted. Thus, expressing the equation system through finite volumes forms a physically intuitive method of achieving a systematic account of the changes in mass, momentum and energy, as fluid crosses the boundaries of discrete spatial volumes within the computational domain. Also, finite volume techniques yield algebraic equations that promote the robustness of the solution process.

3.2. The Generic Equation and Its CFD Approximation

The basic requirement of transfer modelling is to obtain a solution to a set of governing PDE's of the transported variable. As discussed above the transport phenomena are generally described by commercial CFD developers using the finite volume method. The generic conservation convection-diffusion equation, after application of Gauss's theorem to obtain the surface integrals can be described as follows:

$$\int_V \frac{\partial \rho\phi}{\partial t} + \int_A \mathrm{n}.(\rho U\phi)dA = \int_A \mathrm{n}.(\Gamma_\phi \; grad \; \phi)dA + \int_V S_\phi \tag{1}$$

The conservation principle is explicit in Eq. (1), i.e. the rate of increase of φ in the control volume plus the net rate of decrease of φ due to convection is equal to the increase in φ due to diffusion and an increase in φ due to the sources (Versteeg and Malaskeera, 1995). For a numerical solution, both the surface and volume integrals need to be solved on a discrete level, which means numerical interpolation schemes are required. As the convection term, the only non-linear term in the equation, needs to be approximated at each mesh element face, it presents the greatest challenge in allowing the numerical scheme to preserve properties such as the stability, transportiveness, boundedness and accuracy of a solution. For the same reason, numerical schemes are often called convection schemes as the accurate and stable representation of the convection term is a major requirement (Patankar, 1980).

3.3. The Numerical Mesh

The volume mesh in a simulation is a mathematical description of the space or the geometry to be solved. One of the major advances to occur in meshing technology over recent years was the ability for tetrahedral, hexahedral hybrid and even polyhedral meshes to be incorporated into commercial codes. This has allowed mesh to be fit to any arbitrary geometry, thereby enhancing the attainment of solutions for many industrial applications. In addition, some modern commercial modelling codes promote very little interaction between the user and individual mesh elements, with regard to specifying the physics of the problem. Through this decoupling of the physics from the mesh, the user is allowed to concentrate more on the details of the geometry, and the transfer of simulation properties and solutions from one mesh to another is easier when the mesh independence of a simulation needs to be studied or when extra resolution is required.

4. The Equations Governing Food Processes

4.1. The Navier-Stokes Equations

The mathematical formulation of fluid motion has been complete for almost two hundred years, since the emergence of three very important scientists in the field of fluid mechanics. The first one being the Swiss mathematician and physicist Leonhard Euler (1707-1783) who formulated the Euler equations, which describe the motion of an invisid fluid based on the conservation laws of physics, now defined as classical physics; namely the conservation of mass, momentum, and energy. The French engineer and physicist, Claude-Louis Navier (1785-1836), and the Irish mathematician and physicist, George Gabriel Stokes (1819-1903) later introduced viscous transport into the Euler equations by relating the stress tensor to fluid motion. The resulting set of equations now termed the Navier-Stokes equations for Newtonian fluids have formed the basis of modern day CFD (Anon., 2007).

$$\nabla \cdot \vec{v} = 0 \tag{2}$$

$$\rho \frac{\partial v_i}{\partial t} + \rho \vec{v} \cdot \nabla \vec{v_i} = -\nabla p + \mu \nabla^2 \vec{v_i} + \rho g \tag{3}$$

Eq. (2) is a mathematical formulation of the law of conservation of mass (which is also called the continuity equation) for a fluid element where $\vec{v}$ consists of the components of $\vec{v_i}$, the solution for each requires a separate equation. The conservation of mass states that the mass flows entering a fluid element must balance exactly with those leaving for an incompressible fluid. Eq. (3) is the conservation of momentum for a fluid element, i.e. Newton's second law of motion, which states that the sum of the external forces acting on the fluid particle is equal to its rate of change of linear momentum.

Any fluid that does not obey the Newtonian relationship between the shear stress and shear rate is called a non-Newtonian fluid. The shear stress in a Newtonian fluid is represented by the second term on the right hand side of Eq. (2). Many food processing media have non-Newtonian characteristics and the shear thinning or shear thickening behaviour of these fluids greatly affects their thermal-hydraulic performance (Fernandes et al., 2006). Over recent years, computational fluid dynamics (CFD) has provided better understanding of the mixing, heating, cooling and transport processes of non-Newtonian substances.

4.2. The Heat Transfer Equation

The modelling of thermal processes requires that the energy equation governing the heat transfer within a fluid system to be solved. This equation can be written as follows:

$$\rho \frac{\partial \left(c_p \vec{v_i}\right)}{\partial t} + \nabla\left(\vec{v} \cdot T\right) = \lambda \nabla^2 T + s_T \tag{4}$$

The transport of heat in a solid structure is considered in many simulations of food processes, and becomes especially important when conjugate heat transfer is under investigation, in which case it is important to maintain continuity of thermal exchange across the fluid-solid interface (Verboven et al., 2004). The Fourier equation which governs heat transfer in an isotropic solid can be written as:

$$\frac{\partial \left(\rho c_p \vec{v_i}\right)}{\partial t} = \lambda \nabla^2 T + s_T \tag{5}$$

The main difference between the two equations is that the Fourier equation lacks a convective mixing term for temperature, which is incorporated into Eq. (4). For a conjugate heat transfer situation where evaporation at the food surface is considered, and where the heat transfer coefficient is known, the boundary condition for Eq. (5) may be written as

$$h\left(T_{bf} - T_s\right) + \varepsilon\sigma\left(T_{bf}^4 - T_s^4\right) = -\lambda \frac{\partial T}{\partial n} - \alpha \cdot N_s \tag{6}$$

where ε is the emission factor coefficient, σ is the Stefan-Boltzmann constant. Heat transported by radiation and convection from air to food raises the sample temperature and also goes towards evaporating the free water at the surface (Aversa et al., 2007). The solution of Eq. (6) can be used on the food surface to calculate the local heat transfer coefficients (Verboven et al., 2003),

$$h = \frac{-\lambda \left.\frac{\partial T}{\partial n}\right|_{surface}}{\left(T_{bf} - T_s\right)} \tag{7}$$

Eq. (7) can be used provided the surface temperature is assumed independent of the coefficient during calculations.

The equation of state relates the density of the fluid to its thermodynamic state, i.e. its temperature and pressure. The profiles of thermal processing variables, i.e. temperature, water concentration, fluid velocity, are functions of density variations caused by the heating and cooling of fluids. Since fluids flows encountered in thermal processing can be regarded as incompressible there are two means of modelling the density variations that occur due to buoyancy. The first is the well known Boussinesq approximation (Ferziger & Peric, 2002). This has been used successfully in many CFD applications (Abdul Ghani et al., 2001):

$$\rho = \rho_{ref}\left[1 - \beta\left(T - T_{ref}\right)\right] \tag{8}$$

The approximation assumes that the density differentials of the flow are only required in the buoyancy term of the momentum equations. In addition, a linear relationship between temperature and density, with all other extensive fluid properties being constant is also assumed. This relationship only considers a single-component fluid medium however, by using Taylor's expansion theorem the density variation for a multi-component fluid medium can also be derived.

Unfortunately, the Boussinesq approximation is not sufficiently accurate at large temperature differentials (Ferziger & Peric, 2002). Therefore, in such cases another method of achieving the coupling of the temperature and velocity fields is necessary. This can be done by expressing the density difference by means of the ideal gas equation:

$$\rho = \frac{p_{ref} M}{RT} \tag{9}$$

This method can model density variations in weakly compressible flows, meaning that the density of the fluid is dependent on temperature and composition but small pressure fluctuations have no influence.

4.3. The Equation for Mass Transfer

In general, mass transfer in food products depends on local water concentration and is governed by Fick's law of diffusion of the form,

$$\frac{\partial C}{\partial t} = D_s \nabla^2 C + s_c \tag{10}$$

where C is water concentration in food (ppm or mol/m^3 etc.), is the effective diffusion coefficient of water in food (m^2/s). If the food is highly porous and dehydrated then water vapour diffusion may be significant. However, for products with a void fraction lower than 0.3 (May and Perrè, 2002) vapour diffusion can be neglected. For mass transfer in the fluid medium, a passive scalar can represent the diffusion of mass and be written as

$$\frac{\partial C}{\partial t} + \nabla \cdot \left(\vec{v} C \right) = D_s \nabla^2 C + s_c \tag{11}$$

In flows involving gases mixing with air and chemicals in water, diffusion coefficients can be found in the literature but for liquid foods a reasonable approach is to make an educated guess and conduct a sensitivity analysis (Verboven et al., 2004). Using a passive scalar is only valid in low concentrations, and becomes invalid when particulate sizes of about 1 μm are present, which can influence the flow properties.

Simulations of spray dryers, solid-gas flows, or non-homogenous liquid foods exemplify applications where solid and fluid phases coexist and interact with each other. For this modelling, two approaches are used, i.e. either the Eulerian-Lagrangian or the Eulerian-Eulerian. Using the Eulerian-Lagrangian framework, the bulk fluid is modelled as a continuum carrying discrete solid particles (Nijdam et al., 2006; Anandharamakrishnan et al., 2010). The Lagrangian equations of mass and momentum are then solved to determine trajectory of each particle within the continuum, which has resulted in this technique often being referred to as "particle tracking" (Kuriakose and Anandharamakrishnan, 2010). As discussed by Verboven et al. (2004), in turbulent flow simulations, the turbulent velocity fluctuations have been averaged out by the RANS models and the turbulent dispersion of the particles require modelling by mimicking these fluctuations. This is done by extracting random numbers from a Gaussian distribution with a computed mean and variance (Harral and Burfoot, 2005). When the particle passes through any arbitrary mesh element, the energy, mass and momentum transferred to the fluid continuum are calculated and added to the source terms of that element. In this way all trajectories are calculated one by one. In the next iteration the fluid flow is solved using these source terms, and the calculation loop is repeated until sufficient convergence is achieved. For a complete description of how the Eurlerian-Lagrangian framework can be used to determine particle distribution, and as subsequently predict particle history, particle residence time and particle collisions, in fluid-food systems the reviewer is referred to the work of Kuriakose and Anandharamakrishnan (2010).

Whilst the Eulerian-Lagrangian approach provides a direct physical interpretation of the particle-fluid, particle-particle, and particle-wall interactions, computational times can become excessive, depending on the number of particles to be solved in the system (Szfran and Kmiec, 2004). This means that applications of the Eulerian-Lagrangian approach are still limited to small scale computations. In the Eulerian approach, the fluid and particle phases are treated as interacting and interpenetrating continua (Nijdam et al., 2006). The governing equations for each phase are convection-diffusion equations, which contain extra source terms to account for the turbulent dispersion of the particles. Coupling of the phases is achieved through pressure and interphase exchange coefficients. The equations appropriate to the Eulerian-Eulerian approach are presented by Verboven et al. (2004) and Nijdam et al. (2006).

4.5. Radiation Models

In recent years the number of radiation models incorporated into simulations has increased, and some have been designed by the code developers themselves. The most common radiation models used in thermal processing simulations include discrete ordinate (DO) or surface-to-surface (S2S) models. The DO model takes into account media participation (Modest, 1992). The S2S model considers the radiation heat exchange between two surfaces only (Siegel and Howell, 1992), and the amount of radiation received and emitted by each surface is defined by the surface's view factors and the thermal boundary conditions. In contrast to a S2S solution, the solution for the DO model is coupled to the flow solution and energy is exchanged between the fluid and the radiation field. Therefore, solution times for S2S model can be almost half those of the DO model (Mistry et al., 2006). More recently, Chhanwal et al. (2010) compared three radiation models namely discrete transfer radiation model (DTRM), surface to surface (S2S) and discrete ordinates (DO) for an electric heating oven and found that they all predicted similar temperature evolution in the baked product.

5. Optimising Conventional Food Processes to Ensure Food Safety with Numerical Transfer Modelling

Over the years many of the conventional process such as sterilisation, drying and cooking have been designed and optimised with using numerical modelling in order to ensure safety and quality of food. Some of the studies conducted thus far are reviewed in this section. Given that fluid motion was considered is most of these studies, CFD was the main modelling tool employed.

5.1. Sterilisation and Pasteurisation

5.1.1. Canned Foods

During the sterilisation process, rapid and uniform heating is desirable to achieve a suitable level of sterility with minimum destruction of the colour, texture and nutrients of food products (Tattiyakul et al., 2001). CFD has been used to investigate the heat and mass transfer phenomena occurring in canned products such as soup, carboxyl-methyl cellulose (CMC) or corn starch undergoing sterilisation. As can be seen in Fig. 1, these investigations,

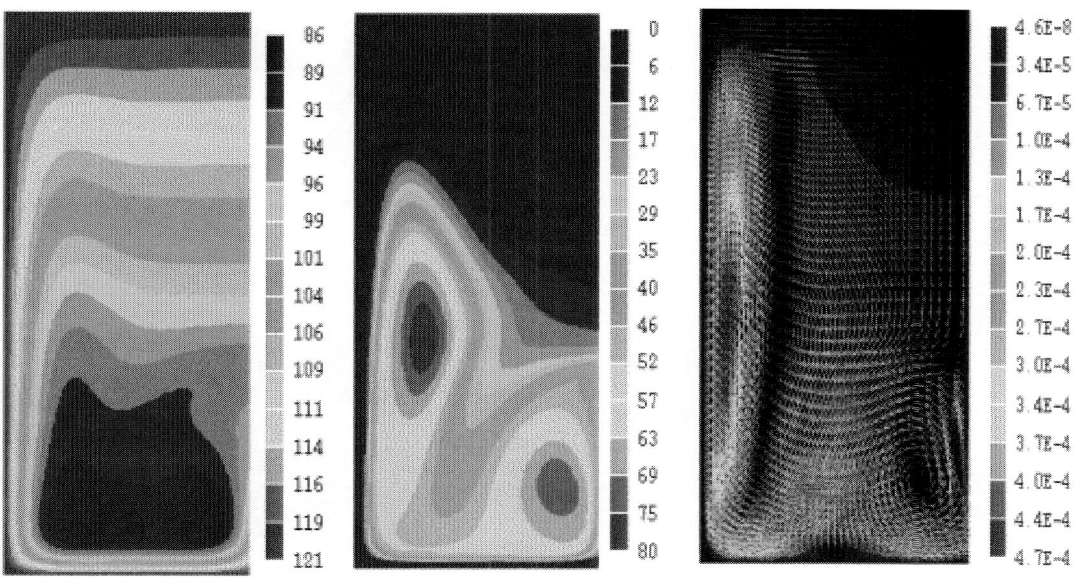

Figure 1. Temperature, bacteria deactivation and flow pattern profiles in a can filled with CMC and heated by condensing steam after 19 min respectively (Abdul Ghani et al. (1999) with permission by Elsevier).

CFD has highlighted the transient nature of the slowest heated zone (SHZ), quantified the amount of time needed for heat to be transferred fully throughout food, as well as illustrating the sharp heterogeneity in temperature profile of the product when no agitation is applied to the can (Abdul Ghani et al., 1999). Abdul Ghani et al. (2002) conducted CFD studies of both natural and forced convection (via can rotation) sterilisation processes of viscous soup, and showed that forced convection was about four times more efficient. Further simulation studies of a starch solution undergoing transient gelatinisation showed that uniform heating could be obtained by rotating the can intermittently during the sterilisation process (Tattiyakul et al., 2001). CFD simulations have also been used to generate the data required for the development of a correlation to predict sterilisation time (Farid and Abdul Ghani, 2003). A more recent study simulated the sterilisation process of a solid-fluid mixture in a can, and showed that the position of the food in the can has a large influence on the sterilisation times experienced (Abdul Ghani and Farid, 2006).

CFD has recently been used to study the effect of container shape on the efficiency of the sterilisation process (Varma and Kannan, 2005; 2006). Conical shaped vessels pointing upwards were found to reach appropriate sterilisation temperature the quickest (Varma and Kannan, 2005). Varma and Kannan (2006) investigated the effects of modifying the geometry of the conventional cylindrical food can as well as the cans orientation on the rate of thermal sterilization of pseudoplastic food from CFD simulations. A cone pointing upwards was found to heat faster than the cylinder of equal volume while the cone pointing downward heated the slowest. In a more recent study, Abdul Ghani (2006) has also analyzed the cooling cycle of the food sterilization process using CFD. The most recent recent CFD study on food sterilisation of canned products has been conducted by Kannan et al. (2008), in which the authors provide insight on the thermal sterilization process in relatively long food cans of different aspect ratios. CFD studies were used to determine regimes during the heating cycle, which led to the development of more realistic heat transfer correlations, which can be used to estimate the heat transfer fluxes as a function of time, and subsequently upon integration can lead to estimation of energy consumption.

5.1.2. Foods in Pouches

In recent years, CFD has provided a rigorous analysis of the sterilization of three-dimensional pouches containing liquid foods (Abdul Ghani and Farid, 2002a, 2002b).

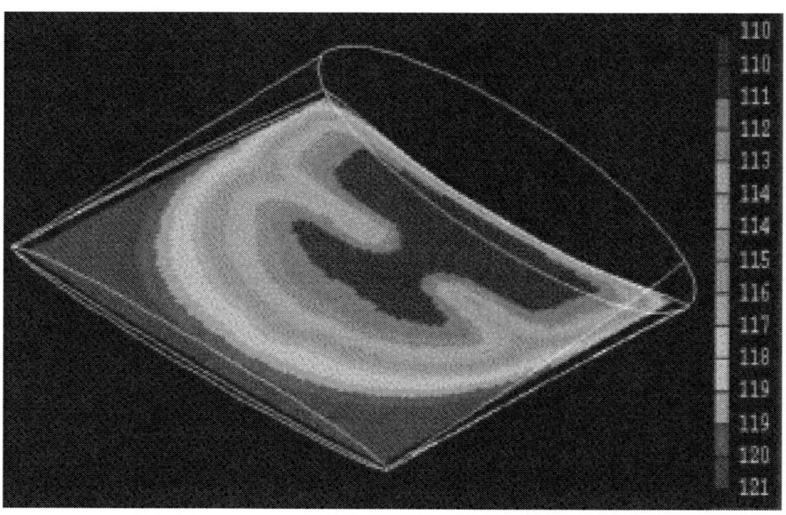

Figure 2. Temperature profiles of *B. stearothermophilus* at 30% height from the bottom of pouch filled with beef–vegetable soup and heated by condensing steam (Abdul Ghani et al. (2002) with permission by Elsevier).

Coupling first-order bacteria and vitamin inactivation models with the fluid flow has allowed transient temperature, velocity and concentration profiles of both bacteria and ascorbic acid to be predicted during natural convection. The concentrations of bacteria and ascorbic acid after heat treatment of pouches filled with the liquid food were measured, and found close agreement with the numerical predictions. The SHZ was found to migrate during sterilisation until eventually resting in a position about 30% from the top of the pouch. As expected, the bacterial and ascorbic acid destruction was seen to depend on both temperature distribution and flow pattern (Fig. 2).

5.1.3. Intact Eggs

Denys et al. (2003, 2004, 2005) has used CFD to predict the transient temperature and velocity profiles in intact eggs during the pasteurisation process with the aim of making the process more effective. Owing to its ability to account for complex geometries, heterogeneous initial temperature distributions, transient boundary conditions and non-linear thermophysical properties, CFD has permitted a comprehensive understanding of this thermal process (Denys et al., 2003). Such an analysis has allowed the gap in the knowledge of this area to be filled, as up to recently little information on the correct processing temperatures and times for safe pasteurisation, without loss of functional properties, was available (Denys et al., 2004). In the series of papers published on this topic by Denys et al. (2003, 2004, 2005), a procedure to determine the surface heat transfer coefficient using CFD simulations of eggs filled with a conductive material of known thermal properties was first developed. After this, conductive and convective heating processes in the egg were modelled (Denys et al., 2004). From this it was revealed that, similar to the phenomena noted by Abdul Ghani et al. (1999) in canned food, the cold spot was found to move during the process towards the bottom of the egg. Moreover, again similar to the findings of Ghani et al. (1999), a cold-zone as opposed to a cold-spot was predicted. The location of the cold zone in the yolk was predicted to lay below its geometrical centre, even for the case where the yolk was positioned at the top of the egg. It was concluded that no convective heating takes place in the egg yolk during processing.

In the final paper of the series Denys et al. (2005) coupled a first order kinetic approach to predict the rate of thermal microbial inactivation for *Salmonella Enteritidis* (SE) during the heating process. Firstly, SE was observed to be less heat sensitive in the yolk when compared to egg-white. Then they found that processing at a temperature of 57°C enabled 5 log reductions of SE within 40 minutes which was considered a reasonable process time after balancing the desire for efficient processing with the desire to retain raw egg properties. The results of the CFD analysis were then presented graphically with an aim to help egg processors to optimise the pasteurizing process.

Recent studies have focused on the fact that inadequate cooling of eggs before storage as improper cooling can result in cooling of chicken eggs can lead to growth of pathogenic bacteria such as Salmonella Enteritidis (SE). The cooling rate of an egg, which depends on several factors, such as cooling air temperature, initial egg temperature, air velocity inside the chiller, type of packaging material, and location of the egg inside the chamber, has been recently investigated using CFD (Kumar et al., 2009). During the research work the authors (Kumar et al., 2009) compared simulated and experimental values of egg centre temperature at various boundary conditions, and found good agreement with root mean squared error ranging from 0.23°C to 0.37°C. It is assumed that future work will use predicted egg temperatures from the heat transfer model as an input to a microbial growth model to estimate the potential growth of Salmonella spp. during egg cooling for assessing food safety. CFD modelling has also been used to predict the decontamination period for intact eggs contaminated by *Salmonella* spp. Results from the model, together with the microbial results on experimentally inoculated eggs, suggests the application of the hot air treatment on eggs before packaging is useful in reaching approximately 90% reduction of *S.Enteritidis* population (Pasquali et al., 2010).

6. Modelling Emerging Processing Technologies

6.1. High Pressure Thermal Processing

High pressure processing (HPP) is a novel food processing method which has shown great potential in the food industry. During compression/decompression phases the internal energy of a high pressure processing system changes, resulting in heat transfer between the internal system and its boundaries. These thermal-hydraulic characteristics were studied with CFD by Hartmann (2002) the technique was deemed necessary in gaining thorough understanding of the phenomena inherent in HPP, especially when the scale-up phenomena need to be analyzed (e.g. layout and design of high pressure devices, packages, etc.) (Hartman, 2002). Hartmann and Delgado (2002) used CFD and dimensional analyses to determine the time-scales of convection, conduction and bacterial inactivation, the relative values of which contribute to the efficiency and uniformity of conditions during HPP. Conductive and convective time-scales were directly compared to the inactivation time-scale in order to provide a picture of the thermal-hydraulic phenomena in the HP vessel during bacterial inactivation. The results showed that pilot scale systems exhibited a larger convection time-scale than the inactivation time-scale, and that the intensive fluid motion and convective heat transfer resulted in homogenous bacterial inactivation. Furthermore, the simulations of industrial scaled systems showed greater efficiency in bacterial inactivation as the compression heating subsisted for greater time periods when compared to smaller laboratory systems. Other CFD simulations showed that the thermal properties of the HP vessel boundaries have considerable influence on the uniformity of the process, and insulated material promoted the most effective conditions (Hartmann et al., 2004). As well as this, the insulated vessel was found to increase the efficiency of the HPP by 40%. A CFD and dimensionless analysis of the convective heat transfer mechanisms in liquid foods systems under pressure was also performed by Kowalczyk and Delgado (2007) who advised that HP systems with a characteristic dimension of 1 m alongside a low viscous medium should be used to avoid heterogeneous processing of the product.

CFD studies have also provided solutions to the thermal-hydraulic phenomena in HPP systems containing packaged UHT (Ultra High Temperature) milk (Hartmann et al., 2003), packaged enzyme mixture (Hartmann and Delgado, 2003), solid beef fat (Abdul Ghani and Farid, 2007), and solid food analogue material (Otero et al., 2007) (e.g. tylose with similar properties to meat, and agar with similar properties to water). In both investigations

(Hartmann et al., 2003; Hartmann and Delgado, 2003) the most significant results showed strong coupling between concentrations of the surviving micro-organisms and the spatial distribution of the food package in the HP vessel, owing to the inhomogeneous temperature field. A low conductive package material was also found to improve the uniformity of processing by preserving the elevated temperature level within the package throughout the pressurization phase; an average difference of about two log reductions was found per 10 fold increase in the package thermal conductivity. The two-dimensional CFD simulations of Otero et al. (2007) found that the filling ratio of the HP vessel played a major role in process uniformity, with convective currents having least effect on heat transfer when this ratio is large. Otero et al. (2007) also showed that by anticipating the temperature increase resulting from compression heating and by allowing the pressure transmitting medium to supply the appropriate quantity of heat, the uniformity of HPP was enhanced when both large and small sample ratios were used. More recently, Abdul Ghani and Farid (2007) used three-dimensional CFD simulations to illustrate both convective and conductive heat transfer in a HPP system loaded with pieces of solid beef fat. The simulation showed a greater adiabatic heating in the beef fat than the pressure transmitting medium owing to the greater compression heating coefficient used in this case (Fig. 3).

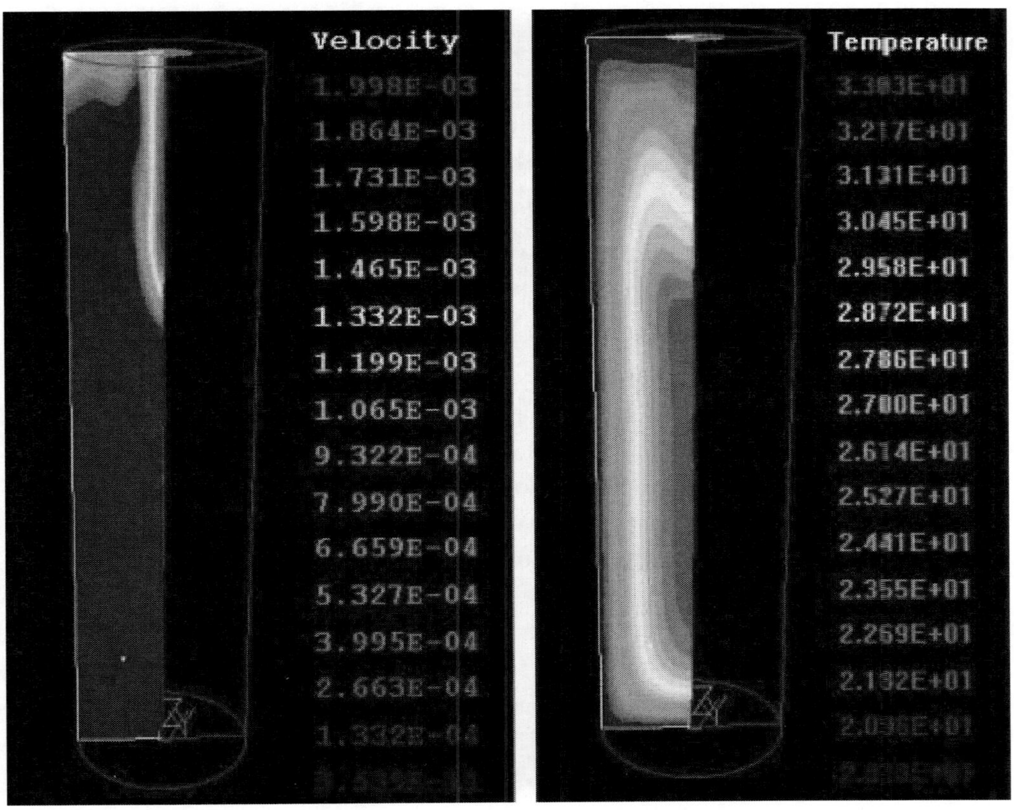

Figure 3. Temperature and velocity profiles of the fluid at the early stage of non-adiabatic compression (t = 24 s, P = 400 MPa). The red arrow shown in the figure is just for a point in the computation process (Abdul Ghani and Farid, (2007) with permission by Elsevier).

6.2. Ohmic Heating

The basic principle of ohmic heating is that electrical energy is converted to thermal energy within a conductor. In food processing, foodstuffs act as the conductor. The main advantage of ohmic heating is that, because heating occurs by internal energy generation within the conductor, this processing method leads itself to even distribution of temperatures within the food and therefore an even kill of potentially harmful micro-organisms throughout the product (Jun and Sastry, 2005). Jun and Sastry (2005) were the first to model the ohmic heating process, with the aim to enhance heating techniques for use by cabin crews during long-term space missions. They developed a two-dimensional transient model, for chicken noodle soup (assumed single phase) and black beans, under the ohmic heating process, by solving the electric field via the Laplace equation. From this solution the internal heat generation was obtained, which was added as a source to the Fourier equation, and was then numerically solved by the CFD code. Electrical conductivity was allowed to vary as a function of temperature. The CFD model was able to predict regions of electric field overshoot in the food, as well as the non-uniformities in the predicted thermal field. Moreover, they noted that as the electrode got wider, the cold zone area developed in the middle of the packaging diminished to a minimum and then appeared and grew at the corners of the packaging, clearly illustrating the existence of a threshold value for electrode size optimisation. They later expanded the model so that a three-dimensional representation of pouched tomato soup could be simulated (Jun and Sastry, 2007), in which they found the electric field strength near the edges of electrodes to overshoot as it got close to the maximum value, as predicted by their first model. On the other hand, the food between the V-shaped electrodes experienced a weak electric field strength, which gave rise to cold zones in the food. Jun and Sastry (2007) recognised that the presence of these cold zones merited further research on pouches via modelling and package redesign.

Marra et al. (2009) developed a CFD model to quantify the ohmic heating effects within a model solid food system, which they then used to optimize heat distribution within the foodstuff in order to evaluate the main parameters affecting the system. Their results showed that heat loss occurs to the cell wall and electrode surfaces could be reduced by improving the system design. It was also found that the temperature gradients that were established in the proximity of the outer curved surface exhibited colder areas at junctions of electrodes with lateral sample surface. However, as noted by Shim et al. (2010) former CFD modelling owever, the majority of CFD models for multiphase food mixture have dealt with a single solid particle system surrounded by liquid fluid, which is over-simplified and far from the reality. Therefore, Shim et al. (2010) conducted CFD simulation of the thermodynamic performances of the multi-component food mixture in an ohmic heating system in order to predict realistic hot and cold spots in the processed food, thereby preventing less conductive food particles from being under-processed. While the model was successfully validated, it was only developed in 2D, therefore the flow regime with moving particle orientation inside the field strength distribution is crucial and could not be interpreted in three-dimensions. The authors proposed that a future 3D model should be developed and should be capable of addressing the particles concentration and orientation in a continuous flow system.

Conclusion

Transfer modelling has played an active part in the design of thermal processes for over a decade now. In recent years simulations have reached higher levels of sophistication. The importance of maintaining a high level of accuracy via circumspect choices made during model development is evident from the reviewed studies, as many studies provide detailed validation exercises. Undoubtedly, with current computing power progressing unrelentingly, it is conceivable that transfer modelling will continue to provide explanations for transport phenomena, leading to better design of thermal processes and ensure safer food in the food industry.

Nomenclature

A= area (m^2)
C = water concentration (mol m^{-3})
C_μ = empirical turbulence model constant
c_p = specific heat capacity (W kg^{-1} K^{-1})
d = width of jet (m)
D = width of cylinder (m)
g = acceleration due to gravity (m s^{-2})
M = Molecular weight (kg $kmol^{-1}$)
Nu = Nusselt number
p = pressure (Pa)
R = gas constant (J $kmol^{-1}$ K^{-1})
Re = Reynolds number
k = turbulent kinetic energy (m^2 s^{-2})
s_T = thermal sink or source (W m^{-3})
s_c = concentration sink or source (mol m^{-3})
T = temperature (K)
t = time (s)
U = velocity component (m s^{-1})
V = volume (m^3)
$\vec{v_i}$ = velocity component (m s^{-1})
x = Cartesian coordinates (m)

Greek Letters

ρ = density (kg m^{-3})
μ = dynamic viscosity (kg m^{-1} s^{-1})
β = thermal expansion coefficient (K^{-1})
λ = thermal conductivity (W m^{-1} K^{-1})
α= water molar latent heat of vaporization, (J mol^{-1})
ε = turbulent dissipation rate (m^2 s^{-3})
φ = the transported quantity

Γ = diffusion coefficient of transported variable

D_s = the diffusivity of the mass component in the fluid ($m^2\ s^{-1}$)

ω = specific dissipation (s^{-1})

μ_t = turbulent viscosity ($kg\ m^{-1}\ s^{-1}$)

Subscripts

i = Cartesian coordinate index

bf = bulk fluid

s = surface

V = mesh element volume

A = area of mesh element

m = mean

0 = with no turbulence

References

Abbott M B, Basco B R (1989). *Computational Fluid Dynamics - an Introduction for Engineers*. Harlow, UK. Longman Scientific and Technical, pp 5-30.

Abdul Ghani A G, Farid M M (2007). Thermal sterilisation of foods using computational fluid dynamics. Chapter 13 in *Computational Fluid Dynamics in Food Processing*, Da-Wen Sun (editor), 28 chapters, CRC Press, USA, pp 347-381.

Abdul Ghani A G, Farid M M (2006). Numerical simulation of solid–liquid food mixture in a high pressure processing unit using computational fluid dynamics. *Journal of Food Engineering*, 80(4), 1031-1042.

Abdul Ghani A.G., (2006). A computer simulation of heating and cooling liquid food during sterilization process using computational fluid dynamics. Association for Computing Machinery New Zealand Bulletin, 2(3), ISSN: 1176-9998.

Abdul Ghani A G, Farid M M, Chen X D (2002b). Theoretical and experimental investigation of the thermal inactivation of Bacillus stearothermophilus in food pouches. *Journal of Food Engineering*, 51(3), 221-228.

Abdul Ghani A G, Farid M M, Chen X D, P Richards (1999). An investigation of deactivation of bacteria in a canned liquid food during sterilization using computational fluid dynamics (CFD). *Journal of Food Engineering*, 42, 207-214.

Abdul Ghani A G, Farid M M, Chen X D, Richards P (2001). Thermal sterilization of canned food in a 3-D pouch using computational fluid dynamics. *Journal of Food Engineering*, 48, 147-156.

Abdul Ghani A G, Farid M M, Zarrouk S J (2003). The effect of can rotation on sterilization of liquid food using computational fluid dynamics. *Journal of Food Engineering*, 57, 9-16.

Abdul Ghani A G, Farid M M. Numerical simulation of solid-liquid food mixture in a high pressure processing unit using computational fluid dynamics. *Journal of Food Engineering*, 80(4), 1031-1042.

Abraham J P, Sparrow E M (2003). Three-dimensional laminar and turbulent natural convection in a continuously/discretely wall-heated enclosure containing a thermal load. *Numerical Heat Transfer Part A: Applications*, 44(2), 105-125.

Anandharamakrishnan, C. (2003). Computational fluid dynamics (CFD) applications for the food industry. *Indian Food Industry*, 22(6), 62-68.

Anandharamakrishnan, C., Gimbun, J., Stapley, A. G. F., Rielly, C. D. (2010). Application of computational fluid dynamic (CFD) simulations to spray-freezing operations. *Drying Technology*, 28, 94-102.

Anon. (2007). *A brief history of computational fluid dynamics*. Internet article, www. fluent.com, Fluent Inc. 10 Cavendish Court, Lebanon, NH 03766, U.S.A.

Ashforth-Frost, S, Jambunathan K (1996). Numerical prediction of semiconfined jet impingement and comparison with experimental data. *International Journal of Numerical Methods in Fluids*, 23, 295–306.

Assarie, M.R., Tabrizi, H.B., Saffar-Avval, M. (2007). Numerical simulation of fluid bed drying based on two-fluid model and experimental validation. *Applied Thermal Engineering*, 27, 422–429.

Aversa M, Curcio S, Calabro V, Iorio G (2007). An analysis of the transport phenomena occurring during food drying process. *Journal of Food Engineering*, 78 (3), 922-932.

Boulet, M. Marcos, B., Dostie, M., Moresoli, C. (2010). CFD modeling of heat transfer and flow field in a bakery pilot oven. *Journal of Food Engineering*, 97(3), 393-402.

Choundhury D (1993). Introduction to the renormalization group method and turbulence modelling. Fluent Inc. Technical Memorandum TM-107.

Chhanwal, N., Anishaparvin, A., Indrani, D., Raghavarao, K.S.M.S. Anandharamakrishnan, C. (2010). Computational fluid dynamics (CFD) modeling of an electrical heating oven for bread-baking process. *Journal of Food Engineering*, (In Press).

Crank J (1975). *The Mathematics of Diffusion*, 2nd ed., Clarendon Press, Oxford, U.K.

De Bonis M V, Ruocco G (2007). Modelling local heat and mass transfer in food slabs due to air jet impingement, *Journal of Food Engineering*, 78(1), 230-237.

De Bonis M.V., Ruocco, G. (2008). A generalized conjugate model for forced convection drying based on an evaporative kinetics. *Journal of Food Engineering*, 89(2), 232-240.

De Bonis, M.V., Ruocco, G. (2009). Conjugate fluid flow and kinetics modelling for heat exchanger fouling simulation. *International Journal of Thermal,* 48(10), 2006-2012.

De Vries U, Velthuis H, Koster K (1995). Baking ovens and product quality - a computer model. *Food Science and Technology Today*. 9(4), 232-234.

Delele, M.A., Schenk, A., Ramon, H., Nicolaï, B.M., Verboven, P. (2009). Evaluation of a chicory root cold store humidification system using computational fluid dynamics. *Journal of Food Engineering*, 94(1), 110-121.

Denys S, Pieters J G, Dewettinck K (2003). Combined CFD and experimental approach for determination of the surface heat transfer coefficient during thermal processing of eggs. *Journal of Food Science*, 68(3), 943-951.

Denys S, Pieters J G, Dewettinck K (2004).Computational fluid dynamics analysis of combined conductive and convective heat transfer in model eggs. *Journal of Food Engineering*, 63(3), 281-290.

Denys S, Pieters J G, Dewettinck K (2005).Computational fluid dynamics analysis for process impact assessment during thermal pasteurization of intact eggs. *Journal of Food Protection*, 68(2), 366-374.

Denys S, Pieters J, Dewettinck K (2007). CFD analysis of thermal processing of eggs. Chapter 14 in *Computational Fluid Dynamics in Food Processing*, Da-Wen Sun (editor), 28 chapters, CRC Press, USA, pp 347-381.

Dincov D D, Parrott K A, Pericleous K A (2002). Coupled 3-D finite difference time domain and finite volume methods for solving microwave heating in porous media. *Lecture Notes in Computer Science*, 2329, 813-822.

Erriguible A, Bernada P, Couture F, Roques M A (2005). Modeling of heat and mass transfer at the boundary between a porous medium and its surroundings. *Drying Technology*, 23(3), 455-472.

Erriguible A, Bernada P, Couture F, Roques M A (2006). Simulation of convective drying of a porous medium with boundary conditions provided by CFD. *Chemical Engineering Research and Design*, 84(A2), 113-123.

Farid M, Abdul Ghani GA (2004). A new computational technique for the estimation of sterilization time in canned food. *Chemical Engineering and Processing*, 43(4), 523-531.

Feng and Tang (1998). Microwave finish drying of diced apples in a spouted bed. *Journal of Food Science*, 63, 679–683.

Fernandes C S, Dias R P, Nobrega J M, Afonso I M, Melo L F, Maia J M (2006). Thermal behaviour of stirred yoghurt during cooling in plate heat exchangers. *Journal of Food Engineering*, 69, 281-290.

Fernandes C S, Dias R, Nóbrega J M, Afonso I M, Melo L F, Maia J M (2005). Simulation of stirred yoghurt processing in plate heat exchangers. *Journal of Food Engineering* 69: 281-290.

Ferry M (2002). New Features of Migal Solver. *The Phoenics Journal*, 14(1), 88-96.

Ferziger J H, Peric M (2002). *Computational Methods for Fluid Dynamics*. Berlin Heidleberg: Springer-Verlag, pp 1-100.

Fletcher D F, Guo B, Harvie D J E, Langrish T A G, Nijdam J J, Williams J (2006). What is important in the simulation of spray dryer performance and how do current CFD models perform? *Applied Mathematical Modelling*, 30(11), 1281-1292.

Friedrich R, Huttl T J, Manhart M, Wagner C (2001). Direct numerical simulation of incompressible turbulent flows. *Computers and Fluids*, 30, 555-579.

Geedipalli S S R, Rakesh V, Datta A K (2007). Modelling the heating uniformity contributed by a rotating turntable in microwave ovens. *Journal of Food Engineering*, 82(3), 359-368.

Ghani AGA, Farid M M, Chen X D (2002b). Theoretical and experimental investigation of the thermal destruction of Vitamin C in food pouches. *Computers and Electronics in Agriculture*, 34(1-3), 129-143.

Gidaspow D, Bezburuah R, Ding J (1992). Hydrodynamics of circulating fluidized beds, kinetic theory approach. In Fluidization VII, *Proceedings of the 7th Engineering Foundation Conference on Fluidization*, Brisbane, Australia, 75–82.

Grijspeerdt K, Vucinic D, Lacor, C (2007). CFD modelling of the hydrodynamics of plate heat exchangers for milk processing. Chapter 19 in *Computational Fluid Dynamics in Food Processing*, Da-Wen Sun (editor), 28 chapters, CRC Press, USA, pp 403-417.

Grijspeerdt K., Hazarika B, Vucinic D (2003). Application of computational fluid dynamics to model the hydrodynamics of plate heat exchangers for milk processing. *Journal Food Engineering*, 57(3), 237-242.

Harral B B, Boon C R (1997). Comparison of predicted and measured airflow patterns in a mechanically ventilated livestock building without animals. *Journal of Agricultural Engineering Research*, 66, 221-228.

Harral, B, Burfoot, D (2005). A comparison of two models for predicting the movements of airborne particles from cleaning operations. *Journal of Food Engineering*, 69, 443-451.

Hartmann C (2002). Numerical simulation of thermodynamic and fluid-dynamic processes during the high-pressure treatment of fluid food systems. *Innovative Food Science & Emerging Technologies*, 3(1), 11-18.

Hartmann C, Delgado A (2003). The influence of transport phenomena during high-pressure processing of packed food on the uniformity of enzyme inactivation. *Biotechnology and Bioengineering*, 82(6), 725-735.

Hartmann C, Delgado A, Szymczyk J (2003). Convective and diffusive transport effects in a high pressure induced inactivation process of packed food. *Journal of Food Engineering*, 59(1), 33-44.

Hartmann C; Delgado A (2002). Numerical simulation of convective and diffusive transport effects on a high-pressure-induced inactivation process. *Biotechnology and Bioengineering*, 79(1), 94-104.

Heldman D R, Singh R P (1981). *Food Process Engineering*, AVI Publishing, Westport, Connecticut, pp 1-10.

Hendrickx M, Silva C, Oliveria F, Tobback P (1993). Generalised (semi-) empirical formulas for optimal sterilisation temperatures of conduction-heated foods with infinite surface heat transfer coefficients. *Journal of Food Engineering*, 19(2), 141-158.

Jamaleddine, T.J., Ray, M.B. (2010). Application of Computational Fluid Dynamics for Simulation of Drying Processes: A Review. *Drying Technology*, 28(2), 120 -154.

Jangam, S.V., Mujumdar, A.S. and Thorat, B.N. (2009). Design of an efficient gas distribution system for a fluidized bed dryer, *Drying Technology*, 27: 11, 1217 — 1228

Jun S, Puri V M (2005). 3D milk-fouling model of plate heat exchangers using computational fluid dynamics, *International Journal of Dairy Technology*, 58(4), 214-224

Jun S, Puri V M (2007). Plate heat exchange thermal and fouling analysis. Chapter 19 in *Computational Fluid Dynamics in Food Processing*, Da-Wen Sun (editor), 28 chapters, CRC Press, USA, pp 417-433.

Jun S, Sastry S (2005). Reusable pouch development for long term space missions: A 3D ohmic model for verification of sterilization efficacy. Journal of Food Engineering, 80(4), 1199-1205.

Jun S, Sastry S (2007). Modeling and optimization of ohmic heating of foods inside a flexible package. *Journal of Food processing Engineering*, 28 (4): 417-436 AUG 2005

Jung A, Fryer P J (1999). Optimising the quality of safe food: computational modelling of a continuous sterilisation process. *Chemical Engineering Science*, 54, 717-730.

Kannan, A., Gourisankar Sandaka, P.Ch. (2008). Heat transfer analysis of canned food sterilization in a still retort. *Journal of Food Engineering*, 88(2), 213-228.

Kaya A, Aydin O, Dincer I (2007). Numerical modeling of forced-convection drying of cylindrical moist objects. Numerical Heat Transfer Part A: Applications, 51(9), 843-854.

Kaya A, Aydin O, Dincer I (2008). Modeling of recirculating flows during air drying of moist objects for various dryer configurations. Numerical Heat Transfer Part A: Applications, 53(1), 18-34.

Kenneth J B (2004). Heat exchanger design for the process industries. *Journal of Heat Transfer*, 126(6), 877-885.

Kocer D, Karwe M V (2005). Thermal transport in a multiple jet impingement oven. *Journal of Food Process Engineering*, 28(4), 378-396.

Kocer D, Nitin N, Karwe M V (2007). Applications of CFD in Jet impingement oven. Chapter 19 in *Computational Fluid Dynamics in Food Processing*, Da-Wen Sun (editor), 28 chapters, CRC Press, USA, pp 469-487

Kondjoyan A (2006). A review on surface heat and mass transfer coefficients during air chilling and storage of food products. *International Journal of Refrigeration*, 29, 6, 863-875.

Kowalczyk W, Hartmann C, Delgado A (2004). Modelling and numerical simulation of convection driven high pressure induced phase changes. *International Journal of Heat and Mass Transfer*, 47(5), 1079-1089.

Kızıltaş, S., Erdoğdu, F., Palazoğlu, T.K. (2010). Simulation of heat transfer for solid–liquid food mixtures in cans and model validation under pasteurization conditions. *Journal of Food Engineering*, 97(4), 449-456.

Kumar, V., Jonnalagadda, D., Subbiah, J., Wee, A. P., Thippareddi, H., Birla, S. (2009). A 3D heat transfer and fluid dlow model for cooling of a single egg under forced convection. *Transactions of ASABE*, 52(5), 1627-1637.

Kuriakose, R. Anandharamakrishnan, C. (2010). Computational fluid dynamics (CFD) applications in spray drying of food products. *Trends in Food Science & Technology*, (In Press).

Langrish T A G, Fletcher D F (2003). Prospects for the modelling and design of spray dryers in the 21st century. *Drying Technology*, 21(2), 197-215.

Launder B E, Spalding D B (1974). The numerical computation of turbulent flows, *Computer Methods in Applied Mechanics and Engineering*, 3, 269–289.

Li, Z., Su, W. , Wu, Z. , Wang, R. and Mujumdar, A. S.(2010). Investigation of flow behaviors and bubble characteristics of a pulse fluidized bed via CFD modeling, *Drying Technology*, 28(1), 78-93.

Madhiyanon T, Soponronnarit S, Tia W (2002). A mathematical model for continuous drying of grains in a spouted bed dryer. *Drying Technology*, 20, 587–614.

Markowski, M., Białobrzewski, I., Modrzewska, A. (2010). Kinetics of spouted-bed drying of barley: diffusivities for sphere and ellipsoid. *Journal of Food Engineering*, 96(3), 380-387.

Marquenie, D., Geeraerd, A.H., Lammertyn, J., Soontjens, C., Van Impe J.F., Michiels C.W., (2003). Combinations of pulsed white light and UV-C or mild heat treatment to inactivate conidia of Botrytis cinerea and Monilia fructigena, International *Journal of Food Microbiology*, 85 (2(1-2), 185–196.

Marra F, Lyng J, Romano V, McKenna B (2007). Radio-frequency heating of foodstuff: solution and validation of a mathematical model. *Journal of Food Engineering*, 79(3), 998-1006.

Mathur, K B, Epstein N (1974). *Spouted Bed*, Academic Press: New York.

Matz S (1989). *Equipment for Bakers*. Elsevier Science Publishers. USA.

Markowski, M., Białobrzewski, I., Modrzewska, A. (2009). Kinetics of spouted-bed drying of barley: Diffusivities for sphere and ellipsoid. *Journal of Food Engineering*, 96(3), 380-387.

Marra, F., Zell, M., Lyng, J.G., Morgan D.J., Cronin, D.A. (2009). Analysis of heat transfer during ohmic processing of a solid food, *Journal of Food Engineering*, 91(1), 56–63.

Marra, F., Valeria M., De Bonis, M., Ruocco, G. (2010). Combined microwaves and convection heating: A conjugate approach. *Journal of Food Engineering*, 97(1), 31-39.

May B K, Perre P (2002). The importance of considering exchange surface area reduction to exhibit a constant drying flux period in foodstuffs. *Journal of Food Engineering*, 54(4), 271-282.

Metcalfe, G., Lester, D. (2009). Mixing and heat transfer of highly viscous food products with a continuous chaotic duct flow. *Journal of Food Engineering*, 95(1), 21-29.

Menter F R (1994). Two-equation eddy-viscosity turbulence models for engineering applications, *AIAA Journal,* 32(8), 1598–1604.

Mistry H, Ganapathi-subbu, Dey S, Bishnoi P, Castillo J L (2006). Modeling of transient natural convection heat transfer in electric ovens. *Applied Thermal Engineering*, 26(17-18), 2448-2456.

Modest M (1992). *Radiative Heat Transfer*, Academic Press, New York, USA.

Mondal, A., Datta, A.K. (2010). Two-dimensional CFD modeling and simulation of crustless bread baking process. *Journal of Food Engineering*, 99(2), 166-174.

Moureh J, Flick D (2005). Airflow characteristics within a slot-ventilated enclosure. *International Journal of Refrigeration*, 26, 12-24.

Moureh, J., Letang, G., Palvadeau, B., Boisson, H. (2009). Numerical and experimental investigations on the use of mist flow process in refrigerated display cabinets. *International Journal of Refrigeration*, 32(2), 203-219.

Mujumdar, A.S., Huang, L-X., Chen, X.D. (2010). An overview of the recent advances in spray-drying. *Dairy Science and Technology*, 90, 211–224.

Nakayama A, Kuwahara F, Xu G, Kato F (2004). An analytical treatment for combined heat transfer by radiation, convection and conduction within a heat insulating wall structure. *Heat and Mass Transfer*, 40(8), 621-626.

Navaneethakrishnan, P., Srinivasan P.S.S. , Dhandapani, S. (2007). Heat transfer and heating rate of food stuffs in commercial shop ovens, *Sadhana-Academy Proceedings in the Engineering Sciences*, 32, 535–544.

Nema, P.K., Datta, A.K. (2006). Improved milk fouling simulation in a helical triple tube heat exchanger, *International Journal of Heat and Mass Transfer*, 49, 3360–3370.

Neto, J. L. Vieira , Duarte, C. R. , Murata, V. V. and Barrozo, M. A. S.(2008). Effect of a draft tube on the fluid dynamics of a spouted bed: experimental and CFD studies, *Drying Technology*, 26(3), 299 -307.

Nicolaï B M, de Baerdemaeker J (1996). Sensitivity analysis with respect to the surface heat transfer coefficient as applied to thermal process calculations. *Journal of Food Engineering*, 28, 21-33.

Nicolaï B M, Verboven P, Scheerlinck N (2001). Modelling and simulation of thermal processes, Chapter 6 in *Thermal Technologies in Food Processing*. Philip Richardson (editor), 14 chapters, Woodhead Publishing LTD, Cambridge, England, pp 91-109.

Nijdam J J, Guo B Y, Fletcher D F, Langrish T A G (2006). Lagrangian and Eulerian models for simulating turbulent dispersion and coalescence of droplets within a spray. *Applied Mathematical Modelling*, 30(11), 1196-1211.

Nitin N, Gadiraju R P, Karwe M V (2006). Conjugate heat transfer associated with a turbulent hot air jet impinging on a cylindrical object. *Journal of Food Processing Engineering*, 29(4), 386-399.

Norton T, Sun D-W (2007). An overview of CFD applications in the food industry. Chapter 1 in *Computational Fluid Dynamics in Food Processing*, Da-Wen Sun (editor), 28 chapters, CRC Press, USA, pp 1-43.

Olsson E E M, Ahrne L M, Tragardh A C (2004). Heat transfer from a slot air jet impinging on a circular cylinder *Journal of Food Engineering*, 63(4), 393-401.

Olsson E E M, Ahrne L M, Tragardh A C (2005). Flow and heat transfer from multiple slot air jets impinging on circular cylinders. *Journal of Food Engineering*, 67(3), 273-280.

Olsson E, Trägårdh C (2007). CFD modelling of jet impingement during heating and cooling of foods. , Chapter 20 in *Computational Fluid Dynamics in Food Processing*, Da-Wen Sun (editor), 28 chapters, CRC Press, USA, pp 478-505.

Otero L, Ramos A M, de Elvira C, Sanz P D (2007). A model to design high-pressure processes towards an uniform temperature distribution. *Journal of Food Engineering*, 78(4), 1463-1470.

Ozalpa, A.A., Dincer, I. (2010). Hydrodynamic-thermal boundary layer development and mass transfer characteristics of a circular cylinder in confined flow. *International Journal of Thermal Sciences*, (in press).

Park K, Choi D-H, Lee K-S (2004). Optimum design of plate heat exchanger with staggered pin arrays. *Numerical Heat Transfer Part A: Applications*, 45(4), 347-361.

Pasquali, F., Fabbri, A., Cevoli,C., Manfreda, G., Franchini, A. (2010). Hot air treatment for surface decontamination of table eggs. *Food Control*, 21(4), 431-435.

Patankar S V (1980). *Numerical heat transfer*. Hemisphere Publishing, Washington.

Patankar S V, Spalding D B (1972). A calculation procedure for heat, mass and momentum transfer in three-dimensional parabolic flows. *International Journal for Heat and Mass Transfer*, 15, 1787-1806.

Perre P, Turner I W (1999). A 3-D version of TransPore: a comprehensive heat and mass transfer computational model for simulating the drying of porous media. *International Journal of Heat and Mass Transfer*, 42 (24), 4501-4521.

Ressing H, Ressing M, Durance T (2007). Modelling the mechanisms of dough puffing during vacuum microwave drying using the finite element method. *Journal of Food Engineering*, 82(4) 498-508.

Richardson P (2001). *Thermal Technologies in Food Processing*. Philip Richardson (editor), 14 chapters, Woodhead Publishing LTD, Cambridge, England, pp 1-3.

Scott G, Richardson P (1997). The application of computational fluid dynamics in the food industry. *Trends in Food Science and Technology*, 8, 119-124.

Seyedein, S.H.; Hasan, M.; Mujumdar, A.S (1994). Modelling of a single confined turbulent slot jet impingement using various k-ε turbulence models. *Applied Mathematical Modelling*, 18, 526–537.

Shih T H, Liou W W, Shabbir A, Zhu J (1995). A new k–ε eddy viscosity model for high Reynolds number turbulent flows—model development and validation, *Computers and Fluids*, 24 (3), 227–238.

Shim, J.-Y., Lee, S.-H., Jun, S. (2010). Modeling of ohmic heating patterns of multiphase food products using computational fluid dynamics codes. *Journal of Food Engineering*, 99(2), 136-141.

Siegel R, Howell J R (1992). *Thermal Radiation Heat Transfer*, Hemisphere Publishing, Co., Washington, USA.

Singh R P, Heldman D R (2001). *Introduction to Food Engineering*. Academic Press, A Harcourt Science and Technology Company, California, USA.

Sobieski, W.(2009). Momentum exchange in solid-fluid system modeling with the Eulerian multiphase model, *Drying Technology*, 27: 5, 653 -671.

Straatsma J, Verdurmen R E M, Verscheuren M, Gunsing M, de Jong P (2007). CFD Simulation of spray drying of food products. Chapter 10 in *Computational Fluid Dynamics in Food Processing*, Da-Wen Sun (editor), 28 chapters, CRC Press, USA, pp 249-287.

Sun D-W (2007). *Computational Fluid Dynamics in Food Processing*, Da-Wen Sun (editor), 28 chapters, CRC Press.

Suresh H N, Narayana P A A, Seetharamu K N, Conjugate mixed convection heat and mass transfer in brick drying, *Heat and Mass Transfer*, 37, 205–213.

Szafran R G, Kmiec A (2004). CFD modeling of heat and mass transfer in a spouted bed dryer. *Industrial and Engineering Chemistry Research*, 43(4): 1113-1124.

Szafran R G, Kmiec A (2005). Point-by-point solution procedure for the computational fluid dynamics modeling of long-time batch drying. *Industrial and Engineering Chemistry Research*, 44(20), 7892-7898.

Tanaka, F., Verboven, P., Scheerlinck, N., Moritaa, K., Iwasakia, K., Nicolaï, B. (2007). Investigation of far infrared radiation heating as an alternative technique for surface decontamination of strawberry. *Journal of Food Engineering*, 79(2), 445-452.

Tattiyakul J, Rao M A, Datta A K (2001). Simulation of heat transfer to a canned corn starch dispersion subjected to axial rotation. *Chemical Engineering and Processing*, 40, 391-399.

Tewkesbury H, Stapley A G F, Fryer P J (2000). Modelling temperature distributions in cooling chocolate moulds. *Chemical Engineering Science*, 55(16), 3123-3132.

Therdthai N, Zhou W B, Adamczak T. (2002). Optimisation of the temperature profile in bread baking. *Journal of Food Engineering*, 55(1), 41-48.

Therdthai N, Zhou W B, Adamczak T (2003). Two-dimensional CFD modelling and simulation of an industrial continuous bread baking oven. *Journal of Food Engineering*, 60(2), 211-217.

Therdthai N, Zhou W B, Adamczak T (2004a). Three-dimensional CFD modelling and simulation of the temperature profiles and airflow patterns during a continuous industrial baking process. *Journal of Food Engineering*, 65(4), 599-608.

Therdthai N, Zhou W B, Adamczak T (2004b). Simulation of starch gelatinisation during baking in a travelling-tray oven by integrating a three-dimensional CFD model with a kinetic model. *Journal of Food Engineering*, 65(4), 543-550.

Turnbull J, Thompson C P (2005). Transient averaging to combine large eddy simulation with Reynolds averaged Navier-Stokes simulations. *Computers and Chemical Engineering*, 29, 379-392.

Varga S, Oliveira J C (2000). Determination of the heat transfer coefficient between bulk medium and packed containers in a batch retort. *Journal of Food Engineering*, 44(4), 191-198.

Varma M N, Kannan A (2005). Enhanced food sterilization through inclination of the container walls and geometry modifications *International Journal of Heat and Mass Transfer* 48, 3753-3762.

Varma M N, Kannan A (2005). CFD studies on natural convective heating of canned food in conical and cylindrical containers. *Journal of Food Engineering*, 77(4), 1027-1036.

Verboven P, Datta A K, Anh N T, Scheerlinck N, Nicolaï B M (2003). Computation of airflow effects on heat and mass transfer in a microwave oven. *Journal of Food Engineering*, 59(2-3), 181-190.

Verboven P, Datta A K, Anh NT, Scheerlinck N, Nicolaï B M (2003). Computation of airflow effects on heat and mass transfer in a microwave oven. *Journal of Food Engineering*, 59(2-3), 181-190.

Verboven P, Datta A K, Nicolaï B M (2007). Computation of airflow effects in microwave and combination heating. Chapter 12 in *Computational Fluid Dynamics in Food Processing*, Da-Wen Sun (editor), 28 chapters, CRC Press, USA, pp 313-331.

Verboven P, de Baerdemaeker J, Nicolaï B M (2004). Using computational fluid dynamics to optimise thermal processes. Chapter 4 in *Improving the Thermal Processing of Foods*. Philip Richardson (editor), 23 chapters, Woodhead Publishing LTD, Cambridge, England, pp 82-102.

Verboven P, Scheerlinck N, De Baerdemaeker J, Nicolaï B M (2000a). Computational fluid dynamics modelling and validation of the isothermal airflow in a forced convection oven. *Journal of Food Engineering*, 43(1), 41-53.

Verboven P, Scheerlinck N, De Baerdemaeker J, Nicolaï B M (2000b). Computational fluid dynamics modelling and validation of the temperature distribution in a forced convection oven. *Journal of Food Engineering*, 43(2), 61-73.

Versteeg H K, Malalsekeera W (1995). *An Introduction to Computational Fluid Dynamics*. Harlow, U.K.: Longman Group Ltd, pp 1-100.

Visser J, Jeurnink T J M (1997). Fouling of heat exchangers in the dairy industry. *Experimental Thermal and Fluid Science*, 14(4), 407-424.

Wang, H. G., Yang, W. Q. Senior, P., Raghavan , R. S., Duncan , S. R. (2008). Investigation of batch fluidized-bed drying by mathematical modeling, CFD simulation and ECT measurement. *Particle Technology and Fluidization*, 54(2), 427-444.

Wang L, Sun D-W (2003). Recent developments in numerical modelling of heating and cooling processes in the food industry - a review. *Trends in Food Science and Technology*, 14, 408-423.

Wang L, Sun D-W (2005). Heat and mass transfer in thermal food processing. Chapter 2 in *Thermal Food Processing*, Da Wen Sun (editor), CRC Press, Taylor and Francis group, New York. 35-71.

Welti-Chanes J, Vergara-Balderas F, Bermúdez-Aguirre D (2005). Transport phenomena in food engineering: basic concepts and advances. Journal of Food Engineering 67, 113-128.

Wilcox D C (1993). Comparison of 2-equation turbulence models for boundary-layer flows with pressure gradient. *AIAA Journal*, 31(8), 1414-1421.

Woo, M.W., Daud, W.R.W., Mujumdar, A.S., Talib, M.Z.M., Wu Z.H., Tasirin, S.M. (2009). Non-swirling steady and transient flow simulations in short-form spray dryers, Chemical Product and Process Modeling, 4(1), Article 20..

Wong S Y, Zhou W B, Hua J S (2006). Robustness analysis of a CFD model to the uncertainties in its physical properties for a bread baking process. *Journal of Food Engineering*, 77(4), 784-791.

Wong S Y, Zhou W B, Hua J S (2007). CFD modeling of an industrial continuous bread-baking process involving U-movement. *Journal of Food Engineering*, 78 (3), 888-896.

Wong S Y, Zhou W B, Hua J S (2007). Designing process controller for a continuous bread baking process based on CFD modelling, *Journal of Food Engineering*, 81(3), 523-534.

Xia B, Sun D-W (2002). Applications of computational fluid dynamics (CFD) in the food industry: a review. *Computers and Electronics in Agriculture* 34, 5-24.

Zhao W N, Kumar K, Mujumdar A S (2004). Flow and heat transfer characteristics of confined noncircular turbulent impinging jets. *Drying Technology*, 22(9), 2027-2049.

Zheng L, Delgado A, Sun D-W (2007). Surface heat transfer coefficients with and without phase change. Chapter 23 in *Food Properties Handbook* (Second Edition). Shafiur Rahman (editor), in editing.

Zhonghua W, Mujumdar A S. (2007). Simulation of the hydrodynamics and drying in a spouted bed dryer. *Drying Technology*, 25, 59–74.

Zhou W, Therdthai N (2007). Three-dimensional CFD modelling of a continuous baking process. Chapter 11 in *Computational Fluid Dynamics in Food Processing* Da-Wen Sun (editor), 28 chapters, CRC Press, USA, pp 287-213.

In: Progress on Quantitative Approaches in Thermal … ISBN: 978-1-62100-842-2
Editors: V. P. Valdramidis and J. F. M. Van Impe © 2012 Nova Science Publishers, Inc.

Chapter 7

PARAMETER IDENTIFICATION UNDER DYNAMIC TEMPERATURE CONDITIONS IN INACTIVATION KINETICS

Kirk D. Dolan[1,2] and Dharmendra K. Mishra[2,3]
[1]Department of Food Science & Human Nutrition,
Michigan State University, East Lansing, MI, US
[2]Department of Biosystems & Agricultural Engineering, MSU, MI, US
[3]Nestle Nutrition, Fremont, MI, US

Abstract

This chapter describes the importance and need for parameter estimation for inactivation kinetics during nonisothermal conditions and how to use the mathematical and statistical tools needed to perform the estimation. The chapter is in three sections: 1. Theoretical background, 2. Examples of dynamic temperature parameter estimation for microbial inactivation using ordinary least squares, and 3. Examples of sequential estimation. Within these sections, the usefulness of plotting scaled sensitivity coefficients prior to or after experimentation is shown. How to choose the proper reference temperature for Arrhenius-type secondary models to avoid unacceptably large rate-constant confidence intervals is described. Examples are given to illustrate how to a) avoid parameter estimation convergence difficulties; b) estimate nonisothermal parameters, their confidence intervals and confidence regions; c) estimate asymptotic confidence bands and prediction bands for log(survival ratio); d) obtain more accurate confidence intervals for both parameters and log(survival) using resampling techniques, e) approximate the parameter distribution function using Monte Carlo techniques, and f) interpret the sequential estimation results as each data point is added.

Introduction

Why estimate kinetic inactivation parameters for dynamic temperature conditions? Two reasons are that virtually all commercial food processes at temperatures above ambient are nonisothermal to some degree; and because experiments run under conditions as close as possible to those of the intended application will have less error in the parameter estimates

(Peleg and others 2003). Canning is an example of a nonisothermal process. Even aseptically processed foods held at constant temperature experience nonisothermal conditions during the rapid initial heating and rapid cooling. However, estimation of nonisothermal parameters requires implicit, nonlinear regression as opposed to explicit computation of linear isothermal parameters. It is only in the last 10-20 years that inexpensive and fast software has become widely available, allowing nonisothermal parameters to be estimated more easily.

But aren't isothermal experimental results typically used for commercial processes, giving satisfactory results? Yes, but with a knowledge that there may be some difference between these parameters and the true dynamic parameters (Claeys and others 2001a) (Claeys and others 2001b), and always with a generous safety factor. Now that food companies want to meet consumer demand for higher quality, processing can be reduced if they have more accurate kinetic parameters and can still ensure food safety requirements. Isothermal parameters must be verified by estimating dynamic parameters from commercially similar processes. Slower heating rates (Stephens and others 1994) (Hassani and others 2005) (Valdramidis and others 2007) (Corradini and Peleg 2009) or longer sublethal history (Stasiewicz and others 2008) (Juneja and Marks 2003) may result in smaller rate constant parameters, perhaps because the microorganisms become more resistant. Therefore, the use of isothermal parameters may incur significant error in prediction of food safety.

Another reason for dynamic temperature experiments is that solid foods heated above 100°C must be in sealed containers to prevent moisture loss. It is technically difficult to seal a container with a sample size small enough to ensure virtually no temperature gradient across the sample. These types of experiments have been conducted by thermobacteriologists in small thermal-death-time cans with thermocouples inserted, and the assumption of no temperature gradient. The validity of this assumption can now be tested using commercial finite-element simulation software (such as Comsol or Ansys) (Halder and others 2007) (Halder and others 2011) that will predict the temperature anywhere in the sample during retorting. However, the situation becomes a "Catch-22" because the simulation software requires knowledge of the solid food thermal diffusivity at temperatures above 100°C. The line-source method is inadequate to estimate food thermal diffusivity because by the time the sample reaches the required isothermal temperature above 100°C, the food has been well overcooked and is no longer in its original state. For this reason, we have found no literature estimating the thermal properties of solid or viscous foods at temperatures above 100°C. Yet these properties are critical for today's aseptic processing of viscous foods, which may soon include particulates. Therefore, not only are dynamic temperature experiments needed to estimate microbial inactivation parameters, but they are also needed for estimation of thermal properties at elevated temperatures (Mishra and others 2008)

An additional reason for dynamic temperature experiments is that they could eliminate the need to perform isothermal experiments, which are technically difficult to perform and whose parameters are only approximate (Peleg and Normand 2004). The ideal case would be if one nonisothermal experiment could be conducted in the desired food at the actual commercial conditions with multiple samples removed at different heating times, resulting in the most accurate estimation of all the parameters simultaneously. This chapter will discuss how to analyze the data from such experiments.

The goal of this chapter is a) to present theory and step-by-step examples on how to estimate dynamic temperature parameters and form confidence bands for log(survival ratio), and b) to make such estimation accessible to a wide audience range. The assumption is that the reader has a need to go beyond isothermal experiments to obtain more accurate parameter estimates, and to determine confidence bands and prediction bands for log(microbial retention). When prediction bands for log(microbial retention) are known, we can be confident that 95% or 99% of the data will fall within the band. This information may be more important than knowing the confidence intervals for parameters (a situation different from the enzymologist desiring to estimate Michaelis-Menten parameters). Therefore, estimation of the microbial inactivation parameters, though important for prediction models, is typically a stepping stone to the $\log(N/N_0)$ prediction intervals.

This chapter will also show how the error in the original log(survival ratio) data is maintained all the way through the estimation process. This procedure is more accurate than the typical process of a two-step isothermal estimation of D and z values, with no confidence intervals, leading to *one* prediction of log(survival ratio) with no error bounds. In other words, the common procedure completely drops the error of the original data during the first estimation step and produces parameters D_r and z as if they had zero error. We strongly recommend rather to do a residual analysis of the log(survival) data and give prediction bands for the predicted log(survival ratio).

All methods in the chapter are independent of the primary model and of the secondary model. However, because a completely generic approach is typically more difficult to learn, we have chosen for our examples the Weibull as the primary model, and two different secondary models, the modified Arrhenius and a log logistic. Since the first-order model (using the "*d*" value) is a subset of the Weibull model, the first-order model is already included in this chapter implicitly.

There has been considerable discussion in the literature about microbial inactivation data that are not "log-linear," and whether the Weibull model is more appropriate than the 80-year-old first-order *d-z* model. "Non-log-linear" means on a semi-log plot at constant temperature, there is curvature of the data, e.g. the exponent in the Weibull model is $\neq 1.0$. The terms "log-linear" and "non-log-linear" by definition have meaning only for isothermal experiments. Because this chapter deals only with nonisothermal experiments, these terms do not apply. All nonisothermal experiments are expected to result in semi-log curves that are concave down, because the constantly rising sample temperature increases the rate of kill, and would increase the negative value of the slope. However, this type of plot cannot be considered "non log-linear" due to the reasons above.

Temperature History Assumptions for This Chapter

1. The dynamic temperature history can be arbitrary, as opposed to requiring analytically integrable functions, such as a linear temperature increase. The software will handle numerically solving differential equations or integrating functions.
2. The dynamic temperature history data *are not* required to be fit to a function for the integral solution method, because the trapezoidal rule can be used to integrate data points. The dynamic temperature history data *are* required to be fit to a function for the differential solution method, because the solution method (typically Runge-Kutta) uses interpolated points in between the observed data.
3. Statistical assumptions are listed later in the chapter.

Theoretical Background

Nonlinear parameter estimation necessarily is iterative, and requires knowing reasonable guesses of the parameters before starting. Nonlinear parameter estimation is also known as an "inverse problem," because the direction is opposite from the forward problem. The forward problem starts with known parameters, and ends by computing y(t). The inverse problem starts with collected y(t) data, and iterates by using various minimization methods, such as ordinary least squares. The end result is estimation of unknown parameters. There are many potential pitfalls, such as lack of convergence, convergence to local minima, over-parameterization, high correlation leading to large parameter confidence intervals, and poor sequential estimation as data are successively added.

Because dynamic temperature experiments require nonlinear parameter estimation, it is wise first to discuss the basics of parameter estimation. We propose the following outline for the estimation procedure:

1) Choice of the model;
2) Choice of the nonlinear parameter estimation method;
3) Choice of the solution method (integral or differential);
4) Scaled sensitivity coefficients.
5) Statistical results for both parameters and the dependent variable;
6) Standard statistical assumptions using residual analysis.

1. Choice of the Model

For purposes of illustration of the generic methods shown in this chapter, the preferred primary model is the Weibull model in this form:

$$\log S(t) = -\left(\frac{1}{d(T)}\right)^n t^n \tag{1}$$

A common secondary model is the *d-z* model:

$$d(T) = d_r 10^{\frac{Tr-T}{z}} \tag{2}$$

$S(t) = N(t)/N_0$, where N = number of microorganisms at time = t;

N_0 = number of microorganisms at time = 0;

T_r is an arbitrary reference temperature (°C or K);

d_r is the characteristic time at T_r, where log$S$(t) = -0.434 at d_r /2.303 for any value of n (van Boekel 2002) at constant temperature T_r;

z is the temperature increase that causes a 10-fold increase in d; (°C or K);

n is the order of the reaction.

Eq. (1) is a preferred form because the units of d are time, and the equation becomes the well-known first-order d-z model when $n = 1$. Another primary model will also be used, despite the undesirable units on b (Mattick and others 2001) (Stasiewicz et al. 2008):

$$\log S(t) = -b(T)t^n \tag{3}$$

In this model, usually b is increasing with temperature. The following three secondary models for $b(T)$ have been proposed in the literature:

$$b(T) = b_r 10^{\frac{T-T_r}{z}} \tag{4}$$

$$b(T) = b_r \exp\left[-\lambda\left(\frac{1}{T(t)} - \frac{1}{T_r}\right)\right] \tag{5}$$

$$b(T) = \ln\{1 + \exp[k(T - T_c)]\} \tag{6}$$

The kinetic degradation parameters to be estimated are b_r, z, n, λ, k, or T_c depending on the model. A total of three parameters will be estimated for any choice of a primary and a secondary model from above.

A model log$S$ is linear with respect to parameter j if the first derivative of the model with respect to the parameter is not a function of the parameter, i.e. d(log$S$)/d(parameter$_j$) $\neq$ f(parameter$_j$) (vanBoekel 1996). The model is considered nonlinear if it is nonlinear with respect to at least one of its parameters (Ratkowsky 1990). Therefore, the model (Eq. (3) and (4)) is a nonlinear model because it is nonlinear with respect to all the parameters except b_r: $\partial \log S(t)/\partial b_r \neq f(b_r)$, $\partial \log S(t)/\partial z = f(z)$. Model Eq. (3) with secondary models Eq. (5) and (6) is a nonlinear model because $\partial \log S(t)/\partial n = f(n)$, $\partial \log S(t)/\partial \lambda = f(\lambda)$, $\partial \log S(t)/\partial k = f(k)$, $\partial \log S(t)/\partial T_c = f(T_c)$. Therefore, nonlinear regression and initial guesses are required. For the d-z model case, Eq. (1) and (2) apply where $n = 1$. The model is nonlinear in both d_r and z.

2. Choice of the nonlinear parameter estimation method

Knowledge of what is happening inside of a nonlinear regression algorithm prevents a black-box approach, and gives insight as to how to minimize estimation difficulties before they arise. Therefore, this section will provide the main statistical details needed before performing parameter estimation.

There are several types of nonlinear regression methods, such as ordinary least squares (OLS), weighted least squares (WLS), maximum likelihood (ML), and maximum a posteriori (MAP). This chapter will focus on OLS in all cases except the sequential estimation section, where MAP will be used. The reader is referred to (Beck and Arnold 1977) (Beck and Arnold 2007) for more detailed discussion of all four methods.

Assume that a given model η is nonlinear in its parameters, has a total of q independent variables, and has a total of p unknown parameters (Dolan 2003):

$$\eta = \eta(x_1, x_2, ..., x_q; \beta_1, \beta_2, ..., \beta_p) \tag{7}$$

or, in matrix notation,

$$\boldsymbol{\eta} = \boldsymbol{\eta}(\mathbf{x}; \boldsymbol{\beta}) \tag{8}$$

The sensitivity coefficient X_i for β_i is

$$X_i = \frac{\partial \eta}{\partial \beta_i} \tag{9}$$

The matrix form of the sum of squares of the errors, **S** (Beck and Arnold 1977), where **b** is the p-by-1 parameter estimate vector for the true parameter estimates β_i is:

$$\mathbf{S} = [\mathbf{Y} - \boldsymbol{\eta}(\mathbf{b})]^{\mathrm{T}} [\mathbf{Y} - \boldsymbol{\eta}(\mathbf{b})] \tag{10}$$

Minimization of Eq. (10) is known as Ordinary Least Squares (OLS) method.

In matrix notation, the generalized recursive relation for parameter estimates is a vector **b** (Himmelblau 1970):

$$\mathbf{b}^{(t+1)} = \mathbf{b}^{(t)} + h^{(t)}\, \mathbf{B}^{(t)} \tag{11}$$

where h is a scaling factor that varies depending on the algorithm used.

The vector $\mathbf{B}^{(t)}$ is calculated as

$$\mathbf{B}^{(t)} = \left[(\mathbf{X}^T \mathbf{w} \mathbf{X})^{(t)}\right]^{-1} (\mathbf{X}^T \mathbf{w} \mathbf{F})^{(t)}; \tag{12}$$

the sensitivity matrix **X** (also known as "the Jacobian") is a $n \times p$ matrix

$$\mathbf{X}^{(t)} = \begin{pmatrix} \left(\frac{\partial \eta_1}{\partial \beta_1}\right)^{(t)} & \cdots & \left(\frac{\partial \eta_1}{\partial \beta_p}\right)^{(t)} \\ \vdots & \ddots & \vdots \\ \left(\frac{\partial \eta_n}{\partial \beta_1}\right)^{(t)} & \cdots & \left(\frac{\partial \eta_n}{\partial \beta_p}\right)^{(t)} \end{pmatrix} = \begin{pmatrix} X_{11}^{(t)} & \cdots & X_{1p}^{(t)} \\ \vdots & \ddots & \vdots \\ X_{n1}^{(t)} & \cdots & X_{np}^{(t)} \end{pmatrix} \tag{13}$$

where **F** is the n x 1 error vector:

$$\mathbf{F}^{(t)} = \mathbf{Y} - \boldsymbol{\eta}^{(t)} \tag{14}$$

Initial guesses $\mathbf{b}^{(0)}$ are made of the true parameters $\boldsymbol{\beta}$. The collected data sets **x** and **Y** are used to calculate the matrices $\mathbf{X}^{(0)}$ and $\mathbf{F}^{(0)}$ for the initial guesses. The vector $\mathbf{B}^{(0)}$ is calculated and substituted into Eq. (11) to provide the new parameter estimates $\mathbf{b}^{(1)}$. The procedure is repeated until **S** is reduced below a pre-determined value, or until there is negligible change in any component of **b**.

One of the more common difficulties in nonlinear estimation is the inability to find a unique minimum **S**. Examination of Eq. (12) reveals that $\mathbf{X}^T\mathbf{w}\mathbf{X}$ can have an inverse only if the determinant is non-zero:

$$\Delta = \left|\mathbf{X}^T\mathbf{X}\right| \neq 0 \tag{15}$$

because the inverse of any matrix requires dividing by the determinant of the matrix (if $|\mathbf{X}^T\mathbf{X}|= 0$, the value of the weighting matrix **w** is irrelevant). If $|X^TX| =0$, there is no unique point at which a minimum **S** exists, and a unique combination of parameters cannot be estimated. Another indicator of this troublesome condition is that the sensitivity coefficients X_i are linearly dependent (Beck and Arnold 1977), that is

$$D_1X_1 + D_2X_2 + \ldots + D_pX_p = 0 \tag{16}$$

is true for all n observations and for not all the D_p values equal to zero. Eqs. (15) and (16) are two descriptions of the same condition. To deal with this difficulty, before beginning parameter estimation, the determinant in Eq. (15) can be evaluated and the model adjusted, if necessary.

If there are only 2 parameters in the model, Eq. (16) collapses to

$$\frac{X_1}{X_2} = \textbf{a constant} \tag{17}$$

If the ratio of the two sensitivity coefficients over all time (or whatever the independent variable is) is a constant, then the sensitivity coefficients are linearly dependent, and it is impossible to estimate the two parameters separately. The parameters may be able to be estimated as a product or ratio of each other. If one desires to know the parameters separately, the model can be adjusted or reparameterized in some manner. An example is to substitute $\log\hat{\beta}$ for β. Scaled sensitivity coefficients can be plotted for the new parameterization. (Ratkowsky 1983) and (Bates and Watts 1988) have examples of reparameterization.

Reference temperature T_r for Arrhenius-type secondary models

Because so many researchers set $T_r = \infty$ (often represented as Y = A_0exp(-kt), and k=$k_0$exp(E/RT)) and still attempt to estimate rate constant and activation energy, the Arrhenius reference temperature deserves special discussion. In most cases, estimating these two parameters when $T_r = \infty$ will result in useless parameter estimates, if the estimate even converges, because the determinant of $|\mathbf{X}^T\mathbf{X}|$ is nearly 0.

The Arrhenius model (Eq (5)) states that b_r, the rate at reference temperature T_r, exponentially increases with reciprocal temperature difference. Therefore, $\exp\left[-\lambda\left(\frac{1}{T(t)}-\frac{1}{T_r}\right)\right]$ is a multiplier. Eqs. (2) and (4) are Arrhenius-type models, because the multiplier is based on temperature difference, which is a good estimate when the temperature range is small (less than 100°C). The two parameters to be estimated are b_r and λ (or d_r and z). The correlation between these two parameters is controlled by the choice of T_r, because a finite T_r value between the minimum and maximum T(t) "centers" the quantity

($1/T - 1/T_r$) (Bates and Watts 1988). The resulting positive and negative values appearing in the λ column of the X matrix stabilizes the $\mathbf{X^TX}$ matrix, and gives a finite value to $|\mathbf{X^TX}|$ (Himmelblau 1970). If T_r is set to ∞, as is often done, the **X** matrix has all small positive values, $|\mathbf{X^TX}| \sim 0$, the correlation between b_r and $\lambda \sim 1.0$, the confidence interval of b_r will be much larger than the b_r estimated (Dolan 2003), and the joint confidence region for b_r and λ will be very long and thin (Himmelblau 1970). The estimates for b_r, and possibly λ, will be statistically meaningless. Therefore, it is recommended that T_r is always set to a finite value, usually but not necessarily, within the temperature range, for parameter estimation purposes.

The term T_r is a "nuisance" parameter, because it is not estimated by minimizing sum of squares, but by minimizing the correlation coefficient. Therefore, T_r should not appear in the normal list of parameters to be determined simultaneously. Instead, T_r is determined after the other parameters have been estimated. This optimal value of T_r depends on the model and the data, and cannot be predicted beforehand. It is the authors' experience that for models of the type in this chapter, the optimal value of T_r is often in the upper quarter of the total temperature range that typically ranges from 25°C to 130°C. Optimal value of T_r is found empirically by running the nonlinear regression at different values of T_r and choosing the value with the lowest correlation coefficient for the two parameters (b_r and λ) (Schwaab and Pinto 2007). The minimized sum of squares is identical at any value of T_r.

Asymptotic Confidence Intervals for Parameters

The symmetrical confidence interval is a linear approximation to the true asymmetric confidence interval for nonlinear parameters. This approximation can be highly inaccurate when sample sizes are small and the model is highly nonlinear. As the sample size increases, this approximation better approaches to the true confidence interval. Approximate confidence intervals for each parameter are estimated using the t test statistic at confidence level 1-0.5α (van Boekel 1996):

$$b_i \pm \sigma_i t_{(1\text{-}0.5\alpha),\, n\text{-}p} \tag{18}$$

where σ_i is the square root of the corresponding diagonal of the symmetric parameter variance-covariance matrix

$$\mathbf{cov(b)} = \mathbf{(X^TX)^{-1}}(MSE) = \begin{pmatrix} \sigma^2_{11} & \sigma_{12} & \sigma_{13} \\ & \sigma^2_{22} & \sigma_{23} \\ \text{symmetric} & & \sigma^2_{33} \end{pmatrix} \tag{19}$$

The correlation coefficient for the ith and jth parameters is $\rho_{ij} = \sigma_{ij} / \sqrt{\sigma^2_{ii}\sigma^2_{jj}}$, and $-1.0 \le \rho \le 1.0$, where higher values of absolute value of ρ indicate less accuracy in the estimation of parameters. Excel assumes a two-tailed t-distribution, so the Excel equation for the t-value for 24 data and 3 parameters would be =TINV(0.05, 21), which gives t = 2.079. The equivalent MATLAB formula is TINV(0.975,21).

Approximate elliptical parameter joint confidence regions

The confidence region is a set of values for the two parameters where the sum of squares is less than a set value. The third parameter is held constant. Approximate inference regions for the estimated parameters using the nonlinear regression were drawn per the following equation (Bates and Watts 1988)

$$(\theta - \hat{\theta})^T \hat{V}_1^T \hat{V}_1 (\theta - \hat{\theta}) \le ps^2 F(p, n-p; \alpha) \tag{20}$$

where, the derivative matrix $\hat{V} = \hat{Q}_1 \hat{R}_1$ and it is calculated at $\hat{\theta}$.

The boundary of the inference region as mentioned above is

$$\left\{ \theta = \hat{\theta} + \sqrt{ps^2 F(p, n-p; \alpha)}\ \hat{\mathbf{R}}_1^{-1}\ \mathbf{d}\ \ \|\mathbf{d}\| = 1 \right\} \tag{21}$$

Where $\hat{\mathbf{R}}^{-1}$ is the inverse of the upper triangular matrix (2-by-2) from the orthogonal-triangular decomposition; θ is the set of boundary values forming the ellipse; $\hat{\vartheta}$ is the single least-squares parameter estimate at the center of the ellipse; p is the number of parameters; s^2 is the mean square error; F is the value of the F distribution at p, n-p degrees of freedom at 1-α level of confidence; and **d** is a 2-by-1 vector of the cosine and sine of the independent variable from 0 to 2*π.

This equation will provide an ellipse which is an approximation of the true contour, and is used for computational efficiency, compared to the iterative method below. Approximate confidence regions are more accurate at lower percent levels (such as a 50% confidence region) than at the more commonly used higher 90-99% confidence regions (Beck and Arnold 1977). The approximate CR is typically smaller than the true confidence region.

3. Likelihood ratio joint confidence regions (Beck and Arnold 1977, p. 380-386)

A more exact joint confidence region for two parameters, β_1 and β_2, is found using a likelihood ratio. This CR is defined as the set of points (b_1,b_2) where the sum-of-squares of errors is less than or equal to a constant value determined by the level of confidence required. Specifically, the constant value is

$$SS_{all-fixed} = SS_{best-fit}\left(\frac{p}{n-p}F(p, n-p) + 1\right) \tag{22}$$

The confidence contour bounding the confidence region was computed per the iterative method of (Motulsky and Christopoulos 2004). The MATLAB function fzero can be used to find the two roots on both sides of b_1,b_2 at $SS_{all\text{-}fixed}$.

Asymptotic confidence intervals for Y

The equations for asymptotic confidence intervals for $Y = \log S(t)$ are found in (Seber and Wild 2003). MATLAB function nlpredci can be used to obtain the asymptotic confidence interval (options = 'curve') or asymptotic prediction interval (options = 'observation.')

4. Choice of the solution method

Either method below can be used, but each method has a disadvantage of requiring a different additional step, discussed below.

Differential Method

The parameters are estimated by numerically solving the differential equation(s) with initial guesses for the parameters, and continuing iteratively until the sum of squares of errors is minimized. A common solution method is the Runge-Kutta (RK) method, available in various software, such as MATLAB and Mathematica. An advantage of this differential method is that no integration need be set up before the estimation, saving a step. The required "extra step" is that the temperature profile must be fit to a function, because the RK method requires determining numerical derivatives at the observed times and at the midpoints of the observed times. There are no observed temperatures at the midpoints, so a linear interpolation curve is needed.

A simple example of the differential solution method is for the first-order equation:

$$\frac{d(\log S)}{dt} = -k \qquad (23)$$

Assuming the secondary model is the Arrhenius function:

$$k = k_r \exp\left[-\lambda\left(\frac{1}{T(t)} - \frac{1}{T_r}\right)\right] \qquad (24)$$

The parameters to be estimated are k_r (time^{-1}), λ (K), and $\log S(0)$, the initial condition. Often $\log S(0)$ is assumed to be fixed = 0, which is reasonable if there is small variance of the replicate measurements. Otherwise, it is recommend to keep $\log S(0)$ as a parameter (vanBoekel 1996). The reference temperature, T_r, is assumed fixed for this example. The independent and dependent variables are time and $\log S$, respectively. The extra step is that $T(t)$ must be converted from individual points to a fitted function before iteratively solving the differential Eq. (23).

Integral Method

For the integral solution method, the "extra step" is that Eq. (23) must be set up manually as an integral first:

$$\int_{\log S(0)}^{\log S} d(\log S) = -\int_0^t k\, dt \qquad (25)$$

Substituting Eq. (24) into Eq. (25) yields

$$\log S(t) = -k_r \int_0^t \exp\left[-\lambda\left(\frac{1}{T(t)} - \frac{1}{T_r}\right)\right] dt + \log S(0) \tag{26}$$

The parameters, independent and dependent variables are all the same as in the differential method. However, *T(t)* can be input as individual observed data because the right side of Eq. (26) can be evaluated using the trapezoidal rule. If *T(t)* is collected at evenly spaced time intervals, increased accuracy can obtained using Simpson's rule. Gauss-Legendre integration, which produces the highest accuracy, can be used only if *T(t)* is fit to a function. Gauss-Legendre integration is not recommended unless extreme accuracy is required.

This model (Eq. (26)) is a nonlinear model because it is nonlinear in the λ parameter.

5. Scaled sensitivity coefficients

Before performing parameter estimation, it is recommended to plot the scaled sensitivity coefficients X' versus the independent variable (typically = time), to identify if there is linear dependence (undesirable), and to compare the maximum absolute value of each X' to the total change of the dependent variable.

It is not uncommon that the model η is very complicated, so the derivative may be difficult to evaluate analytically. Especially in these cases, to avoid errors in the analytical derivative, the sensitivity coefficient may be approximated numerically as follows.

One finite difference approximation for the *i*th observation, *j*th parameter, and *l*th dependent variable is the forward difference approximation:

Numerical method to estimate sensitivity coefficient (Beck and Arnold 1977)

$$\frac{\partial \eta_l(i)}{\partial b_j} = X_{lj} \approx \frac{\eta_l(b_1 \ldots, b_j + \delta b_j, \ldots b_p) - \eta_l(b_1 \ldots, b_j, \ldots b_p)}{\delta b_j} \tag{27}$$

Where δb_j is some small quantity, such as

$$\delta b_j = 0.001 b_j \tag{28}$$

If we want to plot the scaled sensitivity coefficient, Eq. (27) becomes

$$\beta_j \frac{\partial \eta_l(i)}{\partial b_j} = X'_{lj} \approx \frac{\eta_l(b_1 \ldots, b_j + \delta b_j, \ldots b_p) - \eta_l(b_1 \ldots, b_j, \ldots b_p)}{0.001} \tag{29}$$

For example, if there are two parameters, d_r and z, the scaled sensitivity coefficients would be

For d_r:

$$X'_{D_r} = \frac{\eta(1.001 * d_r, z, t) - \eta(d_r, z, t)}{0.001} \tag{30}$$

For z:
$$X_z' = \frac{\eta(d_r, 1.001 * z, t) - \eta(d_r, z, t)}{0.001} \tag{31}$$

If the software, such as nlinfit in MATLAB®, generates the Jacobian (same as the sensitivity matrix **X**), the user can alternatively plot the Jacobian column j multiplied by the corresponding j^{th} parameter. The result is identical to Eq. (30) and (31).

The scaled sensitivity coefficient X_j' has the same units as η, allowing direct comparison between X_j' and η. For best estimates, we desire our scaled sensitivity coefficients X_j' to be *large* and *uncorrelated*. The larger the absolute value of X_j', the more easily the jth parameter can be estimated, and the smaller the variance of the parameter estimate. To plot X_j', one must know 1) some reasonable value of the parameter, and 2) some reasonable input of the independent variables. Observed data are not required to compute scaled sensitivity coefficients. Scaled sensitivity coefficients less than ~5% of the total span of Y can be considered candidates to be eliminated or fixed, as they cannot be easily estimated.

6. Statistical results for both parameters and the dependent variable

The following are several statistical tools used in this chapter to produce and evaluate results of the parameter estimation. These will be discussed in the Outline #6) below.

Parameter Confidence Interval

The statistic R^2 is not useful for nonlinear models (Ratkowsky, D.A., 2003. Model fitting and uncertainty. In: McKellar and Lu 2004, pp. 151-196. Chapter 4) (Ratkowsky 1983). Instead, for nonlinear models we will use a group of standard statistical assumptions. Let η be the errorless (ideal) model; β be the parameter vector; ε the error vector.

The following statistical assumptions will be examined for each model:

6.A. Standard Statistical Assumptions using residual analysis and examples (Beck and Arnold 2007, pp.5.28-5.29)

1. $\mathbf{Y} = \eta(\mathbf{X}, \boldsymbol{\beta}) + \varepsilon$.; additive errors in measurements
2. $E(\varepsilon) = 0$; Zero mean measurements
3. Constant variance errors
4. Uncorrelated errors
5. The errors ε have a normal (Gaussian) distribution of errors
6. Known statistical parameters describing ε (such as covariance matrix of the errors)
7. Errorless independent variables
8. $\boldsymbol{\beta}$ is a constant parameter vector and statistics of $\boldsymbol{\beta}$ are unknown (such as covariance).

The estimation problems are generally less difficult when the standard assumptions are valid (Beck and Arnold 2007). We will examine assumptions after the nonlinear regression by doing a *residual analysis* for assumptions #1-5. The validity of each standard assumption is indicated by a "1", and if it is not valid a "0" may be used. For example, if all assumptions except #6 and #8 are valid, the succinct notation "11111010" is used.

When nothing is known regarding the measurement errors, OLS (Eq. (10)) is recommended. For microbial inactivation models, little has been reported on the measurement errors. Therefore, it is appropriate to use OLS in this chapter.

6.B. Resampling Methods

Microbial data are expensive and labor-intensive to obtain. Therefore, any one study will have only a limited number of data from which parameters can be estimated. Although the error distribution can be inferred from the limited data set, the true distribution can be better approximated by using statistical resampling methods to generate more data.

There are two bootstrap resampling methods (Efron and Tibshirani 1998, p. 113) The first bootstrap method is "bootstrapping the pairs," where all the (x, y) data pairs are resampled with replacement to obtain a new set of data. The second bootstrap method is "bootstrapping the residuals," where all the residuals are resampled with replacement, and are then added on to the OLS regression fit (Sahinler and Topuz 2007). (Efron and Tibshirani 1998). A uniform random distribution is used. The other common resampling technique is the jackknife, which systematically recomputes the parameter estimates leaving out one observation at a time from the sample. We will discuss how to do the bootstrap method, which will be used to verify standard statistical assumptions #1-5.

(Efron and Tibshirani 1998) discuss when to use either the "pairs" or the "residuals" bootstrap methods. From the authors' experience, if there are few data (under 100) and the scaled sensitivity coefficients are small for at least half the experiment, the "pairs" method may give distorted confidence intervals. The reason is that some of the bootstrap sets may fall entirely in the half of the experiment where parameters cannot be estimated. Therefore, the "residuals" method is recommended for this case.

We also wish to find the probability distribution of each parameter using Monte Carlo simulation, which requires an accurate estimate of the standard deviation of the errors. The standard deviation from a small sample set (fewer than 100 data), is likely to significantly underestimate the true σ. A better estimate of σ will be given by resampling the original data with replacement, using the bootstrap method. The second reason for resampling is to obtain a better estimate of the parameter confidence intervals.

Bootstrap method: Generate n random numbers with a uniform distribution. For the "pairs" method, randomly choose n pairs of the original data, with replacement, based on the n random numbers (there may be repeats of the pairs). Estimate the parameters for this new set (bootstrap) of data. Repeat this process at least 999 times for confidence intervals (Efron and Tibshirani 1993; Efron and Tibshirani 1998). Now there are 1,000 sets of parameter estimates = 999 sets plus the one set from the original data. An example is

Original Data

x_1	Y_1
x_2	Y_2
x_3	Y_3
x_4	Y_4
x_5	Y_5
x_6	Y_6
x_7	Y_7

Where (x_i, Y_i) are the original observed data.

If the first set of random numbers is (3, 3, 6, 1, 7, 6, 1) then the first bootstrap set would be:

x_3	Y_3
x_3	Y_3
x_6	Y_6
x_1	Y_1
x_7	Y_7
x_6	Y_6
x_1	Y_1

For the "residual" method, the first set of data would be:

x_1	$\hat{Y}_1 + e_3$
x_2	$\hat{Y}_2 + e_3$
x_3	$\hat{Y}_3 + e_6$
x_4	$\hat{Y}_4 + e_1$
x_5	$\hat{Y}_5 + e_7$
x_6	$\hat{Y}_6 + e_6$
x_7	$\hat{Y}_7 + e_1$

Where $\hat{Y}_i$ indicates the ith predicted Y value, based on the estimated parameters from the original data set. Unlike symmetrical asymptotic parameter CIs, nonlinear parameter CIs are asymmetrical. The bootstrap CIs will also be asymmetrical.

Bootstrap confidence intervals for parameters

The 95% bootstrap confidence interval for a parameter is found as follows:

Sort each set of 1,000 bootstrap parameter estimates from lowest to highest. The 95% confidence interval limits are point #25 to point #975, while the 90% confidence interval will be from point #50 to point #950 (Devore and Berk 2007). This procedure gives one bootstrap confidence interval for each parameter.

Bootstrap confidence intervals for Y

For each pair of 1,000 parameter estimates, compute *Y(t)*. For example, if there are 24 heating times, one would have 24 sets of 1,000 Y values. Sort the 1,000 Y_j values at each heating time t_i from lowest to highest. At each heating time, Y_{25} and Y_{975} will be the 95% confidence interval for *Y* at each heating time. The confidence band is constructed by connected all the Y_{25}'s (lower band) and all the Y_{975}'s (upper band). The 95% confidence band is the region within which one would expect 95% of the *regression lines* to fall. Confidence bands are more narrow than prediction bands, because regression lines will vary less than individual data.

Bootstrap prediction intervals for Y

The 95% bootstrap prediction intervals at each time t_i are calculated using (Montgomery and others 2006)

$$\text{prediction width}_i = \sqrt{(\text{confidence width}_i)^2 + (t_{1-\alpha/2,n-p} * RMSE)^2} \quad (32)$$

where 'confidence width' is the upper or lower width of the bootstrap confidence interval for each heating time t_i :

$$\begin{aligned} \text{confidence width}_{\text{upper, i}} &= Y_{CI,upper,i} - \hat{Y}_i \\ \text{confidence width}_{\text{lower, i}} &= \hat{Y}_i - Y_{CI,lower,i} \end{aligned} \quad (33)$$

$$\begin{aligned} Y_{PI,upper,i} &= \hat{Y}_i + \text{prediction width}_{\text{upper, i}} \\ Y_{PI,lower,i} &= \hat{Y}_i - \text{prediction width}_{\text{lower, i}} \end{aligned} \quad (34)$$

6.C. Monte Carlo methods

Unlike the "bootstrapping the pairs", Monte Carlo methods generate *new* data ($Y_{synthetic}$) based on the original data, the original predicted points, the original parameter estimates, and the original error variance = σ^2. Because σ^2 is unknown, we can choose an estimate, in order of increasing accuracy and increasing difficulty, to compute: MSE, σ^2_{orig}, or σ^2_{boot}.

Monte Carlo step: Use the original fitted function and parameters and perturb the original data with random error of some level based on your knowledge of the error (e.g. RMSE of the original data, or the most accurate is the standard deviation ($\sigma_{bootstrap}$) of the 1,000 bootstrapped residuals) (vanBoekel 1996). Use the equation

$$Y_{synthetic,i} = \hat{Y}_i + s * (randn) \quad (35)$$

Where *RMSE* is root mean square error of the original data, *randn* is a normally distributed random number with $\mu = 0$, $\sigma = 1$, and *s* is either RMSE, or σ_{orig}, or σ_{boot}.

Use these data to repeat the Monte Carlo step 500 to 1000 times (Ratkowsky 1983). Each time a new $Y_{synthetic}$ is created, these new data will be fit and a new set of parameters will be estimated. Therefore, 500 – 1000 sets of parameters will be obtained. The distribution of each parameter can be plotted in a histogram. We desire the histogram to approach a normal distribution. If not, a reparamaterization of the model can be considered $Y_{synthetic}$

Examples of dynamic temperature parameter estimation for microbial inactivation using ordinary least squares

The following examples will demonstrate how to perform all the forgoing steps.

EXAMPLE #1: Estimating microbial inactivation parameters under dynamic temperature conditions: No temperature gradient within the samples.

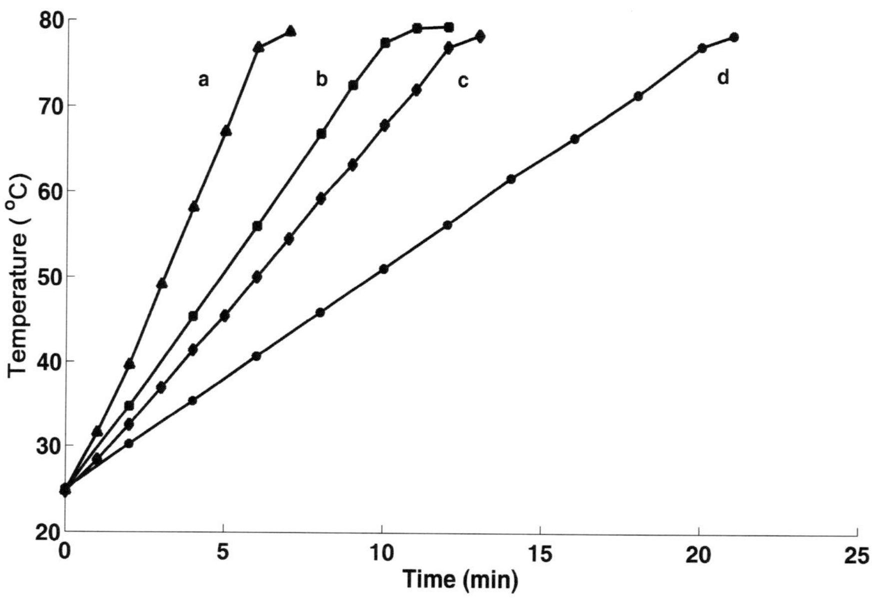

Figure 1. Time-temperature histories of broth with *Salmonella*. Markers are experimental data. Lines are fitted curves for the differential solution method. Data from (Mattick and others 2001).

Four different dynamic temperature profiles of *Salmonella* in a 66% sugar liquid medium are shown in Figure 1 (Mattick et al. 2001). Because the liquid broth for each sample was only 100 microliters in polymerase chain reaction tubes, it was reasonable to assume that there was negligible temperature gradient across the sample.

Step #1. Choose the model.

Primary model: $$\log S(t) = -b(T)t^n \tag{3}$$

Secondary model: $$b(T) = \log\{1 + \exp[k(T - T_c)]\} \tag{6}$$

Independent variables: t (min) and $T(t)$ (°C). The temperature data were input as individual points for each time for each log*S*(t).

Dependent variable: log*S(t)*

Parameters to be estimated: k, T_c, n. (Peleg and Normand 2004) showed that it was reasonable to assume n is independent of temperature.

The model is nonlinear in all 3 parameters, because d(log$S$)/d(parameter$_i$) = f(parameter$_i$).

A schematic view of the secondary model for $b(T)$ is shown in Figure 2 (Peleg and Normand 2004). They used this secondary model because $b(T)$ is very small until the temperature approaches the critical temperature T_c, when which $b(T)$ rises linearly with temperature.

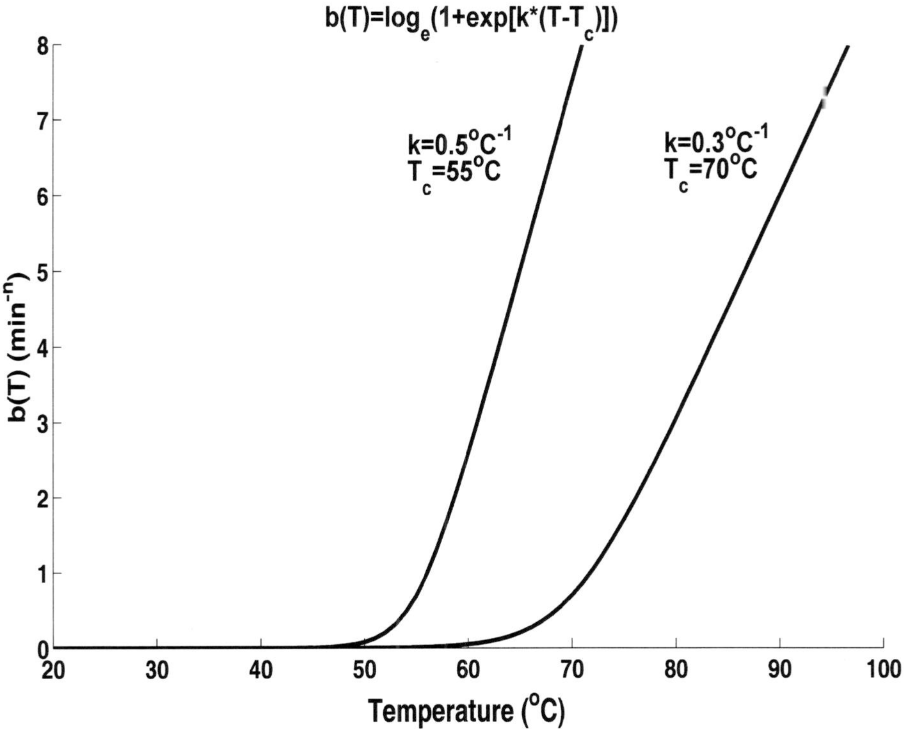

Figure 2. Examples of log logistic model of $b(T)$ vs. temperature per Eq. 6. From (Peleg and Normand 2004).

Step #2. Nonlinear regression method.

Because there is no prior information on the errors, we will use *Ordinary Least Squares*, i.e. the sum of squares will be minimized. If it were known that errors were normally distributed, we could use Maximum Likelihood. If we knew additional information about the parameters, such as the mean, the probability distribution, and the covariance matrix of the parameters, we could use Maximum a posteriori (Beck and Arnold 2007, p. 105).

Step #3. Solution method.

We will use the *integral method*. Therefore, the differential equation must be integrated first. The time derivative at constant temperature for Eq. (3) is

$$\frac{d \log S(t)}{dt} = -b(T)nt^{n-1} \tag{36}$$

Setting up the integral, and substituting Eq. (6) for *b(T)*:

$$\int_{S(0)}^{S(t)} d\log S(t) = -n\int_0^t \ln\{1+\exp[k(T(t)-T_c)]\}t^{n-1}dt \tag{37}$$

The final form to use for running the nonlinear regression was:

$$\log S(t) - \log S(0) = -n\int_0^t \ln\{1+\exp[k(T(t)-T_c)]\}t^{n-1}dt \tag{38}$$

For simplicity, log*S*(0) was fixed = 0 to keep the number of parameters to three. The integral was done by trapezoidal rule (cumtrapz) in matlab.

Step #4. Plot the scaled sensitivity coefficients.

For any nonlinear regression method, some initial estimate of the parameters is necessary. The scaled sensitivity coefficients can be plotted without any observed dependent variable data. All that is needed is the model, independent variable inputs, and a reasonable guess of the parameter values. Figure 3 shows the scaled sensitivity coefficients for the temperature profiles in Figure 1, using fixed parameter estimates k = 0.17 C^{-1}, T_c = 63 °C, n= 0.7 (Peleg and Normand 2004).

From Figure 3, it is clear that T_c and n will be easily estimated, because the absolute values of X'_{Tc} and n are large (10-21 log) compared to the total scale (7 log, shown later). The parameter k will be less easily estimated and have a proportionally larger standard error, because the absolute value of X'_k is small (~1) for all heating rates. Also, the shapes of all three X' are similar, suggesting that there may be high correlation. As a final check, the determinant of X^TX = 3.155 x 10^4, a non-zero result showing that all parameters can be separately estimated.

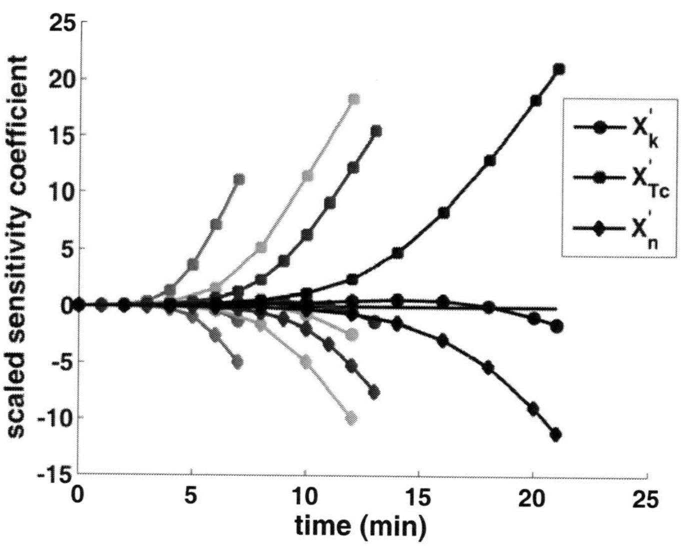

Figure 3. Example 1. Scaled sensitivity coefficients for Figure 1 temperature profiles, primary model Eq. (3) and secondary model Eq. (6), integral method, and initial guesses k = 0.17 C^{-1}, T_c = 63 °C, n=

0.7 (Peleg and Normand 2004). (Red= temperature profile "a" in Figure 1; Blue = "b", green = "c", black = "d".).

Step #5. Results for Example #1, using integral method

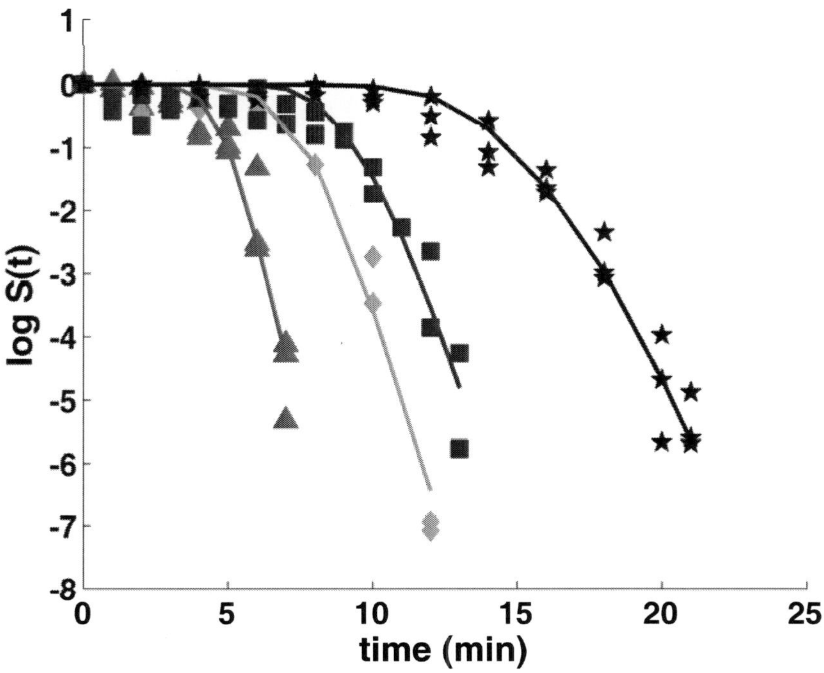

Figure 4. Example 1. Log*S(t)* versus time for Salmonella heated at four different heating rates in 66% sugar broth. Experimental data (markers) and predicted (lines). Data from (Mattick et al. 2001). (Red= temperature profile "a" in Figure 1; Blue = "b", green = "c", black = "d".).

The observed data and predicted lines are shown in Figure 4. The parameter estimates and statistics are shown in Table 1:

Table 1. Example 1. Weibull parameter estimates for *Salmonella* in 66% sugar broth inactivation data (Figure 4) with four temperature profiles (Figure 1), using the integral method. Total number of data = 87

Parameter	estimate	Standard error	95% asymptotic confidence interval	Correlation coefficient, ρ	*SS*
k (°C^{-1})	0.504	0.100	(0.305,0.704)	ρ_{kTc}= -0.237	*SS* = 13.64
T_c (°C)	54.4	0.9479	(52.6,56.3)	ρ_{nTc} = 0.568	RMSE =
n	0.443	0.0474	(0.349,0.5373)	ρ_{kn}= -0.930	0.403 log

As predicted by the small scaled sensitivity coefficient (Figure 3), k was more difficult to estimate than T_c and n, as shown by the proportionally larger standard error of k The standard error of k was ~20% of the estimate, a large value compared to those of T_c and n, which were small standard errors = 1.7% and 11% of the estimates, respectively (Table 1). The confidence intervals are larger around the k estimate than around the T_c and n estimates

(Table 1). Proportionately large standard errors can lead to parameter confidence intervals that are so large that the interval contains zero. In that case, the parameter estimate would be worthless, because it would not be statistically different from zero.

Only the parameters *k* and *n* were more highly correlated (-0.930, Table 1). The RMSE = 0.403 is less than 6% of the total scale of 7, showing that the model fit is good based on sum of squares. However, the appropriateness of a model is evaluated in a residual analysis based on the first six standard statistical assumptions, which we will now discuss.

Step #6. Standard Statistical Assumptions for the integral method.

The residuals (e) are not equal to the error (ε), but the residuals are estimators of the error. The residual scatter plot (Figure 5) shows that additive errors are reasonable (assumption 1), because the residuals are reasonably constant as the absolute value of log*S(t)* becomes larger. The mean of the residuals is slightly negative (-0.119 log, Figure 6), so the second statistical assumption may not be valid. Many of the excessive negative residuals (Figure 5) come disproportionately from temperature profile "c" (Figure 1) during the first 7 minutes, suggesting that the data for this experiment may have experienced something unusual. The third statistical assumption is valid, as shown by a reasonably constant band of residuals (from -1 log to +1 log) over time. The fourth assumption, uncorrelated errors, is true up to ~14 min of heating, after which the slowest heating rate, "d", shows correlation (black stars, Figure 5) as the residuals follow each other as they rise on the positive side with time. The fifth assumption is shown by the fitted Gaussian distribution line (Figure 5). The residuals are reasonably Gaussian, but with the shift to the left, as mentioned. Time can be considered an errorless independent variable (assumption 7). We did not know the parameter covariance matrix of errors (assumption 6), and we did not have prior information on the parameters other than initial guesses (assumption 8). The concise statement for the validity of the standard assumptions for this model for these data is "10101011." Therefore, although the model fit and the parameter estimation is reasonably good, the model for these data could be improved.

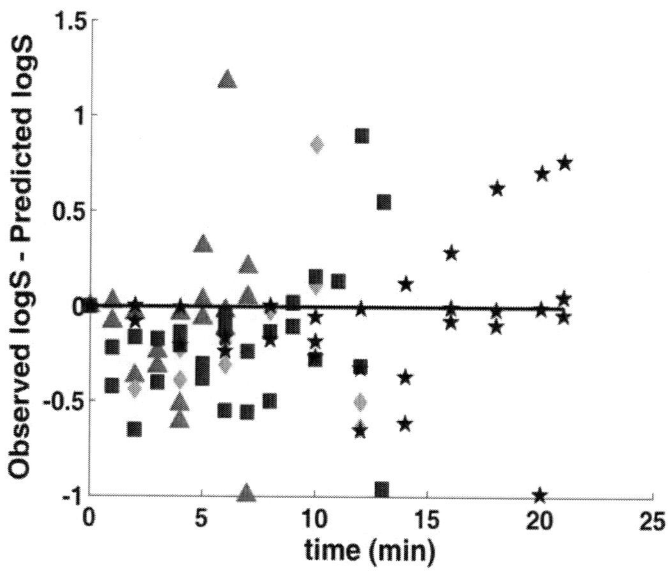

Figure 5. Example 1. Residual scatter plot for *Salmonella* heated in 66% sugar broth for four different heating rates using the integral solution method. Data shown in Figure 4.

The purpose of this residual analysis was to show the reader that examination of how well a model fits requires more than general statements such as, "Figure 4 shows an excellent fit." Residuals can show deficiencies of a model that a plot of observed values and fitted curves cannot. Therefore, researchers are encouraged to plot the residuals and analyze the standard statistical assumptions.

EXAMPLE #2. Repeat Example 1 using the differential method

Steps #1 and #2: Same as in Example #1, integral method, above. The temperature data must be fit to equations.

Temperature data in Figure 1 were fit by (Peleg and Normand 2004):

$$\text{For temperature profile a: } T(t) = 78.1\text{-}\ln\{1+\exp[8.7(6.3\text{-}t)]\} \tag{39}$$

$$\text{For temperature profile b: } T(t) = 78.8\text{-}\ln\{1+\exp[5.2(10.4\text{-}t)]\} \tag{40}$$

$$\text{For temperature profile c: } T(t) = 78.3\text{-}\ln\{1+\exp[4.3(12.4\text{-}t)]\} \tag{41}$$

$$\text{For temperature profile d: } T(t) = 78.5\text{-}\ln\{1+\exp[2.6(20.5\text{-}t)]\} \tag{42}$$

Step #3. Solution method.

We will use the differential method to solve Eq. (36). (Mattick et al. 2001) assumed that the log*S*(t) momentary slope, dlogS(t)/dt, was equal to the isothermal semilogarithmic curve at the momentary temperature T(t), at a time, t*, corresponding to the momentary logarithmic survival ratio log*S*(t). Solving Eq. (3) for this time,

$$t^* = \left(\frac{-\log S(t^*)}{b(T(t^*))}\right)^{\frac{1}{n}} \tag{43}$$

Substituting t^* in Eq. (43) for t in Eq. (36) (Peleg and Normand 2004),

$$\frac{d\log S(t)}{dt} = -b[T(t)]n\left\{\frac{-\log S(t)}{b[T(t)]}\right\}^{\frac{n-1}{n}} \tag{44}$$

assuming negligible change in n with temperature. Eq. (44) was solved in MATLAB using ode45, a Runge-Kutta fourth- and fifth-order solver, nested within nlinfit, the nonlinear regression routine. Starting value was log*S*(0)= -0.0001.

Step #4. The scaled sensitivity coefficients are not plotted because they were very similar to those for the integral method. The same inferences about the parameters can be made.

Step #5. Results for Example #2.

Table 2. Example 2. Weibull parameter estimates for *Salmonella* in 66% sugar broth inactivation data (Figure 4) with four temperature profiles (Figure 1), using the differential method. Total number of data = 87

Parameter	estimate	Standard error	95% asymptotic confidence interval	Correlation coefficient, ρ	*SS*
k (°C^{-1})	0.223	0.0174	(0.189,0.258)	ρ_{kTc} = 0.437	SS = 15.1
T_c (°C)	61.5	0.6426	(60.3, 62.8)	ρ_{Tcn} = 0.309	RMSE =
n	0.438	0.0503	(0.338, 0.538)	ρ_{kn} = -0.675	0.423 log

The statistical results for the parameters for the differential method are shown in Figure 7. Because most of the plotted results were so similar, no plots except the residual histogram are shown for the differential method. Although the differential results gave a slightly worse overall fit than the integral method ($RMSE_{differential}$= 0.423 vs. $RMSE_{integral}$ = 0.403, Table 2 and Table 1), the standard error for k by the differential method was proportionately smaller (7.8%, Table 2) than that for the integral method (20%, Table 1). The estimate of k_{diff} (0.223 °C^{-1}) was less than half the size of $k_{integral}$ (0.504 °C^{-1}) (Figures 8 and 5), and T_c increased from 54.4°C (integral method, Table 1) to 61.5°C (differential method, Table 2). The 95% confidence intervals for k do not overlap for the integral method and the differential method. The same lack of overlap is seen for T_c. Clearly, the integral method and the differential method estimated different values of k and T_c. Conversely, the estimate of n was not dependent on the solution method (Tables 1 and 2), as shown by the overlapping confidence intervals. The integral method was only slightly better than the differential method when considering only how well the prediction line fit (*SS)*. The differential method was superior in terms of lower parameter correlation and smaller standard errors.

Step #6. Standard Statistical Assumptions for the differential method.

The residual histogram for the differential method (Figure 7, mean = -0.115 log) is slightly closer to a zero mean than that for the integral method (Figure 6, mean = -0.119 log). Both residual histograms show a reasonable Gaussian distribution. There was not a large difference in any of the residual results, compared to those for the integral method. Therefore, the validity of the standard assumptions for this model with these data using the differential method is also 10101011. In summary, considering both methods separately, both the integral method and the differential method gave good parameter estimates and good fits, with only marginal differences in RMSE. However, because *both solution methods should give the same parameter results*, and did not for k and T_c, this model (Eq. (44)) or the solution method is inadequate. A strong possibility is that the assumption, Eq. (43), is incorrect, and is unnecessary. Therefore, using both the integral and differential solutions methods to compare estimated parameters is an example of "intrinsic verification," and can be used to disqualify models or assumptions.

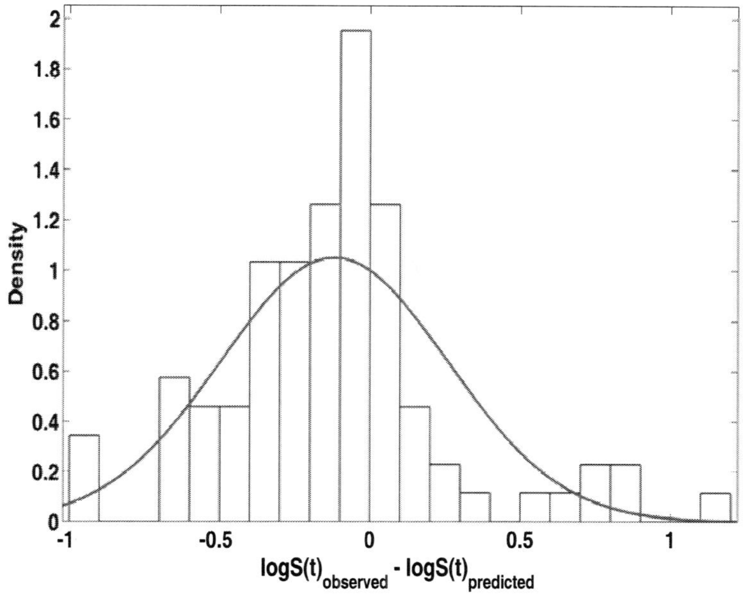

Figure 6. Example 1. Residual histogram for *Salmonella* heated in 66% sugar broth for four different heating rates, based on solution to Weibull model using the integral method. Residuals shown in Figure 6. Fitted line is for a normal (Gaussian) distribution, where mean of residuals = -0.119 log, σ = 0.38 log.

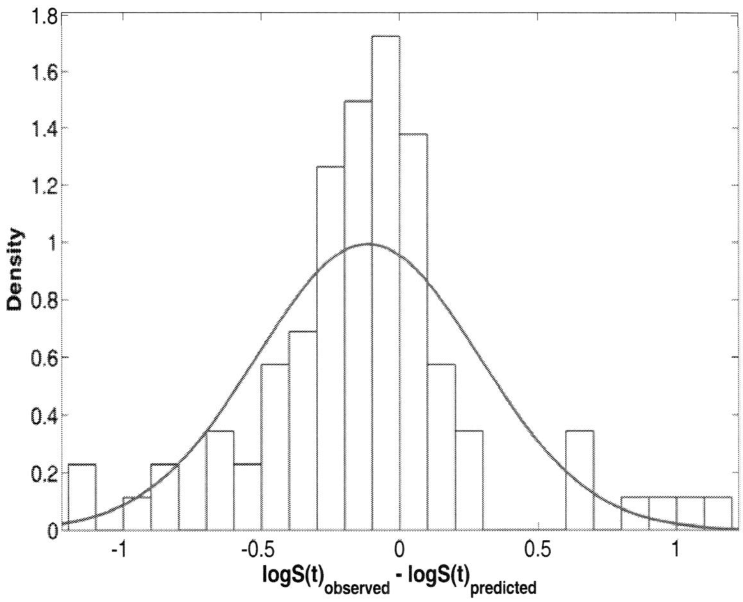

Figure 7. Residual histogram for *Salmonella* heated in 66% sugar broth for four different heating rates based on solution to Weibull model using the differential method. Fitted line is for a normal (Gaussian) distribution, where Mean of residuals = -0.115 log, σ = 0.402 log.

EXAMPLE #3: Estimating microbial inactivation parameters under dynamic temperature conditions: Temperature gradient exists within the samples.

Table 3. Example 3. Retort temperature, heating time, and measured *B. stearothermophilus* spore survival ratios for twenty-four heated cans ((Welt 1996) p. 190-191). Survival ratio for can #2 was > 1.0 because the denominator of survival ratio for all cans was the mean of survivor numbers for cans 1 and 2 (mean = 9 x 10^5 CFU/mL)

Can number	Retort Temperature (°C)	Heating time (min)	Measured survival ratios, N/N_o
1	n/a	0	0.891
2	n/a	0	1.109
3	104.4	50	0.689
4	104.4	50	0.769
5	104.4	90	0.656
6	104.4	90	0.744
7	104.4	180	0.387
8	104.4	180	0.476
9	104.4	180	0.279
10	104.4	180	0.266
11	104.4	240	0.0755
12	104.4	240	0.0697
13	104.4	240	0.0703
14	104.4	240	0.0667
15	104.4	275	0.0418
16	104.4	275	0.0401
17	104.4	305	0.00725
18	104.4	305	0.00855
19	112.8	27.5	0.642
20	112.8	32.5	0.433
21	112.8	44	0.159
22	120.6	19	0.500
23	120.6	24	0.266
24	120.6	27.5	0.093

For cases where the sample size is large (> ~1 gram), or is less than ~80% moisture, or is viscous, there usually is a significant temperature gradient throughout the sample if the heat is applied externally by conventional heating at the surface. Because microorganisms degrade rapidly at higher temperatures, ignoring the temperature gradient may result in large parameter estimate errors. For conduction heating, the temperature gradient at any time can be known by either using an analytical heat transfer solution for a regular geometry, or a numerical solution for any geometry. The problem is further complicated when the thermal properties and heat transfer coefficient are not known, and the former is a function of temperature. (Mishra and others 2009) separated the full problem into two inverse problems: 1) First, estimate temperature-dependent thermal diffusivity by minimizing *SS* of container center temperature errors over time; and second, 2) estimate kinetic nutrient degradation parameters by minimizing *SS* of nutrient concentration over time. The same procedure can be used to estimate kinetic parameters for microbial inactivation. In this example, we will deal only with the second inverse problem, and assume the product thermal properties are known and are temperature-independent.

For Example #2, data from (Welt 1996) were used. They filled cans of pea puree uniformly inoculated with spores of *Bacillus stearothermophilus*, and heated the cans in duplicate at 12 different time-temperature combinations using three different retort temperatures (Welt and others 1997) (Table 3). The number of spores surviving for each can was measured and recorded as a survival ratio = N_i / N_0, where N_0 = 9 (+ 1) x 10^5 CFU/mL. Can size was 0.0602 diameter by 0.0348 m height. Uniform initial temperature = 0°C. Assumed constant thermal properties were: thermal conductivity = 0.682 W/(m°C), density = 958 kg/m^3, specific heat = 4,350 J/(kg °C) (Welt et al. 1997).

Step #1. Choose the model.

Primary model: $$\log S(t) = -\left(\frac{1}{d(T)}\right)^n t^n \qquad (1)$$

Secondary model (modified Arrhenius): $$d(T) = d_r 10^{\left(\frac{Tr-T}{z}\right)} = d_r 10^{\left(\frac{T_\infty - T_i}{z}\right)\left(\frac{Tr-T}{T_\infty - T_i}\right)} \qquad (45)$$

We chose a dimensionless form of the *z* variable to test how much *z* changes with temperature. Previous work (not shown) indicated that there was a modest rise in *z* with temperature.

Step #2. Nonlinear regression method.
Because there is no prior information about the errors, Ordinary Least Squares was used.

Step #3. Solution method.
For this problem, the differential method cannot be used because there are no observed data for a specific point with a known temperature history in the can. Instead, the data are the mass-average logS_a(t)for each can. Therefore, the integral method is required due to the data collection method.

Taking the derivative of Eq. (1) with respect to *t*, we obtain the instantaneous rate of log*S*(t) at any point in the pea puree:

$$\frac{d(\log S(t))}{dt} = -n[d(T)]^{-n} t^{n-1} \qquad (46)$$

Rearranging,

$$\int_{\log S(0)}^{\log S(t)} d(\log S(t)) = -n\int_0^t [d(T)]^{-n} t^{n-1} dt \qquad (47)$$

Integrating and substituting for *d(T)*, we obtain the solution for the log(microbial survival) at an infinitesimally small point in the sample with a known time-temperature history *T(t)*:

$$\log S(t)-\log S(0)=-n\int_0^t [d(T)]^{-n} t^{n-1} dt=-nd_r^{-n}\int_0^t\left[10^{\left(\frac{T_\infty-T_i}{z}\right)\left(\frac{Tr-T(t)}{T_\infty-T_i}\right)}\right]^{-n} t^{n-1} dt=-nd_r^{-n}\psi(t) \quad (48)$$

Where the time-temperature history, $\Psi(t)$ is

$$\psi(t)=\int_0^t\left[10^{\left(\frac{T_\infty-T_i}{z}\right)\left(\frac{Tr-T(t)}{T_\infty-T_i}\right)}\right]^{-n} t^{n-1} dt \quad (49)$$

For pedagogical purposes, the number of parameters to be estimated was reduced by setting $\log S(0) = 0$.

Because the data are the mass-average log survival ratios for each can, we must integrate over space, i.e. over the cylindrical can. First, solve for the survival ratio at one point in the can by taking the anti-log of Eq. (48):

$$S(t)=10^{-nd_r^{-n}\psi(t)} \quad (50)$$

Integrating all survival ratios over the can gives the mass average microbial survival ratio (S_a) for one can at time *t*:

$$S_a(t)=2\iint 10^{-nd_r^{-n}\psi(r,\zeta,t)}\, r\, dr\, d\zeta \quad (51)$$

Where the time-temperature history at a point at radial location *r* and axial distance from the center of the can, ζ,

$$\psi(r,\zeta,t)=\int_0^t\left[10^{\left(\frac{T_\infty-T_i}{z}\right)\left(\frac{Tr-T(r,\zeta,t)}{T_\infty-T_i}\right)}\right]^{-n} t^{n-1} dt \quad (52)$$

r and ζ are the normalized radial and axial coordinates. The integral limits are 0 and 1 for both *r* and ζ. Due to symmetry, we need consider only one-eighth of the can.

Finally, the log of the mass-average survival ratio is

$$\log S_a(t)=\log\left(2\iint 10^{-nd_r^{-n}\psi(r,\zeta,t)}\, r\, dr\, d\zeta\right) \quad (53)$$

Independent variables: *t* (min) and $T(r,\zeta,t)$ (°C).

$T(r,\zeta,t)$ was supplied by computing the analytical heat conduction equation for heating with uniform initial temperature (one equation) and cooling with non-uniform initial temperature (another equation) at constant T_∞ (Dolan and others 2007) using a finite heat-transfer coefficient h=5500 W/m^{2}K. Alternatively, one could compute and supply temperatures via finite-difference or finite-element software, such as Comsol.

Dependent variable: $\log S_a(t)$

Three parameters to be estimated: d_r, $(T_\infty - T_i)/z$, n. For convenience, let $z' = (T_\infty - T_i)/z$

The model is nonlinear in all three parameters. The time integral (Eq. (52)) was computed by trapezoidal rule. The spatial integral (Eq. (53)) was computed by Gauss-Legendre quadrature.

Step #4. Plot the scaled sensitivity coefficients.

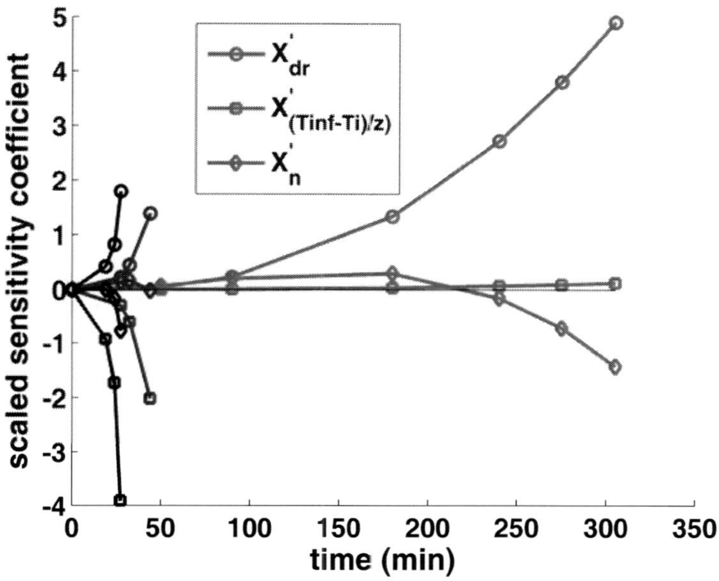

Figure 8. Example 3. Scaled sensitivity coefficients for model in Eq. 34. Parameter values were d_r= 224.7 min, $z' = (T_\infty - T_i)/z = 9.22$, n= 2.45. Red is T_∞ = 104.4°C; blue is T_∞ = 112.8°C; black is T_∞ = 120.6°C.

The maximum scale of observed $\log S$ = -2.1. The absolute value of the scaled sensitivity coefficient X' is largest for d_r compared to the other two parameters (Figure 8). X'_{dr} is very large (up to 5.8) for T_∞ = 104.4°C, and is between 1 and 2 for the two higher T_∞'s (Figure 8). Therefore, d_r will be the most easily estimated parameter with the smallest standard error. The scaled sensitivity coefficient for z' is very different: $X'_{z'}$ is so small for T_∞ = 104.4°C that z' could not be estimated if only these data were used (Figure 8). The information necessary for estimating z' comes from the data at T_∞ = 112.8°C and especially 120.6°C, where the absolute value of $X'_{z'}$ reaches nearly 4.0 (Figure 8). The information for estimating n comes mostly from T_∞ = 104.4°C, and less from T_∞ 120.6°C, as shown by the larger (up to 1.3) absolute values of X'_n (Figure 8). The value of X'_n at 112.8°C was negligibly small, showing that these data did not contribute to the estimation on n. There is similarity in the shapes of all the X', indicating there may be correlation among the parameters.

Step #5. Results for Example #3, using integral method

Parameter results

Table 4. Example 3. Weibull parameter estimates for *B. Stearothermophilus* spores in pea puree in cans (Figure 12) heated at four retort temperatures, using the integral method. Total number of data = 24. The *z* values corresponding to *z′* were 11.3, 12.2, and 13.1 °C, for T_∞ = 104.4, 112.8, and 120.6°C, respectively. The *z* values corresponding to *z′* CIs were (10.7, 12), (11.6,12.9), and (12.4,13.8 °C), for T_∞ = 104.4, 112.8, and 120.6°C, respectively

Parameter	estimate	Standard error	95% asymptotic confidence interval	Correlation coefficient, ρ	*SS*
d_r (min)	224.7	3.76	(216.8, 232.5)	$\rho_{drz'}$ = 4.4 x 10^{-5}	SS = 0.304
$z'=(T_\infty-T_i)/z$	9.22	0.234	(8.73, 9.70)	ρ_{drn} = 0.748	RMSE =
n	2.46	0.182	(2.08, 2.83)	$\rho_{z'n}$ = -0.147	0.121

Figure 9 shows the nonlinear fit of the predicted lines and the observed values for Example #2. The rmse = 0.118 (Table 4), only ~6% of the total scale of 2.1, indicates the fit is relatively good. As predicted by the scaled sensitivity coefficients, the standard error of d_r, the characteristic time at T_r, is small, less than 2% of the estimate = 224.7 min.

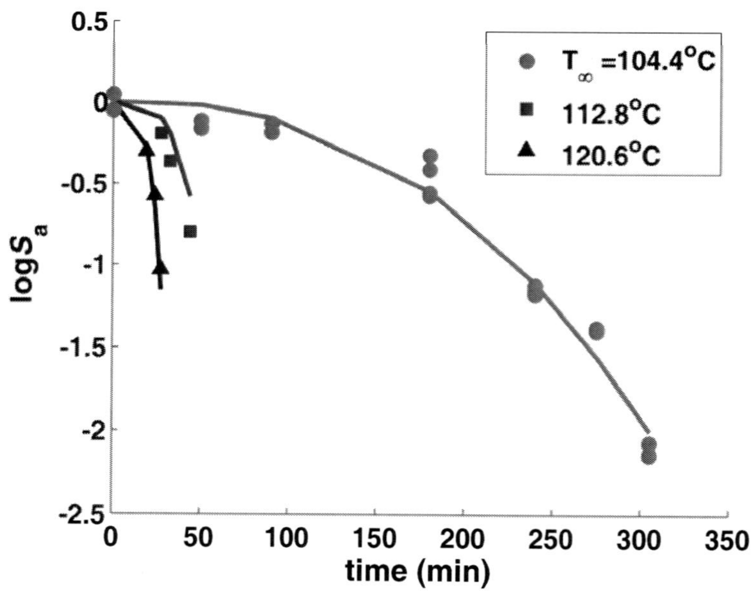

Figure 9. Example 3. Mass-average log survival ratios (logS_a) versus time for *B. Stearothermophilus* spores in pea puree heated in cans at retort temperatures (T_∞) 104.4°C, 112.8°C, and 120.6°C. Markers are observed data listed in Table 2. Lines are predicted values from Weibull model, Eq. (53). T_r was fixed at 104.52°C. Total number of data = 24.

Correspondingly, the d_r CI is small compared to the CIs of z' and n. The CIs for z' and n were still small, being 2.5% and 7.4% of their respective estimates (Table 4). The large estimate of n = 2.46 and 95% CI = (2.08, 2.83) is clear evidence that the commonly used first-order model (n = 1.0) is not appropriate for these data. Correlation was not high for any of the parameters (Table 4), and nearly zero for d_r and z'.

The z values increased with T_∞, from 11.3-13.1 (Table 4, caption). However, the CI for z' was not small enough to draw any statistical conclusion that the z value actually increased with temperature.

Reference temperature T_r for Arrhenius-type secondary models.

T_r was fixed at 104.52°C to minimize correlation between d_r and z' and to minimize error in d_r estimate. Figure 10 shows the plot of correlation coefficient for d_r and z' versus T_r. In our case, d_r was a reciprocal rate, and smaller values of z' represented more sensitivity of the rate to temperature. The standard error of z' was not affected by this correlation for our data, but the standard error of d_r was almost doubled if T_r was not selected correctly. For Example 3, standard errors of d_r at T_r = 110°C and 100°C were 3.2% and 2.9% of the d_r estimates, respectively, versus a 1.7% standard error at T_r = 104.52°C.

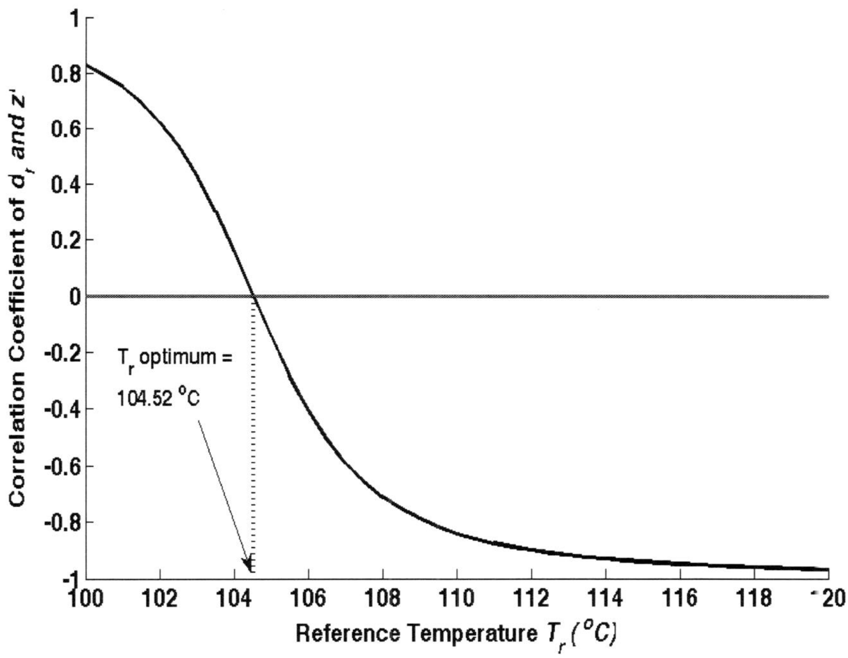

Figure 10. Example 3. Correlation coefficient for d_r and z' versus reference temperature for data *B. Stearothermophilus* in pea puree and Weibull model Eq. 53.

Joint Confidence region

The 95% elliptical approximation to the confidence region was significantly smaller than the more exact likelihood ratio confidence region, (Figure 11). This is a common occurrence for nonlinear models, so the researcher should be aware that the true confidence regions are probably larger than the ellipse (Beck and Arnold 1977, p. 384, Figure 7.12.). Figure 11 has only a slight "slant" and is approaching a circle due to the near-zero correlation = 4.4e-5 (Table 4).

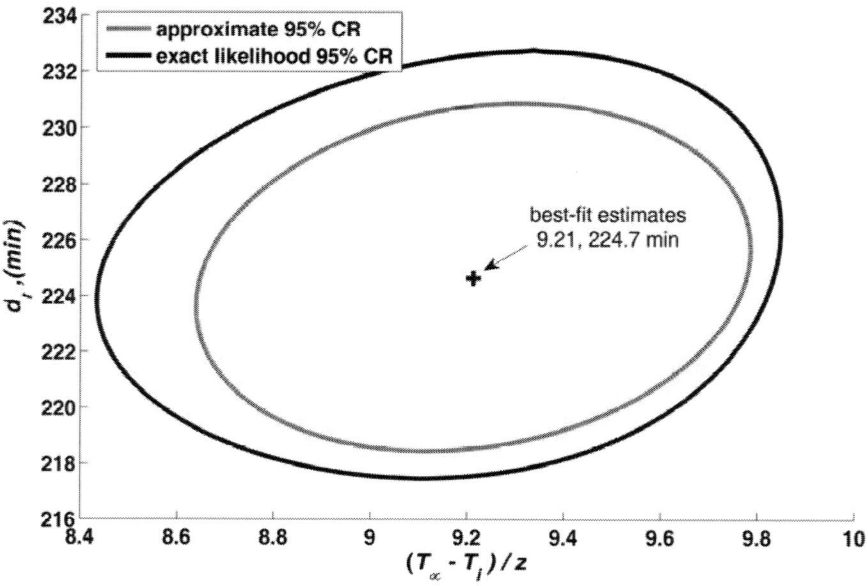

Figure 11. Example 3. 95% Joint confidence regions for d_r and z'. Approximate CR is from Eq. (21). Exact likelihood ratio CR is from Eq. (22).

Step #6. Standard Statistical Assumptions for Example #2.

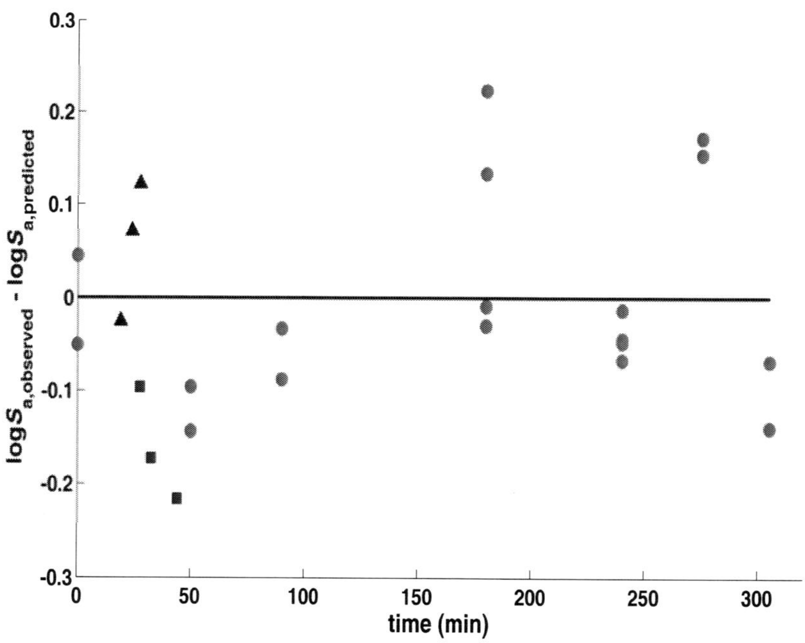

Figure 12. Example 3. Residual scatter plot for *B. Stearothermophilus* spores in pea puree heated in cans at three different retort temperatures. Data shown in Figure 9.

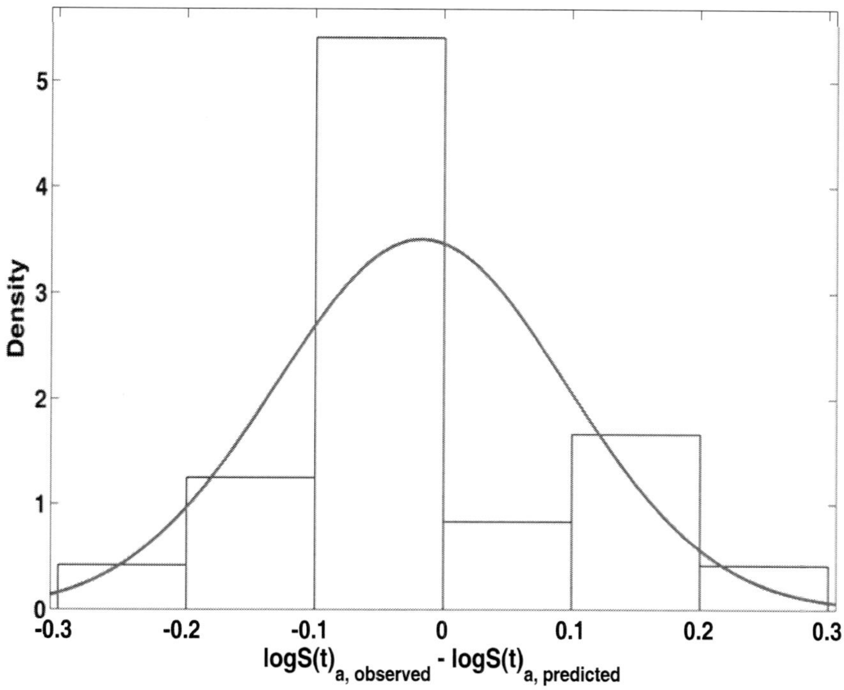

Figure 13. Residual histogram for *B. Stearothermophilus* spores in pea puree heated in cans at three different retort temperatures. Data shown in Figure 9, based on solution to Weibull model using the integral method. Mean of residuals = -0.0175 log, σ = 0.113 log.

The residual scatter plot (Figure 16) shows that assumption 1 is valid (additive errors). The residual histogram (Figure 13) reports the mean of the residuals = -0.0175 log, showing that mean is nearly zero (assumption 2). There is a nearly constant band of residuals from -0.2 to 0.2 (Figure 12), validating assumption 3. Residuals for T_∞ = 104.4°C indicate correlation, as the circle markers (Figure 12) follow a pattern of rising up to a peak at 180 min, and then falling to the negative side at 305 min., so assumption 4 is not valid. The residuals have a Gaussian distribution (Figure 13 and assumption 5). Therefore, the summary of assumptions for this model is 11101011.

6.A. Bootstrap resampling

Resampling by pairs bootstrapping was done 1,000 times. The model was fit each time to obtain 1,000 sets of the 3 parameters. For each fit, 24 new residuals were obtained, resulting in a total of 24,024 residuals. In the scatter plot (Figure 14), the 24,024 residuals are still correlated, shown as a band of width ~0.8 log having a concave-down pattern, as in Figure 12. Correlation of residuals often cannot be avoided, and is common in conduction-heating temperature data (Beck and Arnold 1977). The total range of residuals increased almost 4 times, from $\pm$ 0.22 log (Figure 10) to $\pm$ 0.8 log. The distribution of 24,024 residuals was Gaussian (Figure 15), nearly zero mean, and the estimate of standard deviation increased by 56% to σ_{boot} = 0.172 log from σ_{orig} = 0.11 log (Figure 13). The bootstrap result (σ_{boot}) is a larger and more accurate estimate of the true distribution of the residuals than the small-sample-set result (σ_{orig}). Using σ_{orig} in Monte Carlo simulations would result in a smaller variance of the parameter distribution, which would lead to an inaccurate and misleadingly smaller distribution in the log$S$(t) values. Therefore, to avoid results that are less accurate and less safe, it is recommended to do resampling *before* Monte Carlo simulations.

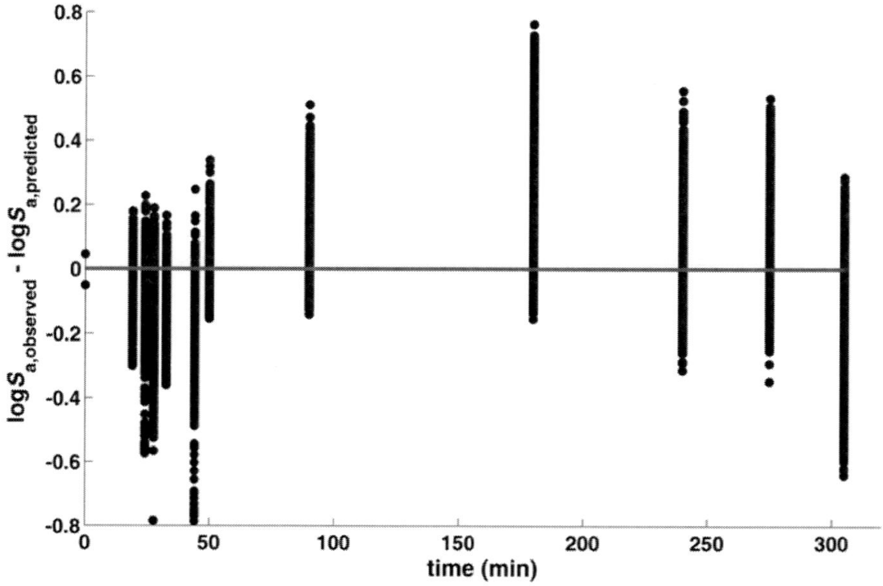

Figure 14. Example 3. Residual scatter plot for 1,000 bootstraps on *B. Stearothermophilus* spores in pea puree data in Figure 9.

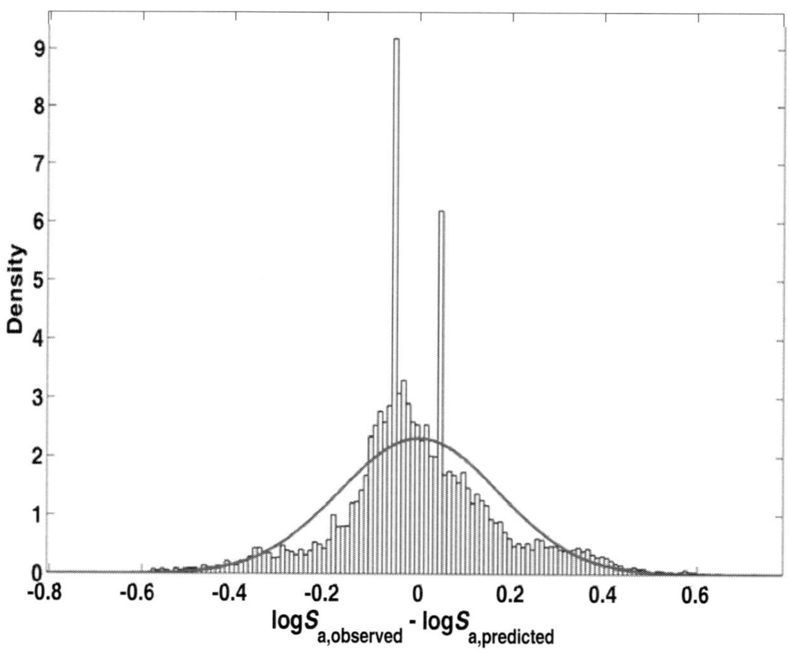

Figure 15. Example 3. Residual histogram for 1,000 bootstraps on *B. Stearothermophilus* spores in pea puree data in Figure 12. Mean = -0.00104 log, σ = 0.1722 log.

The bootstrap 95% confidence intervals for the parameters were:

d_r (200.7, 238.6) min; z' (1.01, 10.6); n (1.32, 2.82)

These boostrap CIs were all larger than the asymptotic CIs (Fig 13) and were asymmetric, being wider on the low side. The bootstrap CI for z' was much larger than the asymptotic, and was highly asymmetric on the lower side (estimate of $z' = 9.22$, Table 4). This difference is caused by most (18 data) of the observed data out of a total of 24 were at $T_\infty = 104.4°C$, where the scaled sensitivity coefficient for z' was negligible (Figure 8). For this reason in this case, the Monte Carlo simulation results (below) and the previously shown exact confidence region (Figure 11) will probably give more realistic confidence intervals for z' than the bootstrap CI.

Confidence bands and prediction bands for dependent variable

In microbial inactivation modeling, knowledge of the distribution of the log$S$ value may be more important than the distribution of the parameters. The reason is that for food safety, one wants to know how much error is in the predicted value of log$S$ if the product is heating for x minutes at T_∞ temperature. Therefore, the most important statistical results may be the 95% prediction band, within which 95% of the data are expected to fall.

To illustrate the procedure, the 95% asymptotic and bootstrap confidence bands (for the regression line) and the 95% asymptotic and bootstrap prediction bands are shown for only the data at 104.4°C (Figure 16). The asymmetric bootstrap results will be closer to the true bands.

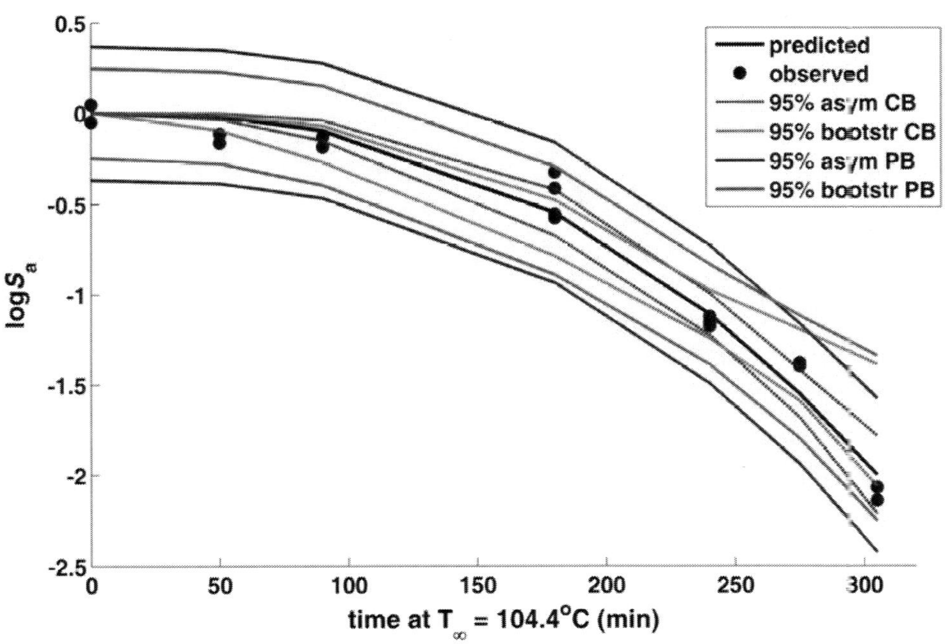

Figure 16. Example 3. Mass-average log(survival ratio) ($\log S_a$) for *B. Stearothermophilus* in pea puree heated in cans versus time at retort temperature = 104.4°C, with 95% confidence bands (asymptotic and bootstrap) and 95% prediction bands (asymptotic and bootstrap). Bootstrap results are based on 1000 bootstrap iterations.

The bootstrap CBs are from 1.2-2.3 times wider than the asymptotic CBs (Figure 16). The bootstrap CBs are more shifted down than symmetric asymptotic CBs, with the predicted line against the top edge of the bootstrap CB (Figure 16). Only at the higher times (280 and 310 min) does the asymptotic CB undershoot the upper bootstrap CB. These CBs show where 95% of all regression lines are expected to fall. In summary, the bootstrap CB shows that the regression lines have more variability, especially on the lower side, than is indicated by the asymptotic CBs.

More importantly, the bootstrap PB is smaller in width than the asymptotic PB, ranging from a ratio PB_{boot}/PB_{asymp} of 0.67 to equal (at 310 minutes, Figure 16). The asymptotic PB is larger than it needs to be, showing that 95% of the data are expected to fall within a larger range than is true, a conservative result. Let's discuss what the PB means by choosing heating at 310 minutes as an example. The asymptotic PB indicates that if we heated a large number of cans for 310 min at T_∞ = 104.4°C, we would expect 95% of the cans to have mass average log reduction between the PB limits of 1.57 to 2.42 log. The other 5% of the cans would be expected to have log reductions above and below these limits. The corresponding results based on the bootstrap PB were 1.34 log to 2.25 log. Other simulation research (Dolan, not shown) has indicated that the bootstrap PBs are very close to the true PBs, while the asymptotic 95% PBs may even wider than the true 99% PBs, showing an over-conservative result. Therefore, thermal processing companies who wish to maintain identical food safety while improving product quality can use the more accurate bootstrap results.

6.C. Monte Carlo simulation to obtain parameter distributions

Once a better estimate for the true σ was obtained, the value σ_{boot} = 0.1722 log was substituted for s into Eq. (35), and 1000 sets of parameters were estimated by fitting the model 1000 times to 1000 different sets of $Y_{synthetic}$.

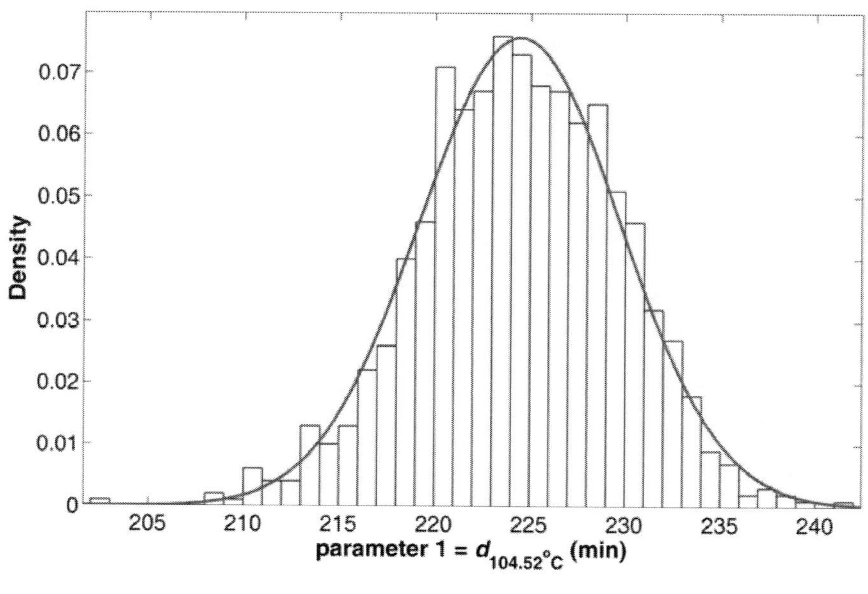

Figure 17. Example 3. Probability distribution histogram for $d_{104.52C}$ in the Weibull model (Eqs. (1) and (30)) for *B. Stearothermophilus* in pea puree heated in cans versus time at retort temperatures (T_∞) 104.4°C, 112.8°C, and 120.6°C, based on 1,000 Monte Carlo simulations. μ = 224.5 min, σ = 5.26 min.

All three parameters showed a clear normal distribution. Figures 17-19 show a distinct normal distribution for d_r, z', and n. These plots are good evidence that the true distribution of the parameters is normal, a desirable characteristic for nonlinear regression.

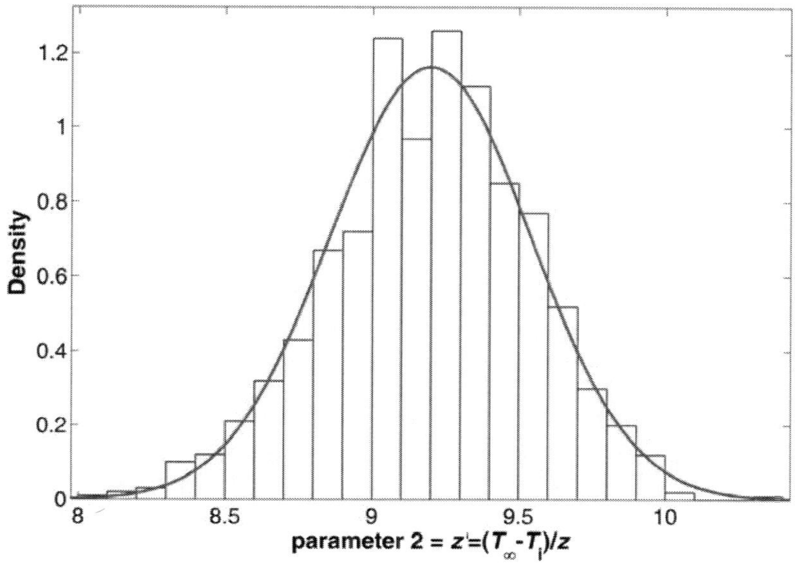

Figure 18. Example 3. Probability distribution histogram for $z' = (T_\infty - T_i)/z$ in the Weibull model (Eqs. (1) and (30)) for *B. Stearothermophilus* in pea puree heated in cans versus time at retort temperatures (T_∞) 104.4°C, 112.8°C, and 120.6°C, based on 1,000 Monte Carlo simulations. $\mu = 9.2$, $\sigma = 0.342$.

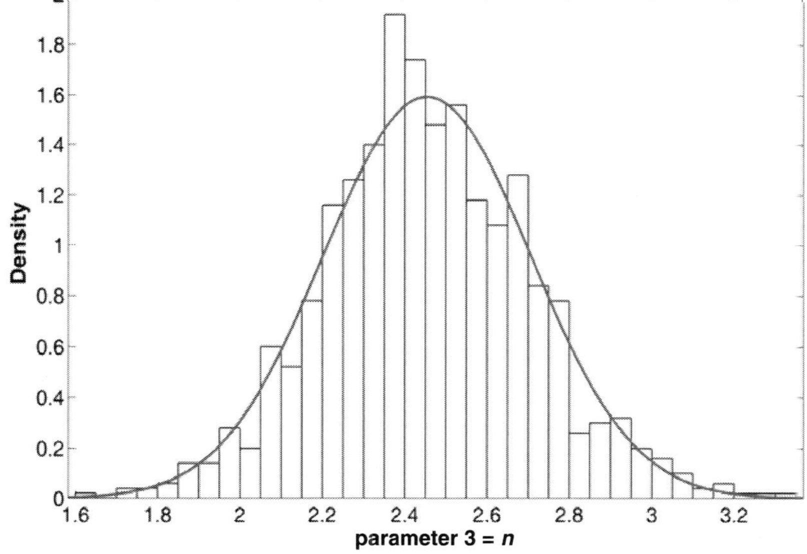

Figure 19. Example 3. Probability distribution histogram for n in the Weibull model (Eqs. (1) and (30)) for *B. Stearothermophilus* in pea puree heated in cans versus time at retort temperatures (T_∞) 104.4°C, 112.8°C, and 120.6°C, based on 1,000 Monte Carlo simulations. $\mu = 2.45$, $\sigma = 0.25$.

Sequential Estimation

We have discussed parameter estimation techniques using all the data at once in earlier sections of this chapter. In this section we are going to introduce you to the sequential estimation of parameters. As the name itself suggests, sequential method of estimation updates parameters as new observations are added. Sequential estimation of parameters in a model provides good insight into building the model and determining uncertainty in parameters. For dynamic processes, this estimation method can be used "on-line" as it accounts for addition of each data point in the model. For example, if parameters come to a constant value after a certain reasonable time then the experiment can be stopped, as further data will not improve the parameter estimate. Prior information of parameters can be used in sequential estimation towards estimation of parameters for a particular experiment.

The mathematical form of non-linear sequential estimation is derived from maximum a posteriori (MAP) estimation. This sequential method uses the Gauss minimization method, and can also be incorporate Box-Kanemasu modification. The minimization function in the Gauss method can be expressed as:

$$S = \left[Y - \hat{Y}(\beta)\right]^T W \left[Y - \hat{Y}(\beta)\right] + [\mu - \beta]^T U [\mu - \beta] \tag{54}$$

Where, Y is the experimental response variable and $\hat{Y}$ is predicted response, μ is prior information of parameter vector β, W is inverse of covariance matrix of errors and U is inverse of covariance matrix of parameters. The extremum of the function given by equation (54) can be evaluated by differentiating it with respect to β. The expression can be given as:

$$\nabla_\beta S = -2\left[\nabla_\beta \hat{Y}(\beta)\right]^T W \left[Y - \hat{Y}(\beta)\right] - 2[I]U[\mu - \beta] \tag{55}$$

The expression for the Gauss method of estimating β can be found by setting equation (55) to zero. As we already have seen the expression for sensitivity matrix, let us assume

$$X(\beta) = \left[\nabla_\beta \hat{Y}(\beta)\right]^T \tag{56}$$

Setting equation (55) equal to zero at $\beta = \hat{\beta}$,

$$X^T\left(\hat{\beta}\right) W \left[Y - \hat{Y}\left(\hat{\beta}\right)\right] + U\left[\mu - \hat{\beta}\right] = 0 \tag{57}$$

As $\hat{\beta}$ is appearing implicitly in $\hat{Y}$ and X as well as explicitly, an iterative procedure is needed to solve for $\hat{\beta}$. Let's assume some notations for the iterative procedure,

$$b^{(k)} = b, \quad b^{(k+1)} = \hat{\beta}, \quad \hat{Y}^{(k)} = \hat{Y}^{(b)}, \quad X^{(k)} = X(b) \tag{58}$$

$$b^{(k+1)} = b^{(k)} + P^{(k)} X^T W \left(Y - \hat{Y}^{(k)}\right) + U\left[\mu - b^{(k)}\right] \tag{59}$$

Where,

$$P^{(k)} = \left[X^{T(k)} W X^{(k)} + U\right]^{-1} \tag{60}$$

For non linear models, iteration of k is required to achieve the estimated parameters. Equation (59) together with equation (60) is a form of MAP equation. For non linear estimation problem, an initial guess of parameters is needed to start the iteration and the system of equations can be continued until convergence is reached. We will provide convergence criteria later in this chapter.

The equation (59) is the basis of sequential procedure. Standard statistical assumptions that allow the use of sequential estimation are; additive errors, zero mean, uncorrelated errors, normally distributed errors, covariance matrix of errors is completely known, no errors in independent variables and subjective prior information of parameters are known

The sequential procedure is developed with above equations using matrix inversion lemma Beck and Arnold 1977, p. 277) based on the Gauss minimization method. The procedure is given as:

$$A_{i+1} = P_i X_{i+1}^T \tag{61}$$

$$\Delta_{i+1} = \phi_{i+1} + X_{i+1} A_{i+1} \tag{62}$$

$$K_{i+1} = A_{i+1} \Delta_{i+1}^{-1} \tag{63}$$

$$e_{i+1} = Y_{i+1} - \hat{Y}_{i+1} \tag{64}$$

$$b_{i+1}^* = b_i^* + K_{i+1}\left[e_{i+1} - X_{i+1}\left(b_i^* - b\right)\right] \tag{65}$$

$$P_{i+1} = P_i - K_{i+1} X_{i+1} P_i \tag{66}$$

In the above equations i represents step for each data addition in the sequential procedure. P is the covariance matrix of parameters, φ is covariance matrix of errors and X is the sensitivity matrix. At this point, it is important to understand that there are two things going on in the above equations. Equations (61) to (66) provide sequential estimation of the parameters by adding one observation at a time and at the end of the final step, the estimated parameters are used as an initial guess for the next iteration. Hence, within iteration data is added one at a time and we call it steps. The parameter vector b* and covariance matrix of parameters P are updated at each step, i. The parameter vector b is the initial parameter vector going into each iteration, and changes only at the end of all the steps before the next iteration, i.e. parameter vector b is constant over the step i. At the end of the sequence (i.e equation(66)), P(k+1), X(k+1) and e(k+1) are evaluated for the next iteration, where k is the index for the number of iterations in the outer loop for the nonlinear regression. Matrix P, X and e are functions of b and not of b*. The stopping criteria for b can be given as;

$$\frac{\left|b_j^{k+1} - b_j^k\right|}{\left|b_j^k\right| + \delta_1} < \delta \tag{67}$$

Where, j is the index for the number of parameters. The magnitude of δ could be 10^{-4}. Another small number is δ_1 which is very small such as 10^{-8}, to avoid the problem when b_j^k tends to zero. In case of well designed experiments and a good model, the number of iterations is very low.

EXAMPLE #4: Plot sequential results for Example #2.

Prior information regarding the parameters and parameter covariance matrix can be used in the model to improve the results. However, in this case where no prior information regarding parameters was known, some good initial guess for parameters and large diagonal components (100) in the parameter covariance matrix was used.

To gain insight into the model, the sequential results are shown for parameters k, T_c, and n for *Salmonella* inactivation in Example 2 using the differential method. The final parameter values for sequential estimation were the same as those for all-data-at-once estimation listed in Table 2.

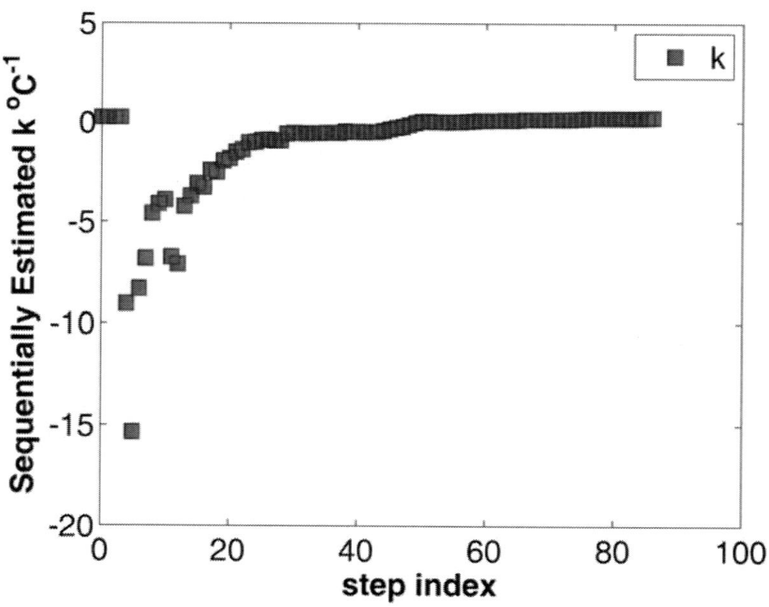

Figure 20. Sequentially estimated parameter k versus step index for Example 2. The number of steps equals the number of data = 87. The step index tracks $-\log S(t)_{predicted}$ from left to right, so step 0 corresponds to $-\log S(t) = 0$, and step 87 corresponds to $-\log S(t) = 5.96$.

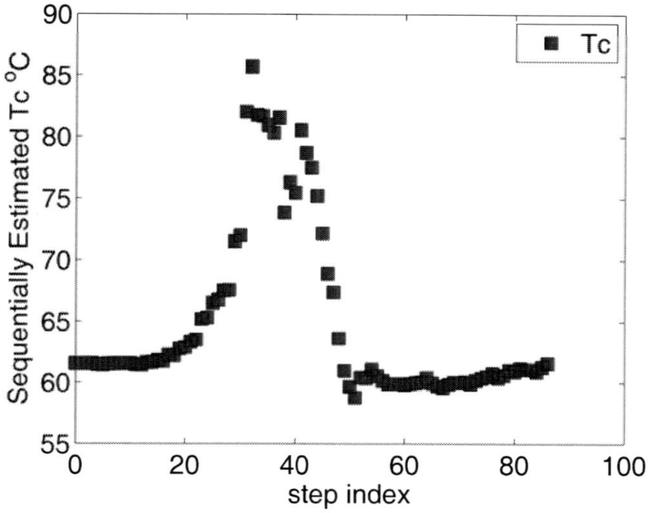

Figure 21. Sequentially estimated parameter T_c (°C) versus step index for Example 2.

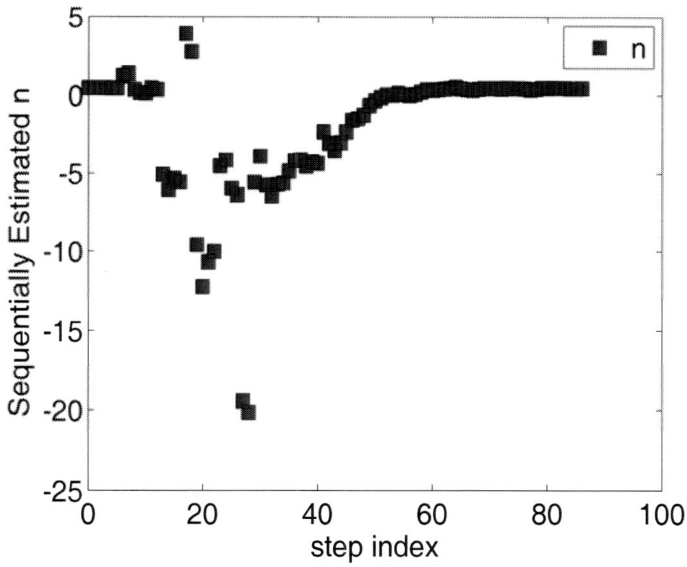

Figure 22. Sequentially estimated parameter n versus step index for Example 2.

The details of the last iteration are shown by plotting the estimate of each parameter versus step index in Figures 20 (k), 21 (T_c), and 22 (n), as each datum was added successively. The order of the data addition was according to –logS(t)$_{predicted}$, so the the first datum was –logS(t)=0, and the last datum was –logS(t)=5.96. We expect each parameter to fluctuate during the early times of the experiment, and desire the parameter to come to nearly a constant value during the last half of the sequential estimation, and before the last datum is added. For all 3 parameters, the values fluctuated during the first ~half of the experiment. Both k and n came to a constant after ~2/3 of the experiment was completed. T_c continued to increase slightly even at the end of the experiment, showing that either the model needs modification, or data need to be collected beyond the time shown. Correspondingly, in the observed data (Figure 4) , log$S$ changed little up to 5 minutes, after which log$S$ began to decrease, and by 13 minutes (~2/3 of the experimental time) the last 3 sets (a, b, and c, Figure 4) were finished, and set d log$S$ had already begun to decrease. In summary, the sequential results show that model was adequate for both k and n, but not for T_c.

EXAMPLE #5: Plot sequential results for Example #3.

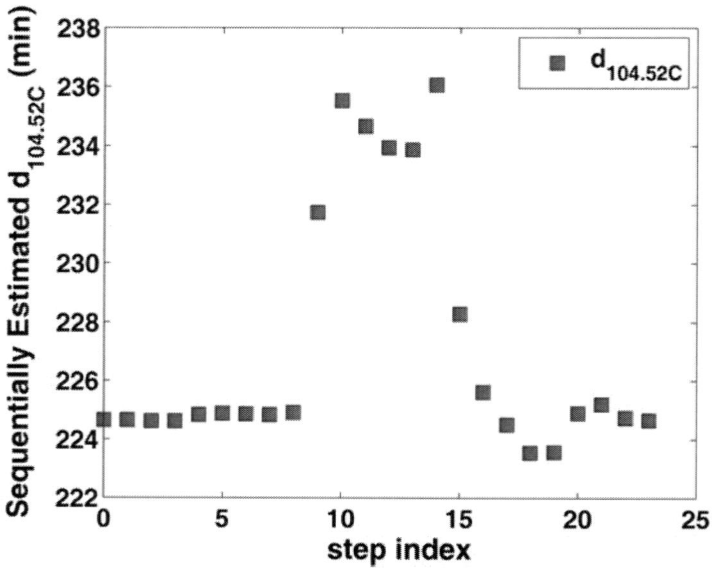

Figure 23. Sequentially estimated parameter d_r versus time for Example 3. The number of steps equals the number of data = 24. The step index tracks –logS(t)$_{predicted}$ from left to right, so step 0 corresponds to -logS(t) = 0, and step 87 corresponds to -logS(t)= 2.1.

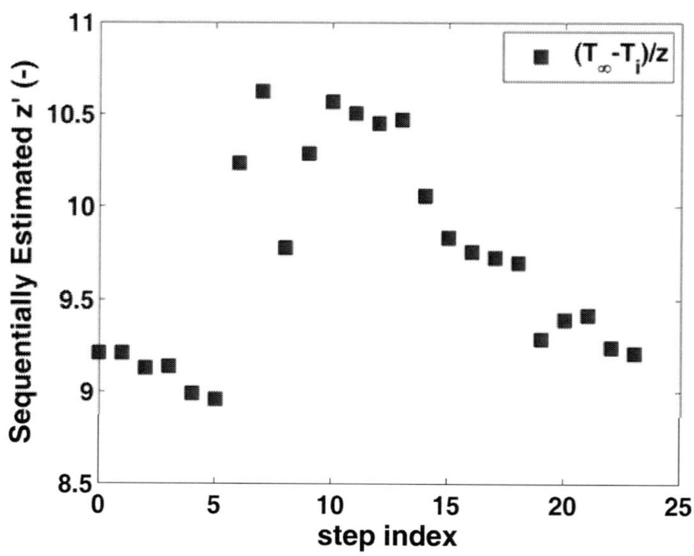

Figure 24. Sequentially estimated parameter $z' = (T_\infty - T_i) / z$ versus time for Example 3.

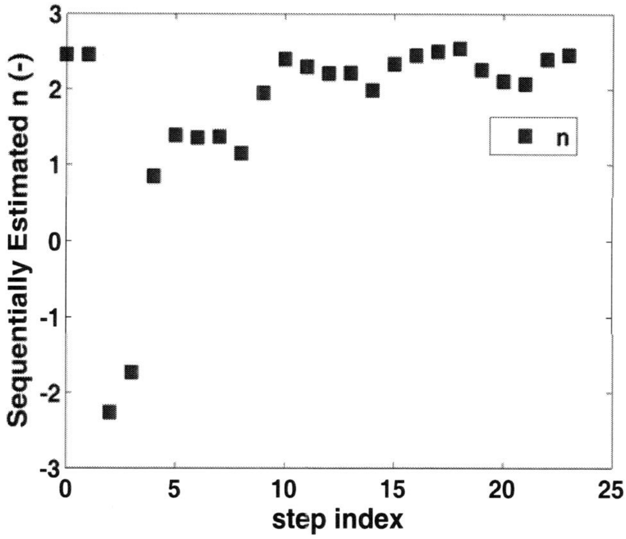

Figure 25. Sequentially estimated parameter n versus time for Example 3. These are the results for the last iteration.

Final sequential parameter estimates were the same as those in Table 2, as expected. Figures 23-25 show the progression with each datum for each parameter. The d_r parameter approached a constant at about step = 16 (Figure 23), corresponding to $\log S_a = -1.1$ (Figure 9) The d_r parameter continued to change slightly until reaching a constant during the addition of the final four data (Figure 23). The parameter z' continued to slope downward and did not approach a constant at the end of the experiment (Figure 24), showing that this part of the model needs improvement. There may be a superior secondary model than the z′ (dimensionless Bigelow model) to represent the effect of temperature on d_r Note also that the slowest heating rate had a very small scaled sensitivity coefficient for z′ (Figure 8), indicating that the data for the slowest heating rate had virtually no effect of estimation of z′. This result makes sense, because the temperature in the pea puree was nearly uniform for the last half of the experiment, when $\log S_a$ for the slow heating rate began to drop. The parameter n continued to vary until the last two data. Since n is an exponent that affects the curvature, the results makes sense because the predicted line for the slowest heating rate continues to curve throughout the experiment (Figure 9). Additionally, the scaled sensitivity coefficient for n is largest for the slowest heating rate, and increases until the last experimental time (Figure 8), showing that n is affected most by the slowest heating rate. The parameter n is also influenced by the fastest heating rate, but is almost completely unaffected by the intermediate heating rate (Figure 8, see the absolute value of the scaled sensitivity coefficients for n). In summary, the sequential results show that the model for Example 3 was adequate for parameters d_r and n, but not for z'.

References

Bates DM & Watts DG. 1988. *Nonlinear regression analysis and its applications*. New York: Wiley.

Beck JV & Arnold KJ. 1977. Parameter *Estimation in Engineering and Science*. New York: John Wiley & Sons, Inc.

Beck JV & Arnold KJ. 2007. *Parameter Estimation in Engineering and Science Revised* Chapter 6. 1935 Danbury West, Okemos MI, 48864: Beck Engineering Consultants Company, jvb@BeckEng.com

Claeys WL, Ludikhuyze LR & Hendrickx ME. 2001a. Formation kinetics of hydroxymethylfurfural, lactulose and furosine in milk heated under isothermal and non-isothermal conditions. *Journal of Dairy Research* 68:287-301.

Claeys WL, Ludikhuyze LR, van Loey AM & Hendrickx ME. 2001b. Inactivation kinetics of alkaline phosphatase and lactoperoxidase, and denaturation kinetics of b-lactoglobulin in raw milk under isothermal and dynamic temperature conditions. *Journal of Dairy Research* 68:95-107.

Corradini MG & Peleg M. 2009. Dynamic Model of Heat Inactivation Kinetics for Bacterial Adaptation. *Appl Environ Microb* 75(8):2590-2597.

Devore JL & Berk KN. 2007. *Modern mathematical statistics with applications.* Belmont, CA: Thomson Brooks/Cole.

Dolan KD. 2003. Estimation of kinetic parameters for nonisothermal food processes. *Journal of Food Science* 68(3):728-741.

Dolan KD, Yang L & Trampel CP. 2007. Nonlinear regression technique to estimate kinetic parameters and confidence intervals in unsteady-state conduction-heated foods. *Journal of Food Engineering* 80(2):581-593.

Efron B & Tibshirani R. 1993. *An introduction to the bootstrap.* New York: Chapman & Hall.

Efron B & Tibshirani R. 1998. *An introduction to the bootstrap.* Boca Raton, FL: Chapman and Hall/CRC Press LLC.

Halder A, Datta AK & Geedipalli SSR. 2007. Uncertainty in thermal process calculations due to variability in first-order and weibull kinetic parameters. *Journal of Food Science* 72(4):E155-E167.

Halder A, Dhall A, Datta AK, Black DG, Davidson PM, Li JJ & Zivanovic S. 2011. A user-friendly general-purpose predictive software package for food safety. *Journal of Food Engineering* 104(2):173-185.

Hassani M, Manas P, Raso J, Condon S & Pagan R. 2005. Predicting heat inactivation of Listeria monocytogenes under nonisothermal treatments. *J Food Protect* 68(4):736-743.

Himmelblau DM. 1970. *Process Analysis by Statistical Methods.* John Wiley & Sons.

Juneja VK & Marks HM. 2003. Characterizing asymptotic D-values for Salmonella spp. subjected to different heating rates in sous-vide cooked beef. *Innov Food Sci Emerg* 4(4):395-402.

Mattick KL, Legan JD, Humphrey TJ & Peleg M. 2001. Calculating Salmonella inactivation in nonisothermal heat treatments from isothermal nonlinear survival curves. *J Food Protect* 64(5):606-613.

McKellar RC & Lu X. 2004. *Modeling microbial responses in foods.* Boca Raton, FL: CRC Press.

Mishra DK, Dolan KD & Yang L. 2008. Confidence intervals for modeling anthocyanin retention in grape pomace during nonisothermal heating. *Journal of Food Science* 73(1):E9-E15.

Mishra DK, Dolan KD & Yang L. 2009. Bootstrap confidence intervals for the kinetic parameters of degradation of anthocyanins in grape pomace. *J Food Process Eng*:no-no.

Montgomery DP, Peck EA & Vining GG. 2006. *Introduction to linear regression analysis,*4th ed.: John Wiley & Sons Inc.

Motulsky H & Christopoulos A. 2004. *Fitting Models to Biological Data Using Linear and Nonlinear Regression*. New York: Oxford University Press.

Peleg M & Normand MD. 2004. Calculating microbial survival parameters and predicting survival curves from non-isothermal inactivation data. *Crit Rev Food Sci* 44(6):409-418.

Peleg M, Normand MD & Campanella OH. 2003. Estimating microbial inactivation parameters from survival curves obtained under varying conditions - The linear case. *B Math Biol* 65(2):219-234.

Ratkowsky DA. 1983. *Nonlinear Regression Modeling*. New York and Basel: Marcel Dekker, Inc.

Ratkowsky DA. 1990. *Handbook of nonlinear regression models*. New York: M. Dekker.

Sahinler S & Topuz D. 2007. Bootstrap and Jackknife Resampling Algorithms for Estimation of Regression Parameters. *Journal of applied Quantitative Methods* 2(2):188-199.

Schwaab M & Pinto JC. 2007. Optimum reference temperature for reparameterization of the Arrhenius equation. Part 1: Problems involving one kinetic constant. *Chem Eng Sci* 62(10):2750-2764.

Seber GAF & Wild CJ. 2003. *Nonlinear Regression*. John Wiley & Sons.

Stasiewicz MJ, Marks BP, Orta-Ramirez A & Smith DM. 2008. Modeling the effect of prior sublethal thermal history on the thermal inactivation rate of Salmonella in ground turkey. *J Food Protect* 71(2):279-285.

Stephens PJ, Cole MB & Jones MV. 1994. Effect of Heating Rate on the Thermal Inactivation of Listeria-Monocytogenes. *J Appl Bacteriol* 77(6):702-708.

Valdramidis VP, Geeraerd AH & Van Impe JF. 2007. Stress-adaptive responses by heat under the microscope of predictive microbiology. *J Appl Microbiol* 103(5):1922-1930.

van Boekel MAJS. 2002. On the use of the Weibull model to describe thermal inactivation of microbial vegetative cells. *International Journal of Food Microbiology* 74(1-2):139-159.

vanBoekel MAJS. 1996. Statistical aspects of kinetic modeling for food science problems. *Journal of Food Science* 61(3):477-&.

Welt BA. 1996. Kinetic parameter estimation in prepackaged foods subjected to dynamic thermal treatments. . *Agricultural & Biological Engineering*. Gainesville: University of Florida.

Welt BA, Teixeira AA, Balaban MO, Smerage GH, Hintinlang DE & Smittle BJ. 1997. Kinetic parameter estimation in conduction heating foods subjected to dynamic thermal treatments. *Journal of Food Science* 62(3):529-&.

In: Progress on Quantitative Approaches in Thermal … ISBN: 978-1-62100-842-2
Editors: V. P. Valdramidis and J. F. M. Van Impe © 2012 Nova Science Publishers, Inc.

Chapter 8

MODELLING THE INDUCED THERMOTOLERANCE DUE TO HEAT, ACID SHOCK AND RISING TEMPERATURES

E. G. Velliou[1], V. K. Juneja[2], J. F. Van Impe[1] and E. Van Derlinden[1]

[1] BioTeC - Chemical and Biochemical Process Technology and Control, Department of Chemical Engineering, Katholieke Universiteit Leuven, Leuven (Heverlee), Belgium

[2]Residue Chemistry and Predictive Microbiology Research Unit, Eastern Regional Research Center, Agricultural Research Service, US Department of Agriculture, Wyndmoor, US

Abstract

The present chapter deals with the mathematical quantification of the microbial heat resistance as a result of exposure to several environmental stressing factors, i.e., heat, acids, rising temperatures and chemical components like osmolytes. More specifically, an overview of the stress phenomenon, the cellular response to stress and the mechanism of stress adaptation is given. Next, the effect of exposure to heat, acids and chemical components on the heat tolerance of several microorganisms at static temperatures is analysed and several mathematical quantification approaches, i.e., via available predictive models, are mentioned. Furthermore, the effect of a dynamic temperature profile on the inactivation kinetics of a variety of microorganisms is being described. Additionally, the cross protection phenomenon, i.e., resistance to an environmental factor after exposure and adaptation to a different king of stress, is explained in detail. Finally, the need for the incorporation of intracellular information related to stress adaptation for an improvement and broader application of the already available predictive mathematical models is being discussed.

Introduction

Nowadays, consumers demand food products with high quality sensory characteristics, i.e., flavour, taste and texture. Therefore recent food processing procedures tend towards less aggressive techniques, e.g., high hydrostatic pressure, mild heat treatment and pulsed electric field processing. In contrast to the more traditional processes like pasteurization and sterilization, these new processing techniques maintain the textural and sensory characteristics of fresh food products. However, as the applied conditions are less harsh, microbial survival and even stress adaptation might not be prevented.

Apart from the processing procedures, food pathogens may get stress adapted in other stages throughout the food chain. For example, inappropriate temperature conditions or specific acid procedures during storage and distribution could be a microbial stress factor. At the consumption stage, temperature during reheating and cooking, or acidic additives, i.e. vinegar, could lead to an adaptation of the pathogens in the food system.

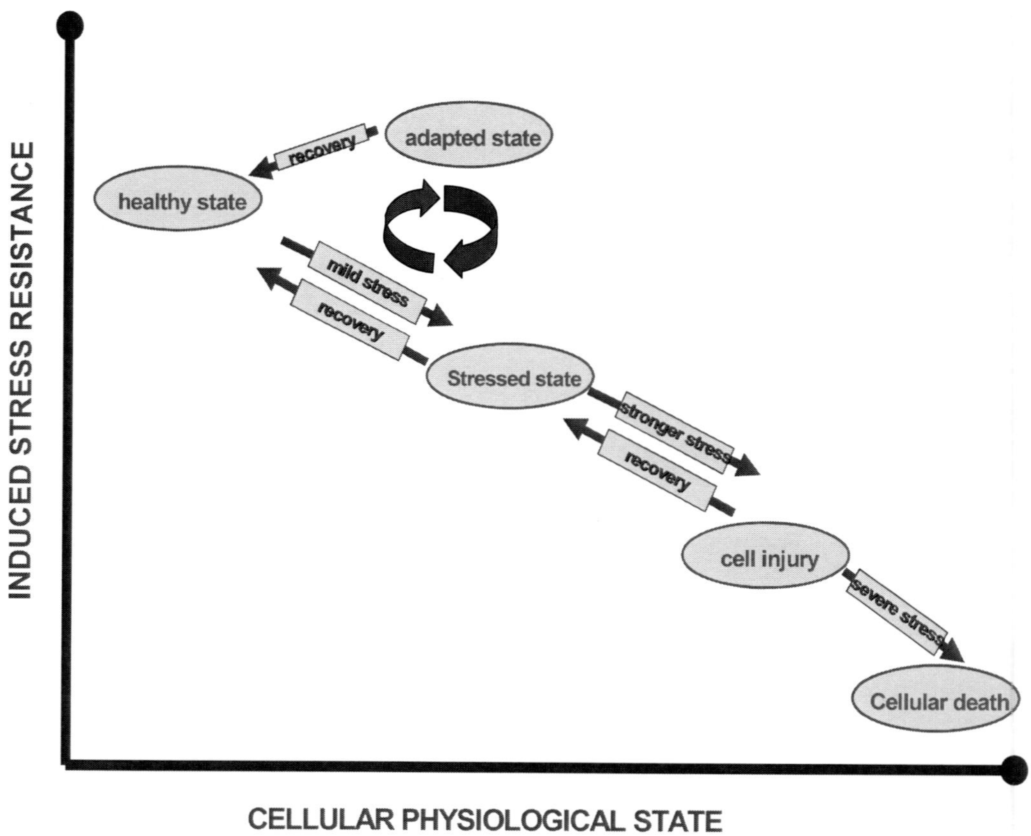

Figure 1. Relations between the physiological state of microbial cells and the level of the stressing environmental factor. (Figure inspired by Yousef and Courtney 2003).

Definition of Stress

Generally, microorganisms may reveal *stress adaptation* which actually is the increase of a microorganism's tolerance to environmental conditions, e.g., temperature, acidity, increased concentration of salt, metabolic products and presence of (toxic) chemical components, which would normally be lethal. This bacterial stress adaptation is usually the result of pre-exposure to a similar or a different stress factor and is dependent on the effect of the stress factor on the physiological state of the cell. Depending on how severe the stress factor may be, the bacterial cells may become *stressed, injured* or might *die*. The *stressed* cells usually become stress adapted, via the *stress adaptive response*. These stress adapted bacterial cells can remain resistant in absence of the stress factor or can return to the initial (healthy) physiological state (see Figure 1) (Yousef and Courtney 2003; Wesche et al. 2009).

Cellular Response to the Stress Factor

When microbial cells are exposed to an environmental stress, they respond in several ways. Most likely, specific genes that are related to the cellular stress response are expressed and/or over-expressed, finally leading to the production of proteins which are known as shock proteins (Figure 2). These shock proteins have as main function to repair the damages caused by the stress factor or eliminate the stress agent. Shock proteins are either protein chaperones -GroEL, DnaK, DnaJ, etc.- which assist folding and assembly of damaged cellular proteins, or they may be ATP-dependent proteases –Lon, ClpAP, etc.- that degrade permanently damaged cellular proteins (Ohtsuka et al. 2007; Yousef and Courtney 2003; Juneja and Novak 2003).

According to the hypothesis of Rowburry (2001), extracellular components might act as stress sensors providing an earlier warning as compared to intracellular sensors. Usually, these extracellular components are proteins, which have as main function the switching on of the stress response. After the detection of stress by the Extracellular Sensing Component(s) (ESCs) which is (are) produced in absence of the stress and is (are) activated by the stress, the production of Extracellular Induction Component(s) (EICs) is switched on. Still according to this conceptual framework, the EICs actually arise from the ESCs (Rowburry et al. 1999). These EICs remain present in the medium in absence of the stress. Since they are extracellular, they act as alarmones inducing tolerance to non-stressed cells or non-stressed regions leading to an intercellular communication, therefore allowing 'cross-talk' between cells which are in different locations (Rowburry et al. 1999; Rowburry 2001; Rowburry 2003).

Another way bacteria protect themselves from an environmental stress factor is via the alternation of the ratio of the fatty acids of the cellular membrane. When the cells experience a stress factor, e.g., an acid, the concentration of saturated fatty acids in the cellular membrane increases and is higher than that of the unsaturated fatty acids. In this case, the membrane fluidity decreases and, therefore, the bacterial cells become more resistant to stress factors such as heat. Brown et al. (1997) observed for five strains of *Escherichia coli*, i.e., MJR, M23, R172, R91 & O157:H7, that when the bacteria were acid shocked (exposed to a low pH of 3) the non-saturated fatty acids were replaced by saturated ones and the bacterial acid resistance increased.

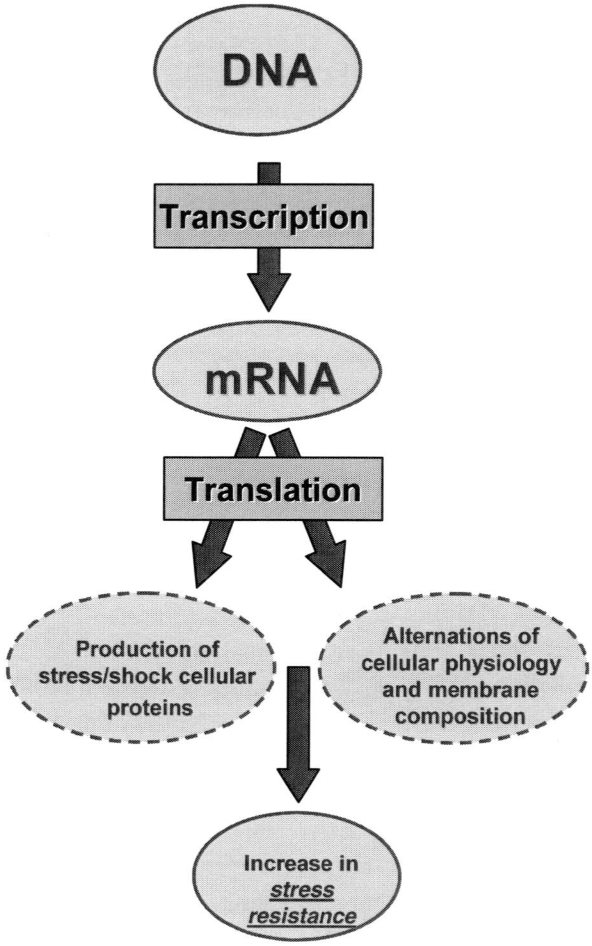

Figure 2. The cellular response to the stress factor.

Heat Stress

Generally, heat stress is exposure to temperatures that are higher than the optimal bacterial growth temperature. Exposure to (mildly) suboptimal temperature conditions may lead to a certain level of stress adaptation and further bacterial tolerance to environmental conditions that would normally be lethal. For example, Cebrian et al. (2007) showed that the heat resistance of *Escherichia coli* cells at 57.5°C was higher when the cells were grown at higher temperatures, i.e., 42°C and 37°C compared to 10°C, 20°C and 30°C. According to Cebrian et al. (2009) the thermotolerance of *Staphylococcus aureus* at 58°C increased after a heat shock of 120 minutes at 45°C. Similar results were obtained by Hassani et al. (2005, 2006) who observed that non-isothermal, i.e., progressive, heat treatment leads to a higher thermotolerance of *Staphylococcus aureus* and *Listeria monocytogenes*. Juneja and Marks (2003a) observed that there is a clear (positive) effect of the heating rate on the heat resistance of *Salmonella* spp at 58°C.

Juneja and Marks (2003) studied the inactivation rates of a cocktail of *Salmonella spp.* in *sous vide* cooked beef for thermal processing from 10°C to 58°C with heating rates that varied from 1 to 3 hours. The inactivation curves, gained at 58°C consisted of a short shoulder phase followed by a linear part. This linear part was modelled by non-zero asymptotic D-values (described also in Juneja et al. (2001)) and an increase of the value of this parameter was observed for slower heating rates, i.e., milder heat pre-treatment, indicating an increased bacterial thermotolerance.

Lee and Kang (2009) studied the combined effect of heat, salt, and acids on the thermal inactivation of *Escherichia coli* O157:H7 in laboratory media at 55°C and quantified the (decreased) bacterial heat resistance in presence of acids via the D-value of the linear microbial inactivation curves. A decrease of the D-value indicates a decreased heat tolerance of *Escherichia coli* O157:H7 in presence of acids at 55°C.

Hassani et al. (2007) studied the effect of heat shocks of various durations up to 60 min on the heat resistance of *Listeria monocytogenes* and *Pseudomonas aeruginosa* at different temperatures between 35 and 45°C in media at pH 4.0, 5.5 and 7.4. All the experimental data, i.e., convex microbial survival curves, were fitted with the Weibull based Mafart model (Mafart et al. 2002) (Equation (1)) (see also Chapter 4 of this book). It should be mentioned that the behaviour of *Listeria monocytogenes* followed a log-linear behaviour and could be equally described by the classical log-linear model as opposed to the case of the inactivation of *Pseudomonas aeruginosa*, during which a slight shoulder was observed and the curves were more convex.

$$N(t) = N(0) \cdot 10^{\left(-\frac{t}{\delta}\right)^{p}} \tag{1}$$

In this model, N [CFU mL^{-1}] is the cell population, $N(0)$ [CFU mL^{-1}] the initial cell population, t [min] the time, δ [min] the time of first decimal reduction and p a shape parameter.

Generally, for both organisms an increased heat resistance was observed at 62°C and 56°C after application of a heat shock. This higher heat tolerance – compared to non heat shocked cultures – was reflected in the model parameters. More specifically, the δ parameter (Figure 3) followed a systematic trend. In order to enable a more efficient comparison at different pH values, the authors represented the relationship between the scale parameter before (δ_0) and after (δ) heat shocks *vs* the shock time (Figure 3). Generally, as it can be seen in Figure 3, there is a clear increase of the δ parameter as the time of the heat shock increases, for all pH values under study. After heat shock the δ parameter increased by 1.1 fold at pH 7.4, 4 fold at pH 5.5 and 2.5 fold at pH 4.0 for *L. monocytogenes*, and 3.7, 3.2 and 1.5 fold respectively for *P. aeruginosa.*

Several researchers have studied the bacterial heat resistance at dynamic temperature profiles, i.e., non-isothermal conditions, as a function of the heating rate. For example, Juneja and Marks (2003) –as previously mentioned- studied the inactivation rates of a cocktail of *Salmonella spp.* in *sous vide* cooked beef for thermal processing under varying time-temperature profiles. Hassani et al. (2005, 2006) investigated and mathematically modelled the heat resistance of *Listeria monocytogenes* and *Staphylococcus aureus*, respectively, under non isothermal conditions.

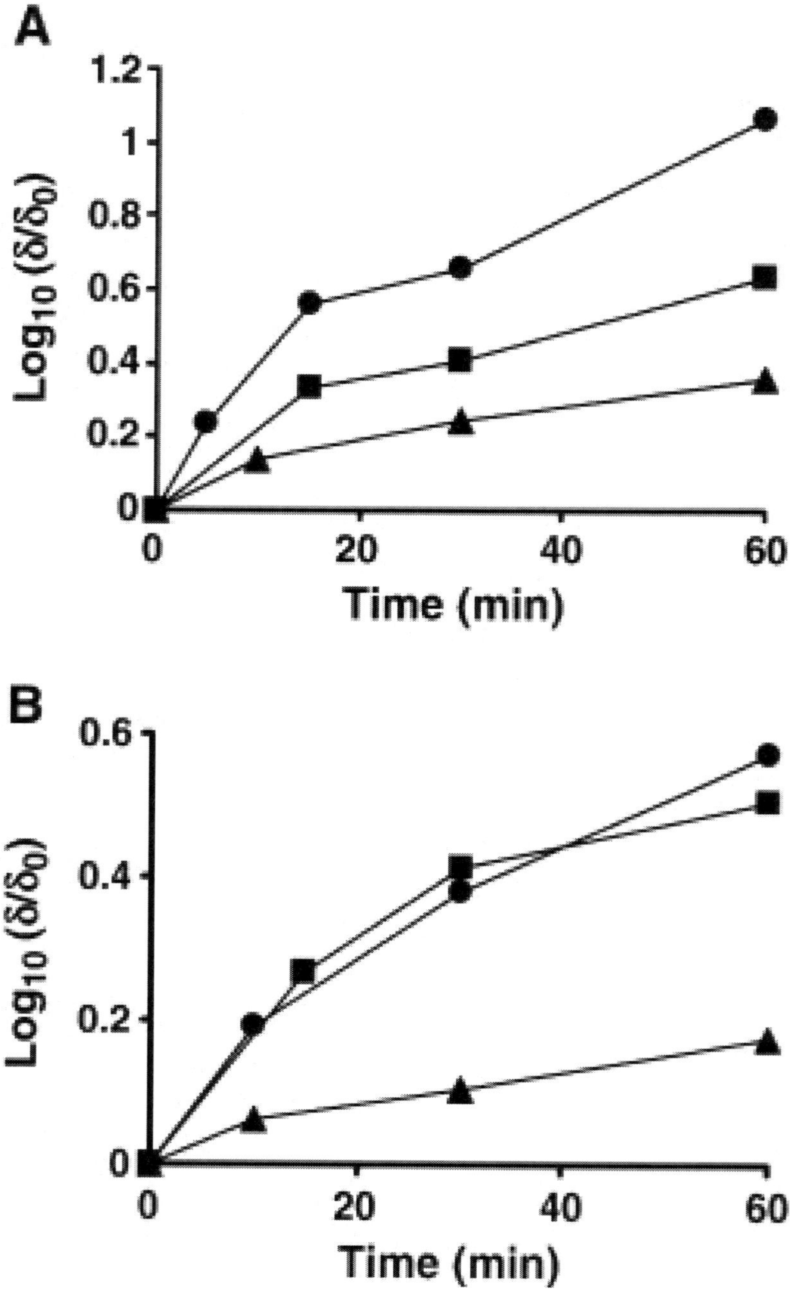

Figure 3. Influence of the heat-shock duration on the log(δ / δ_0) estimated from the fitting of the Mafart equation (Equation 1) the survival curves of *Listeria monocytogenes* (A) and *Pseudomonas aeruginosa* (B) obtained at pH 7.4 (●), 5.5 (■) and 4.0 (▲) (Reprinted with permission from Hassani et al (2007)).

Additionally, Valdramidis et al. (2006) evaluated the microbial dynamics of *Escherichia coli* K12 under dynamic conditions, i.e., heating from 30°C to 55°C with rates of 0.15, 0.20, 0.40, 0.55, 0.82 and 1.64°C/min. More specifically, as it can be seen in Figure 4, from the microbial data, for lower rates the inactivation is also slower indicating an increase of the microbial resistance. The dynamic model of Geeraerd et al. (2000) was used for the description of the microbial inactivation (Chapter 4).

A secondary model related to the microbial physiology under stressful conditions was further developed in the work of Valdramidis et al. (2007), enabling a more efficient description of the inactivation kinetics under non-isothermal conditions, under which bacterial heat adaptation occurs. The introduced factor is the $k_{<phys>}$ and it is related to the microbial physiology under stress conditions (Equation 2).

$$k_{(\langle phys \rangle)} = k_1 \cdot \frac{(dT / dt)}{k_2 + (dT / dt)} \tag{2}$$

In this model, $k_{(phys)}$ is the physiological state of the cells as influenced by the dynamic temperature environment, k_1, k_2 constants and dT/dt the heating rate. From this equation it can be derived that for lower heating rates the physiological state of the cell is lower.

When the microbial heat adaptation is taken into consideration, the inactivation model of Geeraerd et al. (2000) (Chapter 4) could be extended to the following mathematical expression (Equations (3A), (3B)).

$$\frac{dN}{dt} = -k_{max}(T) \cdot \left(\frac{1}{1 + C_c} \right) \cdot \left(1 - \frac{N_{res}}{N} \right) \cdot k_{(\langle phys \rangle)} \cdot N \tag{3A}$$

$$\frac{dC_c}{dt} = -k_{max}(T) \cdot C_c \tag{3B}$$

The above model structure could be summarized by the following equation.

$$\frac{dN_i}{dt} = k(N_i, \langle env \rangle, \langle phys \rangle) \cdot N_i \tag{4}$$

As it can be seen in Figure 4, microbial simulation of the population level taking into account the induced microbial resistance (Equation (3A)) leads to a more accurate description of the microbial dynamics (see thick lines) compared to the simulation without taking into account the induced thermotolerance (see dashed lines). It can, therefore, be concluded that at the level of mathematical modelling it is essential to take into consideration the microbial heat stress adaptation in order to accurately predict the microbial evolution under dynamic time temperature conditions.

Generally, understanding and further quantifying the induced bacterial thermotolerance under rising temperatures is of great importance for microbial safety issues in the food industry, especially since mild heat treatments are progressively replacing the conventional sterilization processes.

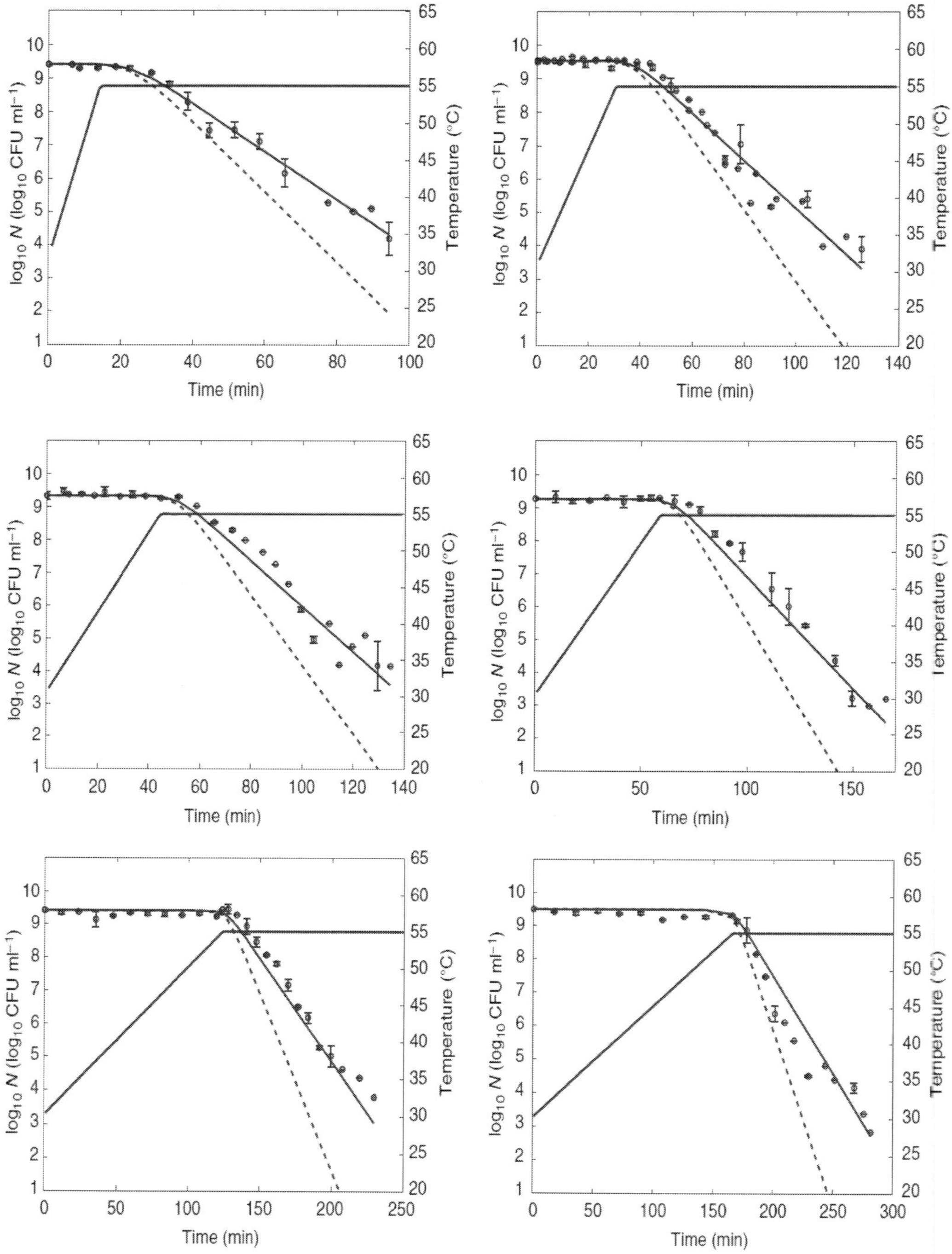

Figure 4. Microbial inactivation experiments *of Escherichia coli* K12 and corresponding temperature profiles for all studied heating rates. From left to right, first row: 1.64, 0.82°C/min, second row: 0.55, 0.40°C/min, third row: 0.20, 0.15 °C/min. Thick line: simulations by incorporating the factor $k_{(<phys>)}$ (see Equation (4)), dashed line predictions without taking into account the $k_{(<phys>)}$ factor (Reprinted with permission from Valdramidis et al. (2007)).

Acid Stress and Cross Protection

Definition of Acid Stress

Acid stress can be defined as the combined biological effect of low pH and/or weak (organic) acids present in the bacterial cellular environment. At lower pH levels *weak (organic) acids* are uncharged and they diffuse via the cellular outer membrane to the internal part of the cell, where they dissociate, leading to a decrease of the pH of the inner part of cell. The lower the external pH is, the higher the influx of organic acids in the cell. *Strong acids*, like HCl, lead to trafficking of the dissociated $[H^{+}]$ in the cell via the membrane, which results in an increase of the internal pH of the cell. Generally, the constant influx of protons in the cell leads to cellular death due to energy depletion (Foster 2004; Bearson et al. 1997).

The bacterial cells respond to acid stress either via the *Acid Tolerance Response* (ATR), i.e., production of acid shock proteins that assist the cellular survival at low pH levels and/or via increasing the proton efflux and/or via increasing the catabolism of amino acids and/or via the induction of enzymes which repair the cellular DNA (Yousef and Courtney 2003). Another way that the microbial cells may react towards the acid stress is by altering the composition of the cellular membrane's fatty acids. More specifically, they increase saturated fatty acids compared to the unsaturated fatty acids finally leading to an alternation of the membrane fluidity. It has been observed that this alternation has a protective role, not only towards the acid but also towards heat stress. For example -as previously mentioned- Brown et al. (1997) observed for five strains of *Escherichia coli* that when the bacteria were exposed to a low pH the non-saturated fatty acids were replaced by saturated ones and there was an increase of the acid resistance. Similarly, Alvarez-Ordonez et al. (2008, 2009) have seen that bacterial growth in an acidified medium leads to an increase of the heat resistance of *Salmonella* Typhimurium and *Salmonella* sentfenberg, respectively. The ratio of the fatty acids also changed as a result of the acid (and further heat) adaptation.

Generally, the extent of the acid tolerance response and the level of the further (induced) resistance is dependent –among other parameters such as the duration of exposure and the rate, on the bacterial strain as well as on the growth phase. For example, Buchanan and Doyle (1997) observed that a non-pathogenic *E. coli* culture is more acid resistant during the stationary phase compared to the log-phase growth phase. Gorden and Small (1993) stated that the pathogenic *E. coli* strains, i.e., enterohaemorrhagic, enteroinvasive and enteropathogenic, have a higher acid resistance compared to the non-pathogenic *E. coli* K12.

Cross Protection

Exposure to a specific environmental stress factor may lead to a certain level of stress adaptation to another kind of environmental stress. This is known as ***cross protection*** (Foster and Specter 1995; Juneja and Novak 2003). Understanding the mechanisms and the relation between heat and other stress factors, like acid stress, and further bacterial adaptation is of great importance for food safety issues, since during a food process or in a food product microorganisms are exposed to a combination of these factors and, therefore, may become stress adapted. For example, acids are used as preservatives in food products or they might be produced in a fermented product. Moreover, mild heat treatments or innovative techniques like high hydrostatic pressure may lead to a progressive heat adaptation of the –possibly present- microorganisms.

Several studies have illustrated that there is a clear relation between exposure to acid (or heat) stress and further heat (or acid) adaptation and resistance. For example, it has been observed that heat (or acid) shocking may increase the resistance of bacteria to heat, acid, ethanol and/or NaCl. More specifically, Leyer and Johnson (1993) studied the relation between acid adaptation and resistance to other environmental stresses for *Salmonella* Typhimurium and observed that acid adapted *S.* Typhimurium was resistant towards heat and salt. Carmelo et al. (1998) studied the effect of rapid acid stress on the modification of the thermotolerance of *Saccharomyces cerevisiae* and observed that loss of viability at 47°C was slower for cells pre-incubated for 10 min in medium of pH 3.5 compared to cells that were pre-incubated in medium of pH 6. Skandamis et al. (2008) indicated that there is an increased heat resistance of *Listeria monocytogenes* at 57°C after pre-exposure to heat or lactic acid. Moreover, the findings of Skandamis et al. (2009) indicated that there is an increase in the acid resistance of *Listeria monocytogenes* after growth in presence of NaCl and/or high acidity. Tetteth and Beuchat (2003) studied the effect of acid stress and heat shock on the acid tolerance of *Shigella flexneri* and concluded that heat shock leads to a higher acid tolerance.

Skandamis et al. (2008) studied and further quantified the adaptive response of *Listeria monocytogenes* to heat (at 52°C, 57°C and 63°C) after exposure to combinations of osmotic (10% NaCl), acidic (pH 5.0 with HCl) and thermal (46 °C) stresses via the Weibull-type model of Mafart et al. (2002). All treatments induced bacterial thermotolerance at 57°C –due to cross protection- and this was revealed in the values of the model parameters, i.e. an increase in the δ parameter of Equation (1).

More recently, Velliou et al. (2011a) investigated the effect of exposure in an acidified broth –shortly before and during heat inactivation- on the heat resistance of *E. coli* K12 MG1655 cells at 54°C and 58°C and concluded that the presence of acids triggered the bacterial cells leading to an increase of their thermal resistance. The acidification of the broth took place with 3 different acids, i.e., acetic, lactic and hydrochloric. For all types of acids under study, different levels of acidification of the medium were tested, i.e. pH 6, 5.5 and 5. A typical microbial curve of that study included a shoulder phase and a linear part. Therefore, the inactivation model of Geeraerd et al. (2000) was fitted to the experimental data (Figure 5).

$$N(t) = \left(N(0) - N_{res}\right) \cdot \exp(-k_{\max} \cdot t) \cdot \left(\frac{\exp(k_{\max} \cdot S\iota)}{1 + \left(\exp\left(k_{\max} \cdot S\iota\right) - 1\right) \cdot \exp\left(-k_{\max} \cdot t\right)} \right) + N_{res} \tag{5}$$

In this model, N [CFU/mL] is the cell population, $N(0)$ [CFU/mL] the initial cell population, N_{res} [CFU/mL] the stress resistant subpopulation, k_{max} [1/min] the maximum specific inactivation rate, Sι [min] the shoulder period and t [min] the time (see also Chapter 4).

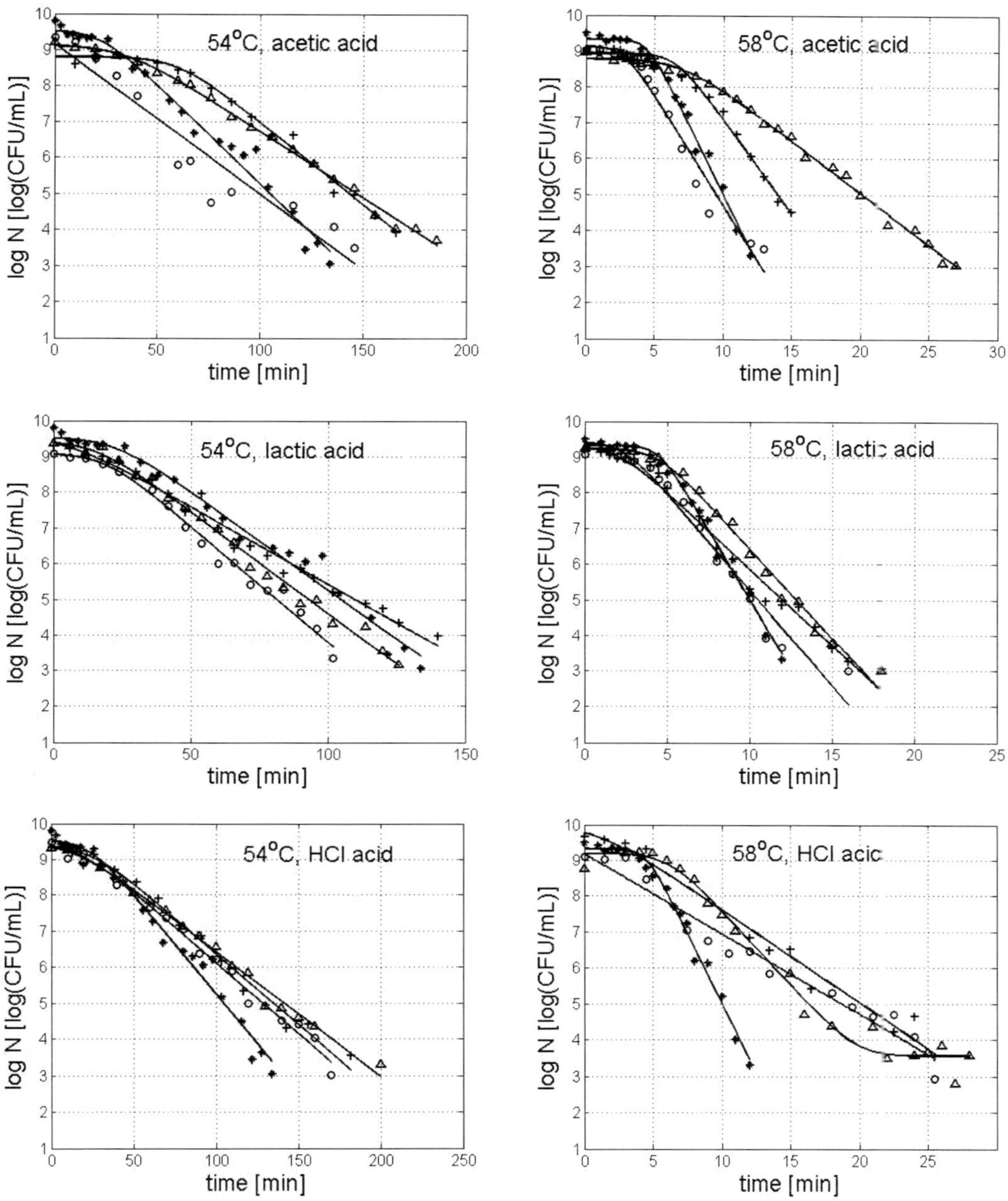

Figure 5. Fits of Geeraerd et al. (2000) inactivation model on the thermal inactivation data for *Escherichia coli* K12 at 54°C and 58°C for acetic, lactic and HCl acid: (*) unmodified BHI broth(pH≈7.5) ,(Δ) pH 6.0, (+) pH 5.5, and (o) pH 5.0 (Reprinted with permission from Velliou et al. 2011a).

Heat resistance was defined as a prolongation of the shoulder (S_l) and/or a reduction of the inactivation rate (k_{max}) and/or the occurrence of a tail after the log-linear part of the inactivation. A change in one or more of these parameters is reflected as an increase/decrease of the time for the fourth decimal reduction (t_{4D}) which was also calculated in order to enable a more efficient comparison between different treatments. Generally, in this study it is observed that acid treatment (before and during heat inactivation) leads to an increase of *E*.

coli heat resistance. The level of the resistance depends on the type of acid used, i.e., weak organic acids induced less heat resistance compared to the strong organic acid and on the quantity added, i.e., different levels of acidification (different pH values of the broth) lead to a different level of heat resistance.

Buchanan et al. (1993) studied and quantitatively determined the effect of pH and increased concentrations of acetic and lactic acid on the heat resistance of *Listeria monocytogenes*. The modelling approach allows the differentiation of the effects of pH and acid concentration. The inactivation curves consisted of a shoulder and a log-linear phase and for the quantification of the inactivation kinetics Equation (6) was applied.

$$Y = Y_0 \qquad [t<t_L]$$

$$Y = Y_0 + s(t - t_L) \qquad [t\geq t_L] \tag{6}$$

In Equation (6), Y [log(CFU/mL)] is the log count of bacteria at time t, Y_o [log(CFU/mL)] the log cont of bacteria at time 0, s [(log(CFU/mL))/h] the slope of the survival curve, t the time [h] and t_L [h] the duration of the lag period (shoulder) before the initiation of the inactivation.

Increased or decreased thermotolerance of *Listeria* could be defined via the value of the shoulder period (t_L), the D-value as well as the value of the time for four decimal reductions (t_{4D}) (Equation 7).

$$t_{4D} = t_L + (4 * D) \tag{7}$$

with t_{4D} [h] the tine for a 4-D(99.99%) inactivation.

Generally, it was concluded that, at a specific pH level, increase of the concentrations of lactic or acetic acid leads to a reduced thermotolerance of *Listeria*, i.e., lower values of the t_L, t_{4D} and D-value. For the pure pH effect, it was shown that, when decreasing the pH of the medium from 7 to 3 there is a progressive decrease of the bacterial heat resistance, i.e. lower values of the t_L, t_{4D} and D-value.

Miller et al. (2009) studied the influence of pH on the thermal inactivation of *Listeria innocua* and quantified mathematically the combined effect of pH and temperature on the bacterial thermotolerance. The effect of combinations of these environmental factors on the heat resistance of *Listeria innocua* has been modelled, based on a Bigelow type equation by Mafart and Leguerinel (1998) and Gaillard et al. (1998) (Equation (8))

$$\log D_{(T,pH)} = \log D_{(T^*,pH^*)} - \frac{T - T^*}{Z_T} - \frac{(pH - pH^*)^2}{Z^2_{pH}} \tag{8}$$

with T^* the reference temperature, pH^* the pH of maximal heat resistance, Z_T the conventional thermal Z-value, Z_{pH} the distance of pH from pH^* which leads to a 10 fold reduction of D-value, and $D_{(T^*, pH^*)}$ the D-value at T^* and pH^*.

It was concluded that the combination of pH and temperature crucially affects the loglinear part, i.e., the inactivation rate or D-value, as well as the shoulder phase of the inactivation.

The Role of Osmolytes on the Bacterial Heat Resistance

Osmolytes

As previously mentioned, heat adaptation and further tolerance may result from exposure to sub(lethal) temperatures. This exposure leads to the production of Heat Shock Proteins (HSP), which on the one hand repair damaged cell proteins and on the other hand digest permanently damaged proteins (Ohtsuka et al. 2007; Juneja and Novak 2003).

Several studies show the presence of a synergistic relation between the function of HSPs and *osmolytes*, i.e., small molecules like Trimethylamine, glycerol and trehalose, that act as chemical chaperones. More specifically, *osmolytes* may have a regulatory effect on the activity of HSP by assisting local folding in chaperone-bound polypeptides and by stabilizing the native end product of the reaction, i.e., the re-folded proteins, via the solvophobic/osmophobic effect. As a result, the presence of osmolytes may improve the intracellular protein stability which leads to increased microbial (heat) stress resistance (Diamant et al. 2001; Caldas et al. 1999).

The Effect of Trimethylamive-N-oxide (TMAO) on the Bacterial Heat Resistance

Trimethylamine-N-Oxide (TMAO) is a well known osmolyte that is naturally produced by fish (Debevere et al. 2001) and it is (bio)synthesized via the conversion of Trimethylamine to the N-Oxide form (Baker et al. (1962). Several studies have illustrated that TMAO is one of the most effective osmolytes, regarding its contribution to protein stabilization. Unfolded proteins are forced to fold by the TMAO due to its solvophobic effect on the peptide backbone which is exposed in the unfolded state. As a consequence, the unfolded state of the protein is very unfavourable compared to the folded state, which finally leads to protein folding (Bolen et al. 2004; Zou et al. 2002; Perlmutter 2002).

Gur et al. (2002) and Van Derlinden et al. (2008) studied and observed the positive effect of TMAO on the growth dynamics of *Escherichia coli* K12. More specifically, Gur et al. (2002) observed that the addition of TMAO (0.3 mol l^{-1}) resulted in growth of *E. coli* at 46°C whereas plain Brain Heart Infusion broth (BHI) did not support growth. Van Derlinden et al. (2008) observed disturbed growth curves at temperatures close to the maximum growth temperature, i.e, 45°C and 46°C, in plain BHI. The addition of 0.3 mol l^{-1} TMAO resulted in approximately sigmoid growth curves. However, adding a higher concentration (0.6 mol l^{-1}) had an inhibitory effect.

Velliou et al. (2011b) studied and quantified the influence of TMAO on the heat resistance of *Escherichia coli* K12 MG1655 cells at static temperatures. For this purpose, stationary phase bacterial cells were inactivated in BHI which contained different concentrations of TMAO, i.e., 0.5, 1, 2 mol l^{-1}, at 52°C, 54°C and 58°C. The mathematical model of Geeraerd et al. (2000) (Equation (5)) was fitted to the experimental data. Heat resistance was quantified as reduction of the inactivation rate k_{max}, and/or an increase of the time for one decimal reduction D and/or an increase of the time for the fourth decimal reduction t_{4D} (Equation (7)). Generally, it was concluded that the bacterial heat tolerance increased at all studied temperatures, due to the protective effect of the osmolyte TMAO.

Stress and Single Cell Variability and Viability

Primary and secondary predictive models are usually developed regarding the microbial populations as homogeneous systems. Due to this assumption, every cell is defined by approximately equal kinetic characteristics, which are reflected in the overall population behavior. This *black box* approach enables the accurate description of growth and inactivation in liquid systems under moderate environmental conditions, e.g., by the dynamic growth model of Geeraerd et al. (2000). However, when microorganisms are exposed to a moderate stress factor, e.g., heat, acid, salts the cell density curves can deviate from the expected sigmoid patterns and the classical models and modelling approaches are no longer valid. The deviation of these curves may be a result of the microbial population heterogeneity. This heterogeneity is induced by the stressing environment.

In the field of microbial inactivation, the co-occurrence of two subpopulations was claimed to explain the tailing in inactivation curves with respect to, e.g., temperature (Humpheson et al. 1998), antimicrobial agents (Dhar et al. 2007, Nikolaou et al. 2006, Olofsson et al. 2005), osmotic stress (den Besten et al. 2007) and pressure (Noma et al. 2006). After an initial shoulder period, in which the bacteria are possibly affected but not yet killed, inactivation starts. The largest fraction of the inoculated cells is severely damaged and inactivates rather fast. A smaller proportion, however, is more resistant and is not inactivated fast, or at a much lower rate. Tailing in inactivation curves results from the survival of the latter, more stress resistant subpopulation. Based on the assumption of the two co-existing subpopulations, Whiting et al. (1993) developed a logistic-based inactivation model that could describe inactivation curves with shoulder and tailing phenomena. Analogous heterogenous modelling approaches were applied by, e.g., Fujikawa and Itoh (1996) and Geeraerd et al. (2000).

Although recognized to be significant, only few attempts have been made to include population heterogeneity in the modelling of microbial inactivation kinetics. For example, Nikolaou et al. (2006) constructed a mathematical framework in order to efficiently describe the effect of antimicrobial agents on heterogeneous microbial populations of distributed antimicrobial resistance

$$\frac{dN(t)}{dt} = \underbrace{k_g \cdot N_{(t)}}_{\substack{\text{physiological}\\ \text{growth rate}}} - \underbrace{r\left(C_{(t)}\right)N_{(t)}}_{\substack{\text{kill rate}\\ \text{due to agent}}},\ N_{(0)} = N_0 \tag{9}$$

with N(t) the cell count, K_g the growth rate and $r(C_{(t)})$ the kill rate due to the antimicrobial concentration $C(t)$.

Tam et al. (2007) mathematically modelled the microbial response of *Pseudomonas aeruginosa* to antimicrobial agents via an equation which determined the rate of change of drug concentration, microbial susceptibility and microbial burden of the surviving (heterogeneous) population over time.

$$\frac{dN(t)}{dt} = k_g \cdot \left[1 - \frac{N(t)}{N_{max}}\right] \cdot N(t) - \left(\frac{C(t)^H \cdot K_k}{C(t)^H + \left[C_{50keff}(t)\right]^H}\right) \cdot N(t) \qquad (10)$$

with $C(t)$, the concentration of the antimicrobial agent, C_{50keff}, the concentration of the antimicrobial agent essential to achieve 50% of maximal kill rate of the bacterial population, H, the sigmoidicity constant for the bacterial population and K_k, the maximum kill rate constant for bacterial population.

Conclusion

Monitoring the survival of microorganisms in food systems is of great importance for food safety issues. Generally, microbial response and adaptation to one or more environmental factors (via cross protection) may occur on the one hand during emerging industrial food processes at which conditions are less harsh and, on the other hand at several points in the food chain, i.e., during distribution, transportation, storage. Therefore, understanding the mechanisms of cellular stress adaptation and further tolerance to stresses/hurdles usually applied in food processing, i.e., temperature, acidification, exposure to salt or chemical preservatives, etc., is crucial for constructing efficient predictive models that can be directly applicable in the food industry. As illustrated in this chapter, several researches have studied and mathematically quantified the effect of pre-exposure to a specific environmental factor, i.e., heat, acids, presence of chemical components such as osmolytes, on the heat resistance of a myriad of microorganisms. In most of these studies the main factor measured is the microbial survival via viable cell counts. However, measurements of the alternation of specific intracellular components, i.e., cellular proteins, during exposure and adaptation to the stressing environmental factor is essential for an in depth understanding of the mechanisms of microbial stress adaptation. Moreover, the incorporation of intracellular information and/or intracellular mechanisms in predictive models in a top-down systems biology approach may lead to mechanistic models that are even more generally applicable.

References

Alvarez-Ordonez, A., Fernandez, A., Lopez, M., Arenas, R., and Bernardo, A. (2009) Relationship between membrane fatty acid composition and heat resistance of acid and cold stressed *Salmonella senftenberg* CECT 4384. *Food Microbiology* 26, 347-353.

Alvarez-Ordonez, A., Fernandez, A., Lopez, M., Arenas, R., Bernardo, and A. (2008) Modifications in membrane fatty acid composition of *Salmonella* Typhimurium in response to growth conditions and their effect on heat resistance. *Food Microbiology* 123, 212-219.

Baker, J.R. and Chaykin, S. (1962) The biosynthesis of Trimethylamine-N-Oxide. *Journal of Biological Chemistry* 237, 1309-1313.

Bearson, S., Bearson, B., and Foster, J.W. (1997) Acid stress responses in enterobacteria. *FEMS Microbiology Lettters* 147, 173-180.

Bigelow, W.D. (1921) The logarithmic nature of thermal death time curves. *Journal of Infectious Diseases* 27, 528–536.

Bigelow, W.D., and Esty, J.R. (1920) The thermal death point in relation to typical thermophylic organisms. *Journal of Infectious Diseases 27, 602-607.*

Bolen, D.W. (2004) Effects of naturally occurring osmolytes on protein stability and solubility: issues important in protein crystallization. *Methods* 34, 312-322.

Brown, J.L., Ross, T., McMeekin, A., and Nichols, P.D. (1997) Acid habituation of *Escherichia coli* and the potential role of cyclopropane fatty acids in low pH tolerance. *International Journal of Food Microbiology* 37, 163-173.

Buchanan, R.L., Golden, M.H., and Whiting, R.C. (1993) Differentiation of the effects of pH and lactic or acetic acid concentration on the kinetics of *Listeria monocytogenes* inactivation. *Journal of Food Protection* 56(6), 474-478.

Buchanan, R.L., and Doyle, M.P. (1997) Foodborne disease significance of *Escherichia coli O157:H7* and other enterohemorrhagic *E.coli. Food Technology* 51, 69-76.

Caldas, T., Demont-Caulet, N., Ghazi, A., and Richarme, G. (1999) Thermoprotection by glycine betaine and choline. *Microbiology* 145, 2543-2548.

Carmelo, V., Santos, R., Viegas, C.A., and Sa-Correia, I. (1998) Modification of Saccharoyces cerevisiae thermotolerance following rapid exposure to acid stress. *International Journal of Food Microbiology* 42, 225-230.

Cebrian, G., Sagarzazu, N., Pagan, R., Condon, S., and Manas, P. (2007) Resistance of *Escherichia coli* grown at different temperatures to various environmental stresses. *Journal of Applied Microbiology* 105, 271-278.

Cebrian, G., Condon, S., and Manas, P. (2009) Heat-adaptation induced thermotolerance in *Staphylococcus aureus*: Influence of the alternative factor sigma(B). *International Jornal of Food Microbiology* 135, 274-280.

Debevere, J., Devlieghere, F., Van Sprundel, P., and De Meulenaer B. (2001) Influence of acetate and CO_2 on the TMAO-reduction reaction by *Shewanella baltica. International Journal of Food Microbiology* 68, 115-123.

den Besten, H.M.W., Ingham, C.J., van Hylckama Vlieg, J.E.T., Beerthuyzen, M.M., Zwietering, M.H., and Abee, T. (2007) Quantitative analysis of population heterogeneity of the adaptive salt stress response and growth capacity of *Bacillus cereus* ATCC 14579. *Applied and Environmental Microbiology,* 73, 4797-4804.

Dhar, N., and McKinney, J.D. (2007) Microbial phenotypic heterogeneity and antibiotic tolerance. *Current Opinion in Microbiology* 10, 30-38

Diamant, S., Eliahu, N., Rosenthal, D., and Goloubinoff, P. (2001) Chemical chaperones regulate molecular chaperones *in vitro* and in cells under combined salt and heat stresses. *Journal of Biological Chemistry* 276, 39586-39591.

Foster, J.W. (2004) *Escherichia coli* acid resistance: Tales of an amateur acidophile, *Nature reviews* 2, 898-907.

Foster, J.W., and Specter, M.P. (1995) How *Salmonella* survive against the odds. *Annual Review of Microbiology* 49, 145-174.

Fujikawa, H., and Itoh, T. (1996) Tailing of thermal inactivation curve of *Aspergillus niger* spores. *Applied and Environmental Microbiology* 62(10), 3745-3749.

Gaillard, S., Lequerinel, I., and Mafart, P. (1998) Modelling combined effects of temperature and pH on the heat resistance of spores of *Bacillus cereus. Food Microbiology* 15, 625-630.

Geeraerd, A.H., Herremans, C.H., and Van Impe, J.F. (2000) Structural model requirements to describe microbial inactivation during a mild heat treatment. *International Journal of Food Microbiology* 59, 185-209.

Gorden, J., and Small P.L.C. (1993) Acid resistance in enteric bacteria. *Infection Immunity* 61, 364-367.

Gur, E., Biran, D., Gazit, E., and Ron, .E.Z. (2002) *In vivo* aggregation of a single enzyme limits growth of *Escherichia coli* at elevated temperatures. *Molecular Microbiology* 46, 1391-1397.

Hassani, M., Manas, P., Pagan, R., and Condon, S. (2007) Effect of a previous shock on the thermal resistance of *Listeria monocytogenes* and *Pseudomonas aeruginosa* at different pHs. *International Journal of Food Microbiology* 116, 228-238.

Hassani, M., Manas, P., Condon, S., and Pagan, R. (2006) Predicting heat inactivation of *Staphylococcus aureus* under nonisothermal treatments at different pH. *Molecular and Nutritional Food Research* 50, 572-580.

Hassani, M., Manas, P., Raso, J., Condon, S., and Pagan, R. (2005) Predicting heat inactivation of *Listeria monocytogenes* under nonisothermal treatments. *Journal of Food Protection,* 68, 736-743.

Humpheson, L., Adams, M.R., Anderson, W.A., and Cole, M.B. (1998) Biphasic thermal inactivation kinetics in *Salmonella enteritidis* PT4. *Applied and Environmental Microbiology* 64(2),459-464.

Juneja, V.K., and Marks, H.M. (2003) Characterizing asymptotic D-values for *Salmonella* spp. subjected to different heating rates in sous-vide cooked beef. *Innovative Food Science and Emerging Technologies* 4, 395-402.

Juneja, V.K., and Novak, J.S. (2003) Adaptation of foodborne pathogens to stress from exposure to physical intervention strategies. In: Yousef, A.E., and Juneja, V.K. (eds.) *Microbial Stress Adaptation and Food Safety*, p.159-211, CRC Press, Boca Raton.

Juneja, V.K., Eblen, B.S., and Marks, H.M. (2001). Modelling nonlinear survival curves to calculate thermal inactivation of *Salmonella* in poultry of different fat levels. *International Journal of Food Microbiology* 70, 37 –51

Lee, S.Y., and Kang, D.H. (2009) Combined effects of heat, acetic acid and salt for inactivating *Escherichia coli* O157:H7 in laboratory media. *Food Control* 20, 1006-1012.

Leyer, G.J., and Johnson, E.A. (1993) Acid adaptation induces cross-protection against environmental stresses in *Salmonella* Typhimurium. *Applied and Environmental Microbiology* 59, 1842-1847.

Mafart, P., and Leguerinel, I. (1998) Modelling combined effects of temperature and pH on the heat resistance of spores by a linear- Bigelow equation. *Journal of Food Science* 63, 6-8.

Mafart, P., Couvert, O., Gaillard, S., and Leguerinel, I. (2002) On calculating sterility in thermal preservation methods : application of the Weibull frequency distribution model. *International Journal of Food Microbiology* 72, 107-113.

Miller, F.A., Ramos, B., Gil, M.M., Brandao, T.R.S., Teixeira, P., and Silva, C.L.M. (2009) Influence of pH, type of acid and recovery media on the thermal inactivation of *Listeria innocua*. *International Journal of Food Microbiology* 133, 121-128.

Nikolaou, M., and Tam, V. (2006) A new modelling approach to the effect of antimicrobial agents on heterogeneous microbial populations. *Journal of Mathematical Biology* 52, 154-182.

Noma, S., Kajiyhama, D., Noriyuki, I., Shimoda, M. and Hayakawa, I. (2006) Mechanisms behind tailing in the pressure inactivation curve of a clinical isolate of *Escherichia coli* O157:H7. *International Journal of Food Microbiology* 109, 103-108.

Ohtsuka, K., Kawashima, D. and Asai, M. (2007) Dual functions of heat shock proteins: molecular chaperones inside the cell and danger signals outside of cells. *Thermal Medicine* 23, 11-22.

Olofsson, S.K., Geli, P., Andersson, D. I., and Cars, O. (2005) Pharmacodynamic model to describe the concentration-dependent selection of cefotaxime-resistant *Escherichia coli* antimicrobial agents. *Chemotherapy* 49, 5081-5091.

Perlmutter, D.H. (2002) Chemical Chaperones: A pharmacological strategy for disorders of protein folding and traffcking, *Pediatric Research* 52(2), 832-836.

Rowburry, R.J., and Goodson, M. (1999) An extracellular stress-sensing protein is activated by heat and u.v. irradiation as well as by mild acidity, the activation producing an acid tolerance-inducing protein. *Letters of Applied Microbiology* 29, 10-14.

Rowburry, R.J. (2001) Cross- talk involving extracellular sensors and extracellular alarmones gives early warning to unstressed *Escherichia coli* of impending lethal chemical stress and leads to induction of tolerance responses. *Journal of Applied Microbiology* 90, 677-696.

Rowburry, R.J. (2003) Physiology and molecular basis of stress adaptation, with particular reference to the subversion of stress adaptation and to the involvement of extracellular components in adaptation. In: Yousef, A.E. and Juneja, V.K. (eds.) *Microbial Stress Adaptation and Food Safety*, p.159-211, CRC Press, Boca Raton.

Skandamis, P. N., Yoon, Y., Stopforth, D., Kendall, P. A., and Sofos, J. N. (2008) Heat and acid tolerance of *Listeria monocytogenes* after exposure to single and multiple sublethal stresses. *Food Microbiology* 25, 294-303.

Skandamis, P. N., Stopforth, J. D., Yoon, Y., Kendall, P. A., Sofos, J. N. (2009). Heat and acid tolerance responses of *Listeria monocytogenes* as affected by sequential exposure to hurdles during growth. *Journal of Food Protection* 72, 1412-1418.

Tam, V., Schilling, A.N., Poole, K., and Nikolaou, M. (2007) Mathematical modelling response of *Pseudomonas aeruginosa* to meropenem. *Journal of Antimicrobial Chemotherapy* 60, 1302-1309.

Tetteth, G.L., and Beuchat, R. (2003) Exposure of *Shigella flexneri* to acid stress and heat shock enhances acid tolerance. *Food Microbiology* 20, 179-185.

Yousef, A. E., and Courtney, P.D. (2003) Basics of stress adaptation and implications in new-generation foods. In: Yousef, A.E. and Juneja, V.K. (eds.) *Microbial Stress Adaptation and Food Safety*, p.1-30, CRC Press, Boca Raton.

Valdramidis, V.P., Geeraerd, A.H., Bernaerts, K., and Van Impe, J. F. (2006) Microbial dynamics versus mathematical model dynamics: The case of microbial heat resistance induction. *Innovative Food Science.& Emerging Technologies* 7, 80-87.

Valdramidis, V.P., Geeraerd, A.H., and Van Impe, J.F. (2007) Stress-adaptive responses by heat under the microscope of predictive microbiology. *Journal of Applied Microbiology* 103, 1922-1930.

Van Derlinden, E., Bernaerts, K., and Van Impe, J.F. (2008) Dynamics of *Escherichia coli* at elevated temperatures: effect of temperature history and medium. *Journal of Applied Microbiology* 104, 430-453.

Velliou, E.G., Van Derlinden, E., Cappuyns, A.M., Nikolaidou, E., Geeraerd, A.H., Devlieghere, F., Van Impe, J. (2011a) Towards the quantification of the effect of acid treatment on the heat tolerance of *Escherichia coli* K12 at lethal temperatures. *Food Microbiology* 28, 702-711.

Velliou, E.G., Van Derlinden, E., Cappuyns, A.M., Aerts, D., Nikolaidou, E., Geeraerd, A.H., Devlieghere, F., and Van Impe, J.F. (2011b). Quantification of the influence of Trimethylamine-N-Oxide (TMAO) on the heat resistance of *Escherichia coli* K12 at lethal temperatures. *Letters in Applied Microbiology* 52(2), 116-122.

Wesche, A.M., Gurtler, J.B., Marks, B.P., and Ryser, E.T. (2009) Stress, sublethal injury, resuscitation and virulence of bacterial foodborne pathogens. *Journal of Food Protection* 72(5), 1121-1138.

Whiting, R. (1993) Modelling bacterial survival in unfavorable environments. *Journal of Industrial Microbiology* 12, 240-246.

Yousef, A.E., and Courtney., P.D. (2003) Basics of stress adaptation and implications in new-generation foods. In: Yousef, A.E., and Juneja, V.K. (eds.) *Microbial Stress Adaptation and Food Safety*, p.159-211, CRC Press, Boca Raton.

Zou, Q., Bennion, B.J., Daggett, V., and Murphy, K.P. (2002) The molecular mechanism of stabilization of proteins by TMAO and its ability to counteract the effects of urea. *Journal of the American Chemical Society* 124(7), 1 192-1202.

In: Progress on Quantitative Approaches in Thermal … ISBN: 978-1-62100-842-2
Editors: V. P. Valdramidis and J. F. M. Van Impe © 2012 Nova Science Publishers, Inc.

Chapter 9

MODELS OF THERMAL INACTIVATION ACCOUNTING FOR MEMORY EFFECTS

Nirupama Vaidya*[1]*, Carlos M. Corvalan*[1]*,
***Georgios N. Yannakakis*[2] *and Vasilis P. Valdramidis*[3]**
[1]Transport Phenomena Laboratory, Department of Food Science,
Purdue University, West Lafayette IN, US
[2]IT University of Copenhagen, Centre for Computer Games Research,
Rued Langgaards Vej 7, Copenhagen S, Denmark
[3]University of Malta – Faculty of Health Sciences, Department of Food Studies
and Environmental Health, Mater Dei Hospital, Msida MSD 2080, Malta

Abstract

This chapter presents different memory models as a useful approach for describing both isothermal and nonisothermal inactivation dynamics in systems that are governed by non-exponential inactivation patterns. The memory models are either built based on the generalization of the conventional inactivation kinetics with the generalization process described by a memory integral or on the construction of Artificial Neural Network (ANN) structures that embed memory with the use of multi-step ahead Feed Forward with delay and autonomous multi-step ahead FFD networks. These generalized approaches may lead to new insights into predictive microbiology with a new appreciation for the importance of considering memory effects.

1. Introduction

This chapter is concerned with the modelling of thermal inactivation processes that deviate from the classical exponential pattern and thus cannot be described by the first-order kinetics conventionally used in food technology. This anomalous non-exponential behaviour is ubiquitous to a diverse number of food related microbial systems, and is detected experimentally by a nonlinear time dependence of the semilogarithmic survival curves (Geeraerd et al., 2000; van Boekel, 2001; Peleg, 2006).

An anomalous inactivation process reflects the breakdown of the Markovian nature of the exponential probabilistic distribution underlying the first-order kinetics caused by broad underlying probabilistic distributions (Peleg and Cole, 1998; Mafart et al., 2002) or by memory effects. By memory effects we refer here to non-local temporal phenomena, such as cumulative cell damage or enhanced cell tolerance, induced by past thermal stresses. This suggests that models to describe non-exponential thermal inactivation may benefit from explicitly taking memory into account.

In what follows, memory models are presented as a useful approach for the description of both isothermal and nonisothermal inactivation dynamics in systems governed by non-exponential (anomalous) inactivation patterns. First we show a family of models that arises naturally as straightforward generalization of the conventional inactivation kinetics with the generalization process described by a memory integral. Then, Artificial Neural Network (ANN) structures that can embed memory are discussed as alternative models for performing predictions under non-isothermal temperature environments. These generalized approaches may lead to new insights into predictive microbiology with a new appreciation for the importance of considering memory effects.

The material in this chapter updates and extends previous work by the present authors (Vaidya and Corvalan, 2009; Valdramidis et al., 2009) in several ways, including (i) an expanded discussion of the stretched exponential inactivation, its generalization as the Mittag-Leffler inactivation, and estimation of memory parameters from nonisothermal inactivation data for the memory-integral approach, and (ii) presenting different structures of ANN capable of embedding memory, assessing the effect of time dependency for predicting microbial populations and proposing the use of autonomous ANN structures for off-line microbial studies for the ANN approach. Most of the motivation and theory is discussed in Sections 2 and 3, the memory-integral model is discussed in Sections 4, while a structure based on the development of ANN is presented in Section 5.

2. Conventional Kinetics and Lack of Memory Property

The conventional first-order inactivation kinetics describes thermal processes with constant risk of inactivation accurately. However, it fails to produce more than weak approximations when the cumulative effect of past thermal stresses on the risk of inactivation attains a dominating influence as a consequence of memory effects.

Conventional Inactivation Kinetics

The decrease in the survival ratio $S(t)$ of a microbial population during a thermal process is conventionally described in food technology assuming the Bigelow model of first-order kinetics

$$\frac{dS(t)}{dt} = -k\,S(t) \quad , \tag{1}$$

with the initial condition $S(0)=1$ (Frazier and Westhoff, 1988; Adams and Moss, 2000; Lewis and Heppell, 2000). The time-independent rate constant k can be regarded as the magnitude of the thermal stress induced by the process temperature.

In the case of isothermal inactivation, the first-order kinetic model yields the classical exponential Debye pattern in the survival ratio:

$$\log S(t) = -t/\tau \, , \tag{2}$$

where $\tau = 1/k$ provides the natural time scale for the inactivation process.

In the following sections we will discuss modelling approaches that extend the description of thermal inactivation beyond the limits of the Markovian first-order kinetics model of Equation (1) by taking into account memory effects. The motivation for incorporating memory as a unifying concept in the treatment of thermal inactivation begins with the recognition that the exponential distribution of Equation (2) is the only survival distribution with lack of memory.

Lack of Memory Property

There is a rough and a more precise answer regarding what the lack of memory property really means, and how it relates to the first-order kinetics. In the rough answer, we say that a survival distribution lacks memory if the probability of cell inactivation at the present time does not change regardless of how long the cell has been exposed to the thermal treatment. In this sense, the first-order kinetic model lacks memory because it describes microbial populations whose risk of inactivation $h(t) = -\left(dS(t)/dt\right)/S(t)$ depends at any time t solely on the thermal stress acting at that instant only, and not on the elapsed processing time.

For a more precise answer we rely on the same core idea as in the rough answer, but we say that a survival function lacks memory if the probability of surviving an additional treatment time t_2 given survival beyond a processing time t_1 does not depend on the past time t_1, and we formalize this statement with the equation: $P(t_1 + t_2 \mid t_1) = P(t_2)$ (Feller, 1967; Yanushkevichius, 2005).

The lack of memory property $P(t_1 + t_2 \mid t_1) = P(t_2)$ can be brought into the equivalent form $P(t_1 + t_2) = P(t_1)\, P(t_2)$ via the definition of conditional probability $P(t_1 + t_2 \mid t_1) = P(t_1 + t_2)/P(t_1)$ and, since the exponential function $\exp(-t/\tau)$ satisfies:

$$P(t_1 + t_2) = \exp[-(t_1 + t_2)/\tau] = \exp(-t_1/\tau)\exp(-t_2/\tau) = P(t_1)\, P(t_2), \tag{3}$$

it follows that the exponential survival function of Equation (2) is 'memoryless' (Balakrishnan and Basu, 1996; Alho and Spencer, 2005). For a more rigorous presentation of the lack of memory property, see Yanushkevichius (2005).

Moreover, the exponential distribution (2) underlying the first-order kinetics is the only distribution characterized by lack of memory property, in the sense that if a given (continuous) survival function satisfies $P(t_1 + t_2 \mid t_1) = P(t_2)$, then it has the distribution

$\exp(-t/\tau)$ (Huang, 1981). This leads to the simple yet essential conclusion that a unifying memory effect is implicit in *all* but the conventional first-order inactivation model.

3. Thermal History and Risk of Inactivation

We argue here that in the natural world, cells may not be the simple forgetful entities they appear to be under the log-linear survival hypothesis. It seems they have a 'memory', in the sense that after a thermal stress has been applied, their survival depends on the thermal history.

Influence of the Thermal History on Cell Survival

At the microscopic level, cells are complex entities for which thermal inactivation is not a single instantaneous event. Some microorganisms suffer a cascade of heat-induced lesions when exposed to thermal stresses that gradually makes them more susceptible to further stress (Watson, 1990), whereas others seem to have an inherent capacity to repair sublethal damage and become more tolerant to stressful conditions (Farber et al., 1990; Juneja et al., 1998; Arsene et al., 2000). Precisely how this occurs is not completely understood (Tanguay, 1983; Schlesinger, 1986) and may reflect multiple factors including synthesis of specialized proteins, such as the chaperon heat shock proteins that assist in the refolding of denatured proteins (Craig, 1985; Lindquist, 1986; Farber and Brown, 1990; Juneja et al., 1998) or proteases that not only degrade the denatured proteins but are also essential for the activation of a large set of signal transduction (El-Samad et al., 2005). Independent of the microscopic mechanisms involved, the influence of the thermal history on cell survival seems to be universal, and has been found not only in food pathogens (Neidhardt and VanBogelen, 1982; Streips and Polio, 1985; Mackey and Derrick, 1986; Mackey and Derrick, 1987a; Mackey and Derrick, 1987b; Fabisiewicz and Piechowska, 1988; Lema et al., 1988, Farber et al., 1990; Juneja et al., 1998; Arsene et al., 2000) but in nearly every cell type (McAlister and Finkelstein, 1980; Hahn and Li, 1982; Craig, 1985; Lindquist, 1986; Lindquist and Kim, 1996; de Maranon et al., 1999; Valdramidis et al., 2006; Guyot et al., 2010).

Temporal Correlation and Memory at the Macroscale

Microscopic cell mechanisms are frequently fast compared to the macroscopic timescale of the overall thermal inactivation process and thus the influence of the thermal history on cell survival can be lost in the long-time limit. If that is the case, the observable decrease of the survival ratio is accurately described by the memoryless prescription $\log S(t) = -t/\tau$ based on the conventional first-order kinetics.

If there is no clear separation between microscopic and macroscopic timescales, however, the macroscopic behaviour can reflect the influence of the thermal history. For instance, the pioneering works by Mackey and Derrick in the UK provide direct evidence of enhanced cell tolerance resulting from past thermal events such as slowly rising temperatures (Mackey and Derrick, 1987a) or mild heat shocks (Mackey and Derrick, 1986; Mackey and Derrick, 1987b). Similar studies over the past two decades have further confirmed that the risk of inactivation of food borne pathogens, in both broth and food systems, can strongly depend on past thermal stresses (Neidhardt and VanBogelen, 1982; Streips and Polio, 1985; Mackey and Derrick, 1986; Mackey and Derrick, 1987a, Mackey and Derrick, 1987b, Fabisiewicz and Piechowska, 1988; Lema et al., 1988; Bunning et al., 1990; Farber et al., 1990; Linton et al., 1990; Linton et al., 1992; Murano et al., 1992; Juneja et al., 1998; Arsene et al., 2000; Juneja and Novak, 2001; Hassani et al., 2007). Lou and Yousef (1996) have generalized these findings showing that history effects can also be induced by application of a variety of non-thermal stresses (see chapter 8).

Therefore, predictions of microbial lethality based on conventional first-order kinetics may not be accurate if there is a macroscopic manifestation of memory, as has been suggested by several authors (Powers et al., 1965; Mackey et al., 1987a; Tjakko et al., 1999; Shimoni et al., 2000; Humphrey, 2004). Frequently observed deviations from the exponential survival pattern seem to support this idea since has been proposed that the nonlinearity of semilogarithmic survival curves may reflect the occurrence of history dependent cumulative effects. For example, it has been suggested that concave, tailed survival curves may reflect the decreasing risk of inactivation associated with enhanced cell tolerance (Augustin et al., 1998; Peleg et al., 1998; Peleg, 2000; Peleg and Penchina, 2000; Peleg et al., 2001; van Boekel, 2001; Alvarez et al., 2003; Buzrul, 2005; Corradini et al., 2010), while convex, shouldered inactivation curves may reflect the increasing risk of inactivation associated with cumulative cell damage (Peleg et al., 1998; Geeraerd et al., 2000; Peleg, 2000; Peleg and Penchina, 2000; Peleg et al., 2001; van Boekel, 2001; Peleg, 2006; Corradini et al., 2010). In recent studies, differential equations describing microbial adaptation in inactivation kinetics by incorporating physiological adjustments were developed as an independent mathematical building block by Valdramidis et al. (2007).

As discussed above, models describing anomalous inactivation may benefit from explicitly taking into account memory effects, such as those often associated with cumulative cell damage or enhanced cell tolerance (Vaidya and Corvalan, 2009; Valdramidis et al., 2009). With this in mind, we discuss in the next section a straightforward generalization of the conventional inactivation kinetics with the generalization process described by a memory integral and in Section 5 by different structures of ANN.

4. Integral Memory Model

The first-order kinetics model implicitly assumes that the microscopic details of cell inactivation are lost above a given time scale so that memory effects can safely be ignored. When this is the case, inactivation processes with arbitrary temperature $T = T(t)$ are accurately described by simple integration of Equation (1) over time:

$$\log S(t) = -\int_0^t k[T(t')]\,dt'. \tag{4}$$

If, on the other hand, observed deviations from the first-order kinetics originate from the influence of past events, the conventional model must be generalized to include memory effects. Fortunately, the generalization of Equation (4) to exhibit memory on the macroscopic level is straightforward:

$$\log S(t) = -\int_0^t k[T(t')] f(t-t') dt'. \tag{5}$$

The convolution integral in Equation (5) is a memory integral, so called because the influence of a thermal stress at a previous time t' is felt at the current time t in a way that depends on its temporal distance $(t-t')$ in the past.

The inclusion of the kernel $f(t-t')$ in the generalized framework of Equation (5) places memory effects as an active participant in microbial inactivation. For instance, in the isothermal limit Equation (5) describes the deviations from the conventional pattern $\log S(t) = -kt$ through a history dependent cumulative term quantified by the expression between brackets in

$$\log S(t) = -k\left[\int_0^t f(t-t')\, dt'\right],$$

More importantly, since the conventional log-linear kinetics is the only continuous distribution characterized by the lack of memory property, the memory model of Equation (5) reflects, in principle, a unified treatment, one master equation for describing nonlinear thermal inactivation.

Proof of Concept Using a Power-Law Kernel

The generalized memory model of Equation (5) has the advantage of great flexibility when it comes to describing memory effects since different memory kernels $f(t-t')$ can be easily accommodated to describe different modes of microbial response to past thermal stresses. Challenges in establishing specific memory kernels include determining the type of memory to be described and the exact kinetics it leads to. To illustrate these challenges, we analyze below a particularly simple class of memory. This will enable us to illustrate the subtleties that can arise when memory integrals are solved, and to consider the role played by the ideas discussed in the previous sections in a more concrete setting. For examples of more accurate, although more complex, memory kernels see Vaidya and Corvalan (2011).

Power-Law Memory

Motivated by the principle that it is frequently best to start from a simple situation, we discuss here the simplest non-trivial category of memory. These are memory kernels defined by a power-law form (Vaidya and Corvalan, 2009):

$$f(t-t') = \alpha (t-t')^{\alpha-1}, \tag{6}$$

with a single parameter $\alpha > 0$ which determines the extent of memory preservation. When substituted into the generalized memory model, Equation (6) yields the inactivation kinetics

$$\log S(t) = -\int_0^t k\alpha(t-t')^{\alpha-1}\,dt' \quad , \tag{7}$$

which describes a class of systems that exhibit memory of the power-law type at the macroscopic level.

While standard integration techniques for calculating thermal process lethality apply directly to Equation (7), we make a few observations to illustrate the care that may need to be taken with this memory integral. A first observation concerns the existence of solutions that satisfy Equation (7). For a memory parameter $\alpha > 1$ the integrand is bounded and continuous, and therefore the integral exists. Instead, for $0 < \alpha < 1$ the kernel $\alpha(t-t')^{\alpha-1}$ is singular, however the integral is convergent and the singularity is integrable. Finally, in the trivial case $\alpha = 1$, the memory effect breaks out and the conventional inactivation of Equation (4) is recovered. A second major observation is that the power-law memory is a long-term memory kernel. This means that in addition to describing cell inactivation, the power-law kernel is also able to describe cell recovery when the applied temperature returns to levels which were hospitable to the microorganism in the past. Since cell recovery is not common in food-borne pathogens, for simplicity Equation (7) was integrated in all the examples discussed here with the explicit assumption that there is no cell growth or recovery during the thermal treatment, that is, $S(t+dt) \le S(t)$. For an example in which cell recovery is allowed, see Vaidya and Corvalan (2009). A final observation is that, instead of the initial condition $S(t=0)=1$ an initial history needs to be specified for memory models. For simplicity, Equation (5) was written assuming $S(t \le 0) = 1$.

We discuss below a few examples showing that, despite its simplicity, Equation (7) leads to kinetics that describe a large number of anomalous survival curves, and makes predictions for nonisothermal treatments that show no marked deviations from experimental inactivation data. Most of the examples were taken from Vaidya and Corvalan (2009) where more details of the solutions can be found.

Stretched Exponential Inactivation

For certain temperature profiles, Equation (7) can be integrated analytically. Of special interest is the isothermal case, which has the solution

$$S(t) = \exp(-kt^{\alpha}), \tag{8}$$

which generalizes the classical exponential pattern of Equation (2) and is ubiquitous to a diverse number of microbial systems. We mention in particular the recent study by van Boekel (2001) who, after analyzing inactivation data from 55 different microorganisms relevant to food science, concluded that all but two cases were statistically better described with a memory parameter α different than one.

According to Equation (8), a power-law memory is observed at the macroscopic level as a change in the curvature of an otherwise log-linear survival curve, and reflects a risk of inactivation

$$h(t) = k\,\alpha\,t^{\alpha-1}, \tag{9}$$

which depends explicitly on the 'age' of the thermal process (i.e., the elapsed processing time t during which the cells have been exposed to the thermal stress) modulated by the memory preservation α. Indeed, an inactivation path bent upward ($\alpha < 1$) would reflect the decreasing inactivation risk often associated with enhanced thermotolerance, whereas an inactivation path bent downwards ($\alpha > 1$) would reflect the increasing inactivation risk frequently associated with progressive cell damage (Figure 1). Due to this ability to describe both tailed and shouldered survival curves, Equation (8) has frequently been applied to describe anomalous inactivation kinetics (Peleg and Cole, 1998; Fernandez et al., 1999; Mattick et al., 2001; Peleg et al., 2001; Peleg, 2006).

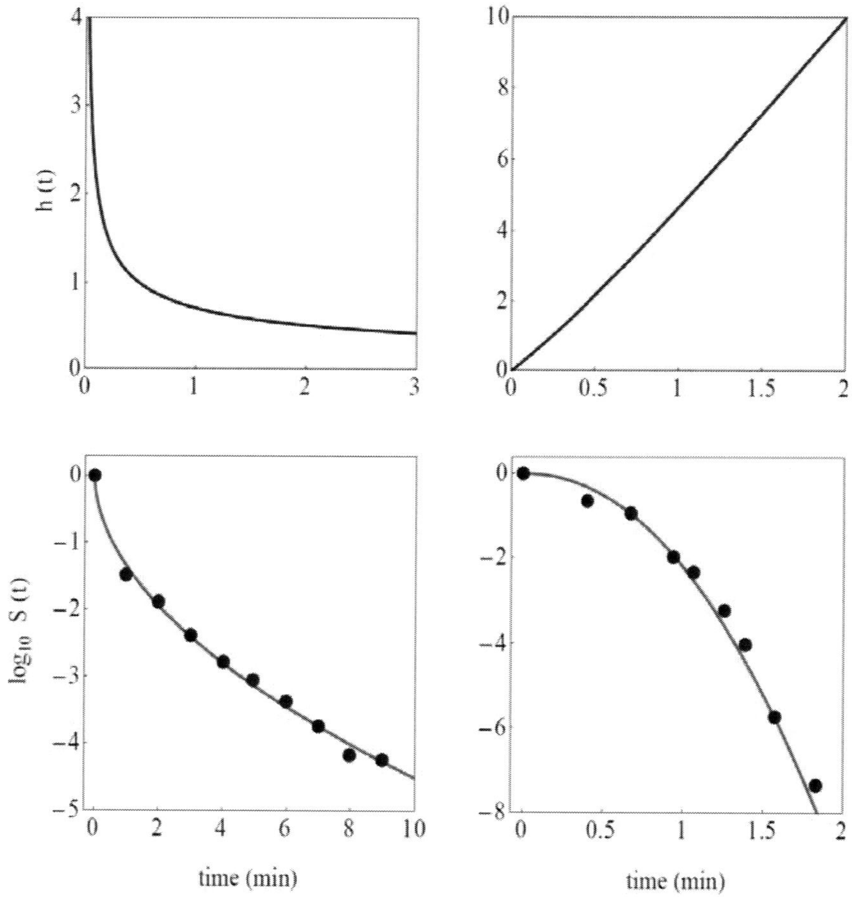

Figure 1. Left. A tailed survival curve (red line) may reflect the decreasing inactivation risk (blue line) associated with enhanced cell thermotolerance. Data from Mattick et al. (2001) for inactivation of Salmonella at 70 °C (dots). Solid lines correspond to Equations 8 and 9 with $k = 1.35$ and $\alpha = 0.52$. Right. An inactivation path bent downwards (red line) may reflect the increasing inactivation risk (blue line) associated with cumulative cell damage. Data from Murphy et al. (2000) for inactivation of Listeria monocytogenes at 70 °C (dots). Solid lines correspond to Equations 8 and 9 with $k = 2.2$ and $\alpha = 2.1$.

Equation (8) also establishes appealing connections with other studies on nonlinear microbial inactivation. In fact, the stretched exponential inactivation can be obtained from several models which correspond to different emphasized mechanisms. For instance, a common interpretation of Equation (8) is that the stretched exponential decline is an artefact resulting from the presence of a mixture of microbial populations, each with a different but constant risk of inactivation. In this interpretation, the stretched exponential kinetics is represented by the Laplace transform

$$\exp(-(kt)^{\alpha}) = \int_0^{\infty} F(k')\exp(-k't)dk' \quad , \tag{10}$$

and interpreted as superposition of infinite exponential decays with constant inactivation risks k' distributed with density $F(k')$. For a critical view of this interpretation see Peleg (2006, p.5).

The close analogy between the interpretation of the anomalous behaviour as resulting from memory effects or from the presence of a mixture of microbial populations, which are apparently very different phenomena, is appealing. Both interpretations can be reconciled if we rewrite Equation (10) in its equivalent form

$$\exp(-(t/\tau)^{\alpha}) = \int_0^{\infty} G(\tau')\exp(-t/\tau')d\tau', \tag{11}$$

Equation (11) shows that the stretched exponential contains information from all timescales τ' distributed continuously with density $G(\tau) = F(1/\tau)/\tau^2$. Therefore, the anomalous evolution of the system is caused by memory effects, that is, by long-time correlations leading to contributions from disparate time scales from the microscale $\tau' \to 0$ to the macroscale $\tau' \to \infty$ (see subsection *Temporal correlation and memory at the macroscale*). It is worth noting that although there are moments of the density $G(\tau)$ there are no moments of the density $F(k)$ which is seen by some statisticians as a formal disadvantage of using Equation (10) for the interpretation of anomalous experimental curves.

For cases where the density of timescales is assumed as originating from a Weibull probabilistic distribution, we find a second common interpretation for the stretched exponential of Equation (8) as the isothermal solution of the well-established differential inactivation model:

$$\frac{d\log_{10} S(t)}{dt} = -k[T(t)]\,\alpha[T(t)]\left\{\frac{-\log_{10} S(t)}{k[T(t)]}\right\}^{\frac{\alpha[T(t)]-1}{\alpha[T(t)]}} , \tag{12}$$

developed by Peleg and co-workers (Peleg and Penchina, 2000; Mattick et al., 2001; Peleg et al., 2003).

Mittag-Leffler Inactivation

We now consider another approach to the problem of identifying a kinetic model that accounts for power-law memory. Instead of generalizing Equation (4) as a memory integral for $\log S(t)$ we will generalize here the conventional Equation (1) in its equivalent integral form

$$S(t) = 1 - \int_0^t k\, S(t')\, dt', \tag{13}$$

With this in mind, we first define a power-law memory integral I^{α}, such that for any suitable function $g(t)$:

$$\mathrm{I}^{\alpha}\, g(t) = \frac{1}{\Gamma(\alpha)} \int_0^t g(t')(t - t')^{\alpha - 1}\, dt'. \tag{14}$$

The memory integral I^{α} is known as a 'fractional integral' (in the sense of Riemann-Liouville) and has been recognized as an important tool in the description of systems with memory ((Podlubny, 1999; Enelund and Olsson, 1999). A complete kinetic theory based on fractional models has been developed by Saxena et al. (2002, 2004) and successfully applied to describe thermal inactivation by Kaur et al. (2008).

Equation (13) can now be generalized as:

$$S(t) = 1 - \mathrm{I}^{\alpha}(k\, S(t)), \tag{16}$$

by straightforward fractional integration (Corvalan et al., 2003; Corvalan et at., 2004; Vaidya et al., 2007; Vaidya and Corvalan, 2009). As in the case of Equation (7), the presence of the power-law kernel in the operator of fractional integration leads to a formulation that is non-local in time and exhibit power-law memory at the macroscopic level. The isothermal solution of Equation (16) is (Podlubny, 1999):

$$S(t) = E_{\alpha}(-k\, t^{\alpha}), \tag{17}$$

which is a generalization of the stretched-exponential inactivation $S(t) = \exp(-k\, t^{\alpha})$. The Mittag-Leffler function $E_{\alpha}(x)$ is defined by

$$E_{\alpha}(x) = \sum_{n=0}^{\infty} \frac{x^n}{\Gamma(\alpha\, n + 1)}. \tag{18}$$

which, in turn, is a generalization of the exponential function $\exp(x)$. Indeed, for the case without memory $\alpha = 1$ Equation (17) becomes

$$S(t) = E_1(-kt) = \exp(-kt), \tag{19}$$

since

$$E_1(x) = \sum_{n=0}^{\infty} \frac{x^n}{\Gamma(n+1)} = \sum_{n=0}^{\infty} \frac{x^n}{n!} = \exp(x). \tag{20}$$

For other values of the memory parameter α , the Mittag-Leffler Equation (17) describes survival curves which, due to memory effects, change more slowly ($\alpha < 1$) or more rapidly ($\alpha > 1$) than the conventional inactivation associated with the first-order ($\alpha = 1$) kinetics (Figure 2).

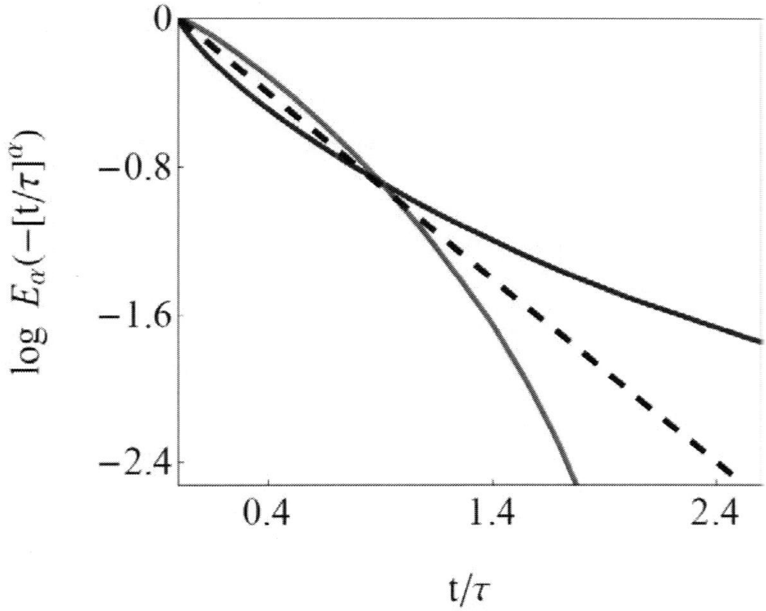

Figure 2. The Mittag-Leffler inactivation kinetics of Equation (17) describes survival curves which change more slowly (blue line, $\alpha = 0.8$) or more rapidly (red line, $\alpha = 1.2$) than the conventional inactivation kinetics (black dashed line, $\alpha = 1$).

Relationship between Mittag-Leffler and Stretched Exponential Kinetics

For times smaller than the characteristic timescale τ , Mittag-Leffler inactivation kinetics asymptotically approaches a stretched exponential form (Podlubny, 1999):

$$E_\alpha(-(t/\tau)^\alpha) \sim \exp\left[\frac{-(t/\tau)^\alpha}{\Gamma(1+\alpha)}\right], \tag{21}$$

as illustrated in Figure 3 for times $t < \tau$.

A consequence of Equation (21) is that experimental data describing a Mittag-Leffler kinetics at short times would also allow for a stretched exponential fit with identical characteristic time and memory parameters. This is exemplified using surrogate data in Figure

4. The upper panels of the figure show that data points generated by adding a small random perturbation to a Mittag-Leffler kinetics with $\alpha = 0.8$ (blue line) can be equally well described by a stretched exponential kinetics with the same value of the memory parameter (red line) for processing times smaller than the characteristic timescale ($\log(t/\tau) < 0$). On the other hand, if the window of available inactivation data becomes wider, the deviations between Mittag-Leffler and stretched exponential kinetics become more pronounced (Figure 3). For such cases, a forced stretched exponential fit to inactivation data following a Mittag-Leffler kinetics may lead to questionably low values of the regression parameters, as illustrated in the lower panels of Figure 4.

This may be the reason for some unusually low values of the stretched exponential parameters reported in literature. Consider, for instance, the isothermal inactivation data for *Salmonella Enteritidis* gathered by Michalski et al. (1999) depicted in Figure 5. van Boekel (2002) shows that this data can be well fitted assuming stretched exponential kinetics with a low memory parameter $\alpha = 0.3$ and an extremely low characteristic time $\tau \sim 150$ milliseconds (red line). In spite of the large number of microorganisms analyzed by van Boekel (2002) and others, such low values of the stretched exponential parameters are uncommon. On the other hand, an alternative analysis of the same experimental data assuming Mittag-Leffler kinetics results in a memory parameter about three times larger $\alpha = 0.93$ and a characteristic time on the order of seconds (blue line). Although this does not necessarily validates a Mittag-Leffler kinetics, the fitting with the Mittag-Leffler model gives more physical values of the parameters than the stretched exponential one.

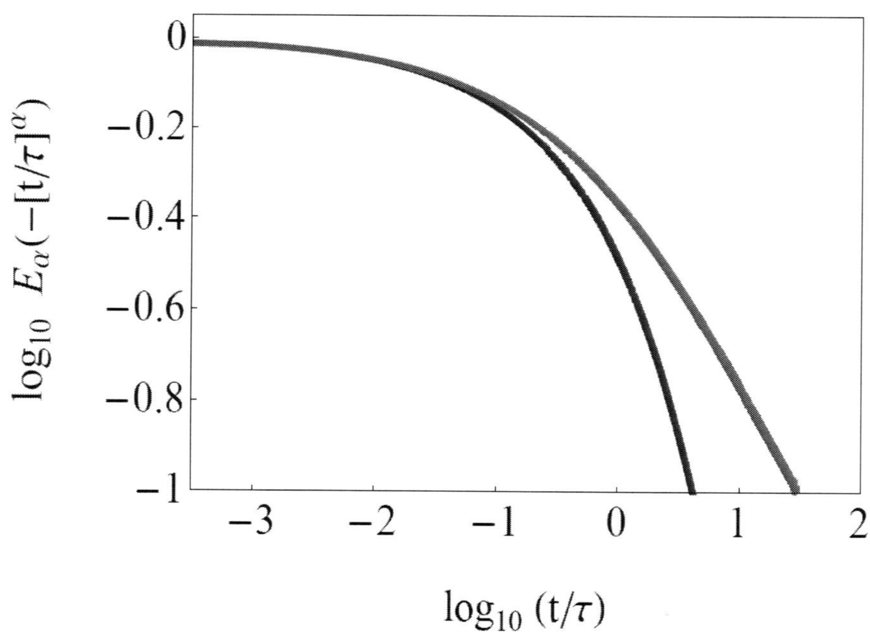

Figure 3. The Mittag-Leffler kinetics $E_\alpha(-(t/\tau)^\alpha)$ describes thermal inactivation processes that exhibit memory at the macroscopic level (red line). For times smaller than the characteristic inactivation time ($\log(t/\tau) < 0$), the Mittag-Leffler kinetics approaches a stretched exponential kinetics (blue line). Both lines correspond to Equation (21) with $\alpha = 1/2$.

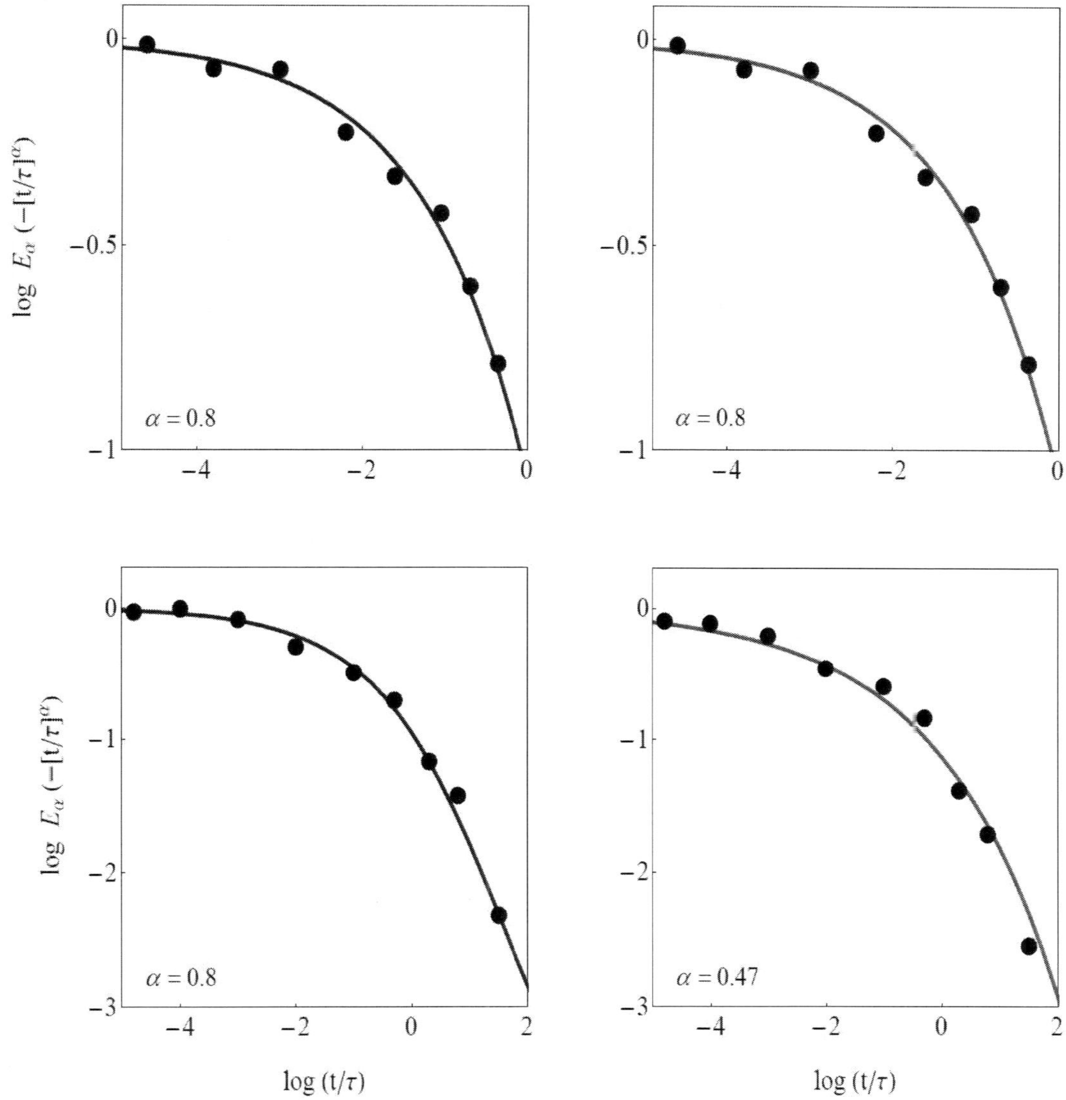

Figure 4. Top row. For times lower than the characteristic time, anomalous survival data can be equally well described by a Mittag-Leffler kinetics (blue line) or a stretched exponential kinetics (red line) with the same value of the memory parameter. The blue line corresponds to Equation (17) and the red line corresponds to Equation (21) both with $\alpha = 0.8$. Bottom row. As the window of available inactivation data becomes wider, assuming stretched exponential kinetics (red line) can result in substantially smaller values of the memory parameter than assuming Mittag-Leffler kinetics (blue line). The blue line corresponds to Equation (17) with $\alpha = 0.8$, and the red line corresponds to Equation (21) with $\alpha = 0.47$.

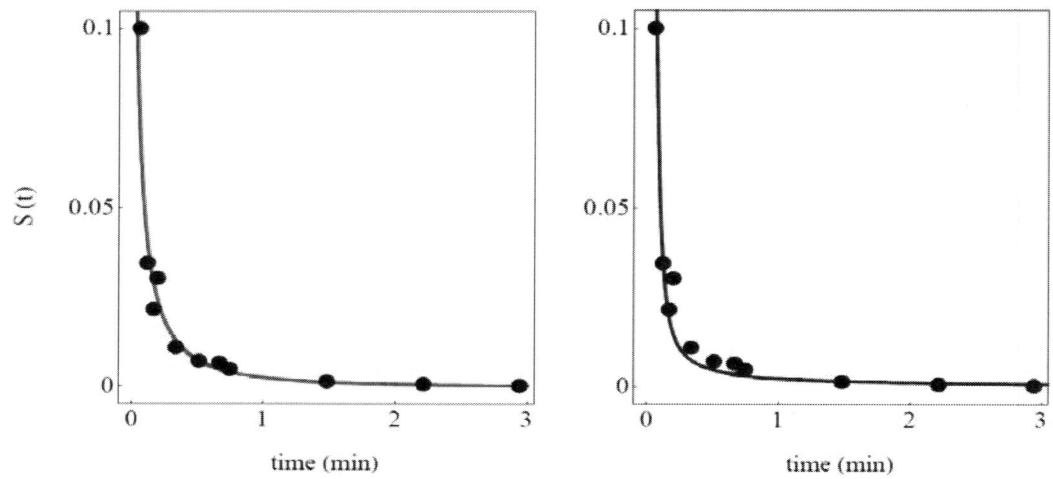

Figure 5. Analysis of survival data from Michalski et al. (1999) for inactivation of Salmonella enteritidis at 64 °C (dots). The red line corresponds to stretched exponential kinetics of Equation (8) with fitting parameter $\alpha = 0.3$ and the blue line corresponds to Mittag-Leffler kinetics of Equation (17) with $\alpha = 0.93$.

Anomalous Inactivation Processes with Linear Temperature Profiles

For certain temperature profiles, Equation (7) can be integrated analytically. An important case corresponds to a linear temperature profile

$$T(t) = T_0 + mt, \tag{22}$$

which increases from the initial value T_0 at a constant heating rate m. On the common assumption that the temperature dependence of the rate constant follows the exponential Arrhenius equation

$$k(T) = \exp[(T - T_c)/b], \tag{23}$$

the integration of the power-law memory model for a linear temperature profile yields the closed solution

$$\log S(t) = -[\Gamma(b) - \Gamma(\alpha, mt/b)] \times \alpha(b/m)^{\alpha} \exp[(T_0 + mt - T_c)/b], \tag{24}$$

where $\Gamma(x, y)$ is the incomplete gamma function, a generalization of the gamma function $\Gamma(x)$ such that $\Gamma(x) = \Gamma(x,0)$ (see Arfken, 1985).

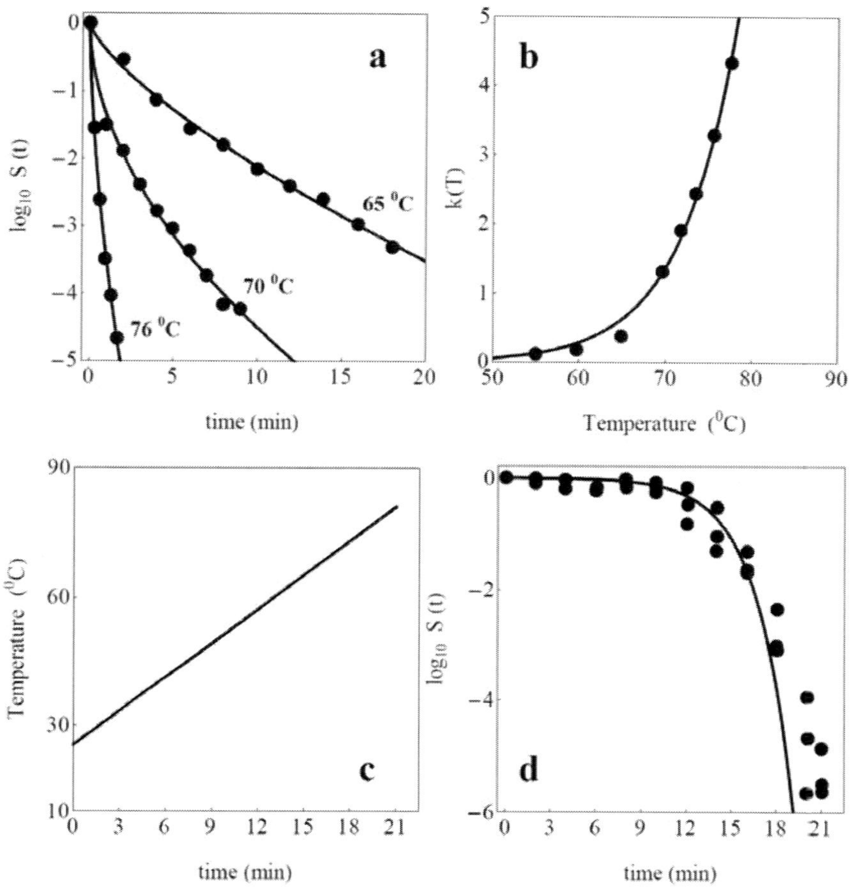

Figure 6. Predicting heat inactivation of Salmonella from anomalous isothermal data. (a) Examples of isothermal survival curves fitted using Equation (8) as a model and (b) the corresponding Arrhenius dependence of the inactivation rate with temperature estimated using Equation (23) as a model ($T_c = 68\,^oC$, $b = 6.4\,^oC$). (c) Linear treatment temperature ($T_0 = 25.4^\circ C, m = 2.7^\circ C\,\mathrm{min}^{-1}$) and (d)the corresponding predictions from the analytical solution Equation (24) ($\alpha = 0.7$). The experimental data points are from Mattick et al. (2001) and the predictions are limited to the linear portion of the heating curves.

The suitability of Equation (24) to predict experimental results was tested using data for *Salmonella enterica* serovar Typhimurium definite type 104 strain 30 reported by Mattick et al. (2001) (Vaidya and Corvalan, 2009). First, a regression analysis of the isothermal survival curves for *S. enterica* cells at temperatures of 55 to 80 °C revealed that the data follow a stretched exponential model with $\alpha < 1$, with regression coefficients of 0.996 to 0.999 (Figure 6a). The regression parameters were then used to determine the Arrhenius dependence of the rate constant using Equation (23) as a model (Figure 6b) and, finally, the nonisothermal survival curve corresponding to the linear temperature profile of Figure 6(c) was predicted using Equation (24). As shown in Figure 6(d) no profound differences were found between the model predictions and the experimental measurements.

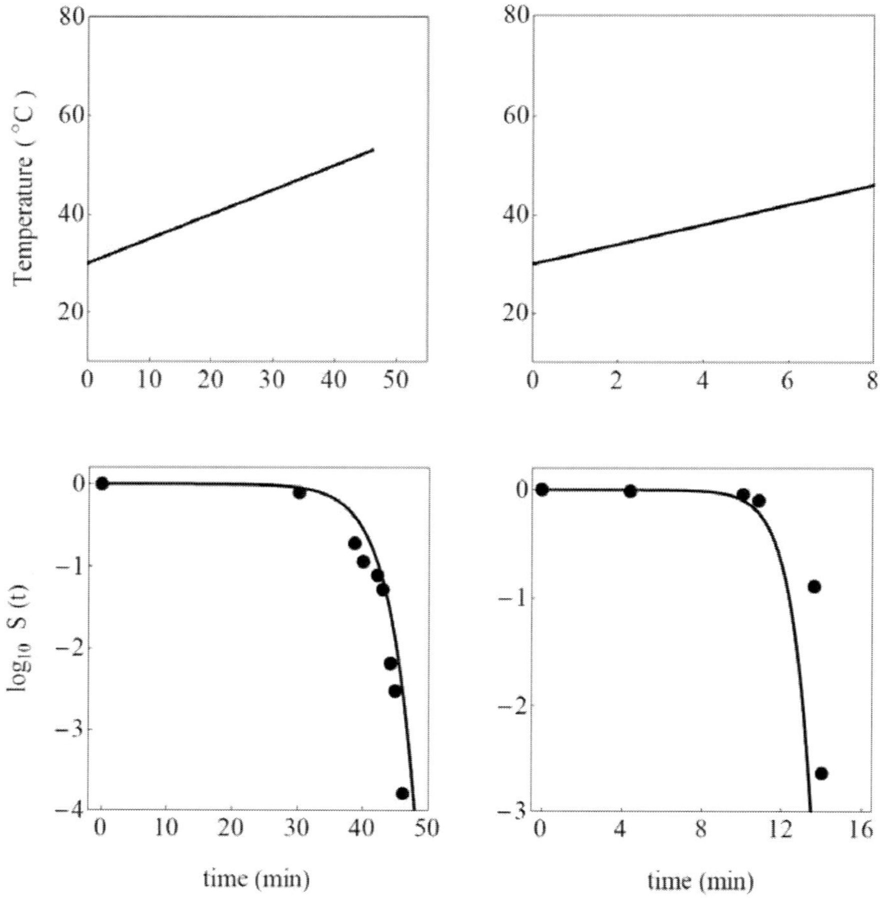

Figure 7. Predicting heat inactivation of Listeria monocytogenes from anomalous isothermal data. The predictions correspond to Equation (24) with $\alpha = 1.27, T_c = 55\ ^{o}C, b = 2\ ^{o}C$ and the linear temperature profiles shown at the top of each figure. Data from Hassani et al. (2005) for inactivation of Listeria monocytogenes (dots) limited to the linear portion of the heating curves. (Adapted from Vaidya and Corvalan (2009). Reprinted with permission from the Journal of Food Protection. Copyright held by the International Association for Food Protection, Des Moines, Iowa, U.S.A.)

In addition to the experimental data for *Salmonella enterica* cells from Mattick et al. (2001) exhibiting memory of the power-law type with $\alpha < 1$, Equation (24) was also tested against experimental data for *Listeria monocytogenes* cells from Hassani et al. (2005) exhibiting power-law memory with $\alpha > 1$. The results are summarized in Figure 7 and show no marked differences between predictions and experiments.

Parameter Determination from Nonisothermal Data

We continue the analysis of the power-law memory model considering the problem of identifying kinetic parameters directly from regression analysis of nonisothermal inactivation data. The problem is not only theoretical, but reflects an experimental necessity. The ability to predict kinetics parameters from nonisothermal data would enable better kinetic characterization of food systems for which the implementation of truly isothermal experiments is challenging, including highly viscous and multiphase food systems (Mizrahi and Karel, 1978; Peleg et al., 2003; Peleg, 2006, Ch. 4).

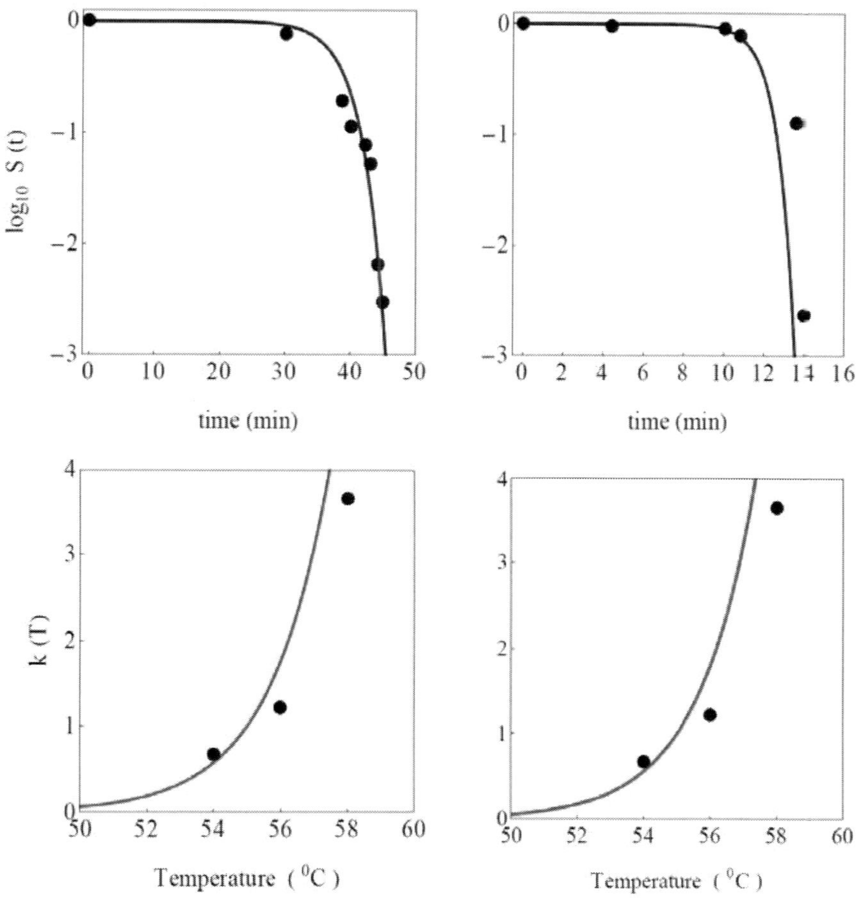

Figure 8. Parameter determination from nonisothermal data. Predictions of the Arrhenius variation of the rate constant k with temperature (red lines) using regression parameters estimated from nonisothermal inactivation data using Equation (24) (blue lines). Data from Hassani et al. (2005) for inactivation of Listeria monocytogenes in TSB buffer (pH = 4) (dots).

With this in mind, we analyzed experimental data for *Listeria monocytogenes* in tryptic soy broth (TSB) buffer (pH = 4) reported by Hassani et al. (2005), in the range in which the microorganism was subjected to nonisothermal treatments with the two linear temperature profiles depicted in Figure 7. The two sets of nonisothermal survival data were first subjected to regression analysis in terms of Equation (24), obtaining the regression values $\alpha = 1.6$, $b = 1.7^{o}C$ for the case with the slower heating rate $m = 0.5$ °C/min, and the regression parameters $\alpha = 1.1$, $b = 1.7^{o}C$ for the case with faster heating rate $m = 2$ °C/min (Figure 8, blue lines). These kinetics parameters estimated from the nonisothermal data were then introduced in Equation (23) to predict the Arrhenius dependence of the rate constant k with temperature. The corresponding predictions (Figure 8, red lines) showed no marked differences with the experimental data gathered by Hassani et al. (2005) from isothermal survival curves at 54, 56 and 58 °C.

Anomalous Inactivation with Nonlinear Temperature Profiles

A fundamental challenge in food science today is the prediction of general nonisothermal inactivation processes for anomalous microbial systems. Thermal inactivation of anomalous microbial systems pose new theoretical challenges as a result of the crucial role played by deviations from the traditional first-order models. Deviations of this kind become more important as processing times progress or as the anomalous character of the microbial system increases, and can lead to substantial departure from the system's expected behavior.

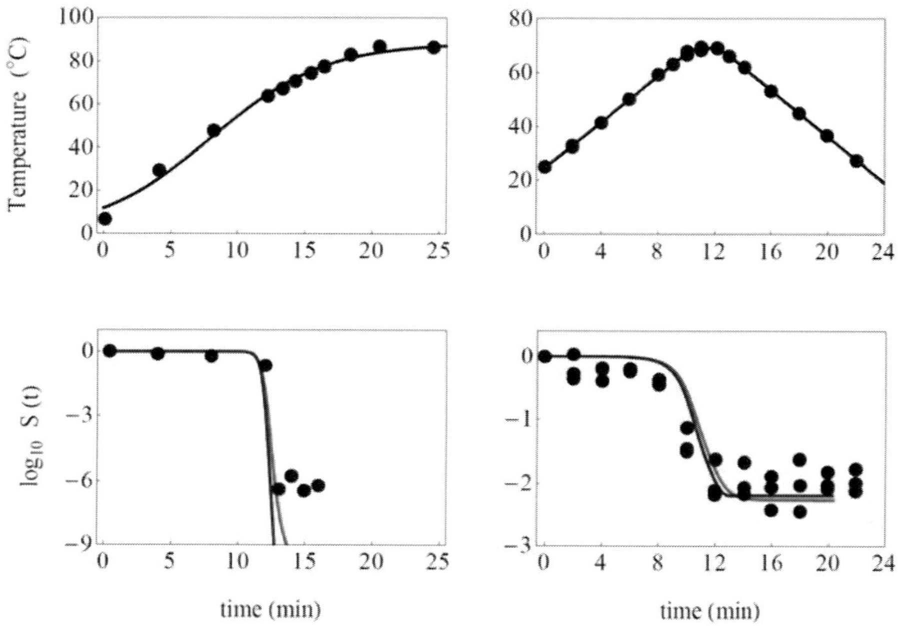

Figure 9. Predictions from the power-law memory model (blue lines) and Peleg et al. (2000) Weibull distribution model (red lines) for the nonlinear temperature profiles shown at the top of each figure (black lines). Left. Data from Coote et al. (1991) for inactivation of Listeria monocytogenes. Right. Data from Mattick et al. (2001) for inactivation of Salmonella. (Adapted from Vaidya and Corvalan (2009). Reprinted with permission from the Journal of Food Protection. Copyright held by the International Association for Food Protection, Des Moines, Iowa, U.S.A.).

Important steps in the modelling of anomalous systems have been taken recently by Peleg and co-workers with the development of the differential Equation (12) for systems following stretched-exponential kinetics (Mattick et al., 2001; Peleg et al., 2001). In Figure 9 we illustrate the potential of the power-law memory model (blue line) by direct comparison against predictions from Peleg's model (red line) and against experimental data reported by Coote et al. (1991) and by Mattick et al. (2001) for microbial systems exhibiting stretched exponential anomalous dynamics with $\alpha < 1$. Results on the left of the figure correspond to inactivation of *Listeria monocytogenes* (Coote et al., 1991) predicted by the memory model using isothermal regression coefficients obtained by Stephens et al. (1994). Similarly, the results on the right side of the figure correspond to nonisothermal survival of *S. enterica* predicted by the memory model using regression coefficients reported by Mattick et al. (2001). For both anomalous systems we found good agreement between the memory model and Peleg's model using identical regression parameters, and we also found good agreement between the memory model and the experimental data.

5. Memory Embedded Structures of Artificial Neural Networks

ANN has received quite some attention in the field of Predictive Microbiology in the last few years (Hajmeer et al. 1997; Geeraerd et al. 1998; Cheroutre-Vialette and Lebert 2000; Garcia-Gimeno et al. 2005; Hajmeer et al. 2006; Panagou 2008; Valdramidis et al. 2009), despite their black box characteristics, i.e., data-driven models purely based on experimental data. In this section their applicability is further evaluated and they are applied as alternative methods to highlight the importance of incorporating microbial past events on the actual microbial responses under dynamic temperature conditions. Considering that the most classical microbiological techniques of predictive modelling in foods account for off-line microbial data for which time intervals for performing microbial predictions are not available to the user beforehand, different structures (i.e., autonomous and non-autonomous) of ANN are also assessed based on microbial inactivation data that are generated at dynamic temperature conditions.

Experimental Data

For illustrating the applicability and efficacy of the ANN structures previously generated, inactivation data of *Escherichia coli* K12 MG1655 under dynamic temperature conditions are used (Valdramidis et al., 2006). These experiments were performed in cell suspensions and at temperature controlled conditions. Six different heating regimes were studied, i.e., dT/dt of 0.15 °C/min, 0.20°C/min, 0.40°C/min, 0.55°C/min, 0.82°C/min, and 1.64°C/min Experiments were performed in sterile glass capillary tubes immersed in a temperature controlled circulating water bath. Colony-forming units of kinetic samples were collected. For further details on the experimental protocol refer to Valdramidis et al. (2006).

All microbial inactivation data are partitioned into training, validation and test sets before being processed by the ANNs. The training sets consist of four experimental sets in which the heating rates of 0.15°C/min, and 1.64°C/min are always included. By including the slowest and the highest heating rates we wanted to ensure that predictions are not performed at temperature rates which are outside the range of the training set. One experimental set is chosen for validation in which an early stopping methodology (Haykin 1994; Bishop 1995) is employed to avoid over fitting.

Model Structures

A multi-step ahead Feed Forward with a delayed artificial neural network (ANN) modelling structure was constructed (Figure 10). In this structure there were three input vectors, namely, time t_k and temperature rate $(dT/dt)_k$, and the microbial load delayed with one time unit, $\log N_{k-1}$. Its architecture consisted of one hidden neuron and one output neuron. The selected transfer functions for the hidden and output neuron were the logistic-sigmoid and the linear functions, respectively. All input values were linearly normalised into [0,1] before they were entered into the network. Initial values that lie within [-1, 1] are picked randomly from a uniform distribution for the ANN's connection weights. The predictive capability of these models was also compared with a Feed Forward (FF) network (with t_k and temperature rate $(dT/dt)_k$ as the input vectors), while an autonomous multi-step ahead FFD network (in which the input vector t_k of the initially developed FFD network is replaced by the temperature T_k) was also considered (Figure 10). The performance of this autonomous structure was also evaluated when the vector of temperature rate $(dT/dt)_k$ was omitted and both in the presence and absence of the bias parameter $\log N_{k-1}$.

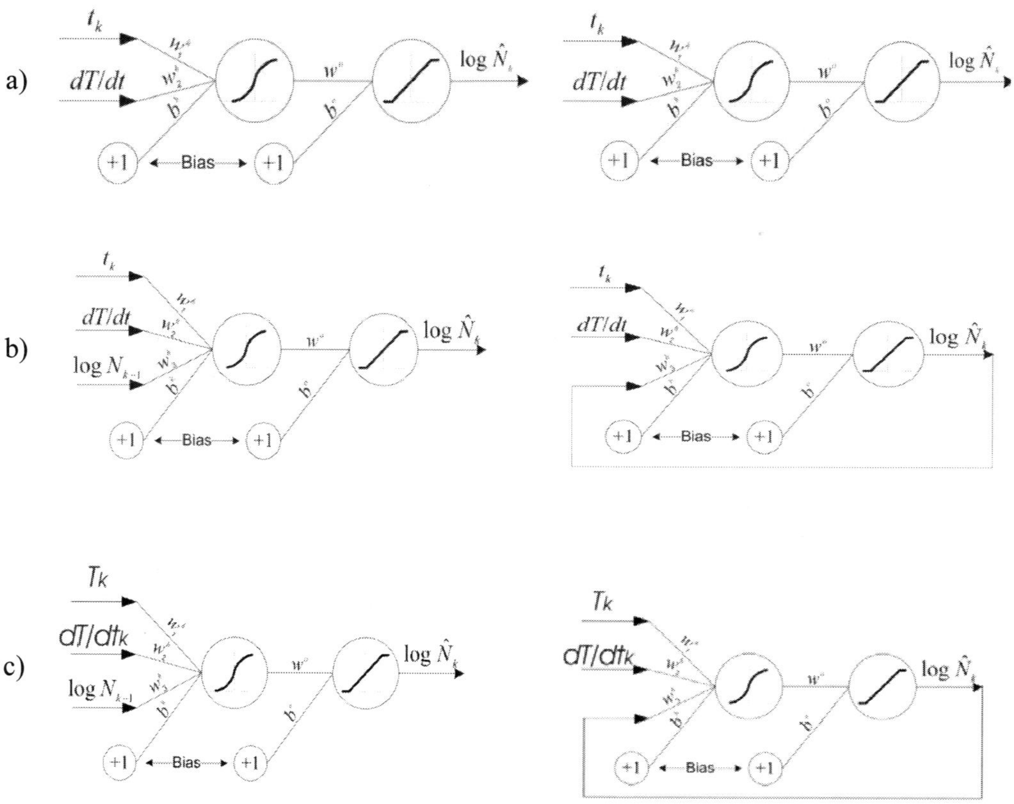

Figure 10. ANN model structures: a) Feed Forward (FF) network, b) multi-step ahead Feed Forward with a delay, c) autonomous multi-step ahead FFD network. Left column: identification, right column: testing.

Parameter Optimisation

The ANNs are trained with the backpropagation methodology by the use of the Levenberg-Marquardt algorithm (Hagan and Menhaj 1994). The algorithm is terminated either when it converges to a good training mean square error value (MSE) or when the MSE on the validation data set increases (i.e., early stopping) or once a predefined large number of epochs (e.g. 5000 epochs) is completed. All backpropagation runs are repeated 100 times (each time with randomly generated connection weight values) and the run with the lowest MSE on the test set is picked. Programs for simulation, optimization, and fitting were written in MatLab® Version 6.5 (The MathWorks, Inc., MA, USA) using the Neural Networks Toolbox Version 6.5.

Modelling the Microbial Dynamics

The increased induction of the heat resistance of *Escherichia coli* K12 under treatments of decreasing heating rates was quantified for all the developed ANN structures. The developed FFD structures appear to have a non-autonomous character due to their time dependency. Consequently, the predictive capability of the ANNs that had t_k as an input vector was influenced by the chosen increments of t_k. This clearly appears in Figure 11. The results show that non-autonomous ANN structures result in different predictions depending on the chosen time increment; this phenomenon is more pronounced for the multi-step ahead FF structure.

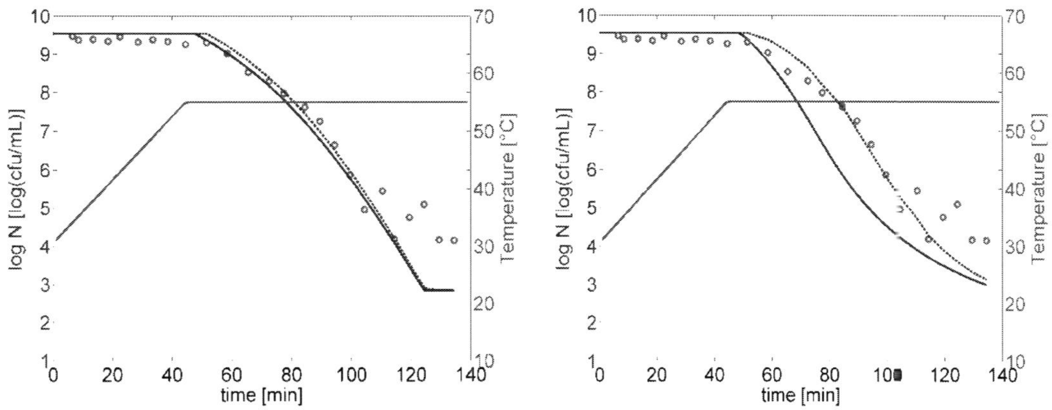

Figure 11. Examples of predictions of Feed Forward (left) and multi-step ahead Feed Forward with a delay (right) ANN architectures for t_k= experimental time of the test set (--) and t_k= 0.5 min increment (-). Microbial inactivation data (used as test set) (o) of E. coli K12 with their corresponding temperature profile at the studied heating rate of 0.55°C/min.

When microbial predictions were performed without presenting time explicitly as an input vector of the ANN structures (i.e., the case of the autonomous multi-step ahead FFD network), it resulted in accurate microbial predictions that incorporated the induced microbial phenomena. Results (Figure 12) show that ANN with two input variables, i.e., T_k, N_{k-1} was sufficient to describe the microbial inactivation kinetics. This means that increasing the parameters of the model did not improve the generalization of the ANN.

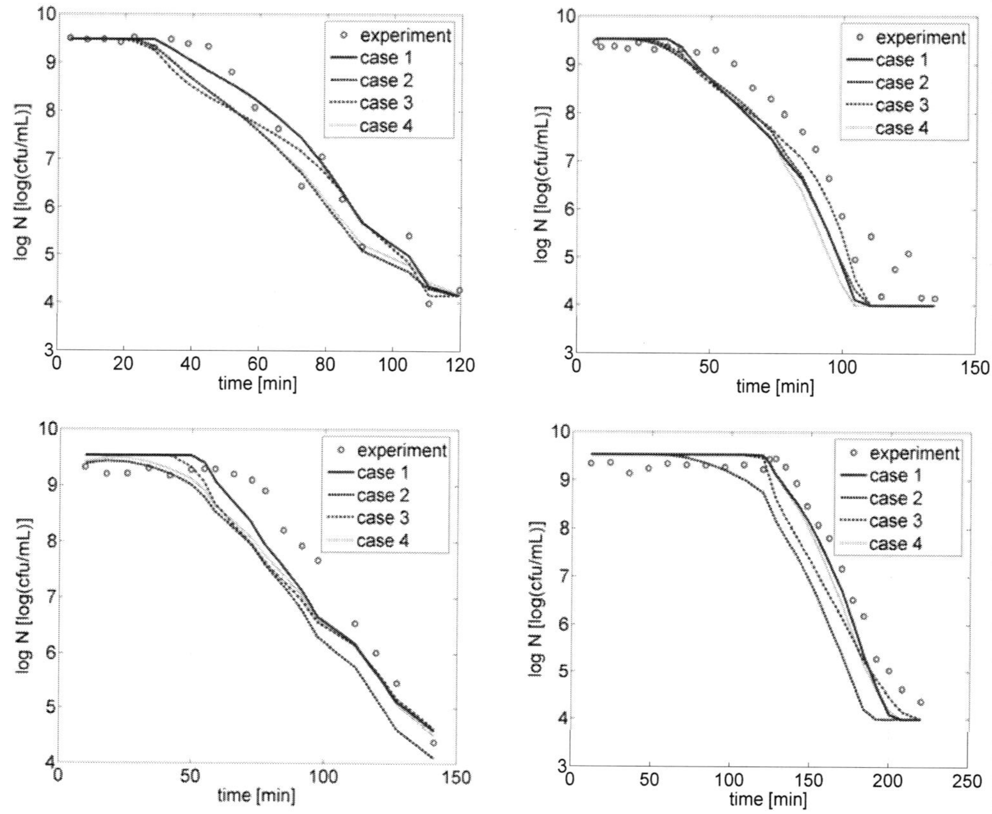

Figure 12. Predictions of autonomous multi-step ahead FFD network at heating treatments of 0.82°C/min (top left), 0.55°C/min (top right), 0.40°C/min (bottom left), 0.20°C/min (bottom right), Experiment: microbial inactivation data (used as test set) (o) of E. coli K12, case 1: input vectors T_k, N_{k-1}, case 2: input vectors T_k, N_{k-1} with bias factor on N_{k-1} case 3: input vectors T_k, N_{k-1}, dT/dt$_k$, case 4: T_k, N_{k-1}, dT/dt$_k$ with bias factor on N_{k-1}.

Although time dependent ANN are a common practise for online control applications, their application in predictive microbiology in which microbial information originates from off-line data, should be reconsidered. For this purpose, an autonomous ANN structure was suggested. The good predictive capability of the FFD network indicates that an input incorporating past events, i.e., N_{k-1}, can improve the modelling performance by encompassing the microbial stress adaptation of the examined microbes due to the slowly increasing temperatures. This input delay can be considered as a dynamic *a priori microbiological knowledge* suitable for extracting the information contained in the microbial experimental data. Further experimental investigation of these adaptations may require studies focusing on the mechanisms influencing the microbial physiology.

Conclusion

We will conclude by recapitulating what we think are the principal concepts to be extracted from what we have tried to describe in this chapter. The first is the concept that the stochastic process underlying the first-order inactivation kinetics lacks memory. This both limits its applicability to microbial systems characterized by constant risk of inactivation and suggests that approaches to describe nonlinear inactivation may benefit from incorporating memory effects, an aspect that has received limited attention in quantitative microbiology. The second, closely related concept is that the memory-integral formalism can make the description of thermal inactivation directly compatible with memory effects Generalized memory-integral models are, in principle, capable of describing all type of nonlinear inactivation processes within a single, all-encompassing, coherent framework. No such general principles apply however when it comes to writing down explicit memory kernels to generate a particular observed behaviour, and in this chapter we concentrated, as a proof of concept, on a simple power-law kernel. Results suggest that memory-integral models provide a promising mathematical framework for describing anomalous microbial inactivation. In its actual form, however, the power-law kernel should be considered a theoretical construct to be developed and optimized further by a combination of analytical and experimental investigations. More important, further studies should propose and test alternative memory kernels to describe data beyond the reach of the power-law memory, and find ways of interpreting those data that may suggest a mechanistic basis on which a larger theory can be founded.

Quantification of adaptive microbial responses can also be tackled by memory embedded modelling structures which could then accurately assess the efficacy of a heating process. The need of novel models that can incorporate the cell's history appears to be imperative for other types of food processes as well, e.g., cooling storage (Juneja et al. 2008).

In conclusion, incorporation of past events seems to be an essential input for taking into account the observed induced microbial heat resistance and can be assessed by the development of memory embedded modelling structures. Non-autonomous ANNs have implications on the prediction of momentary microbial population. Their application should be restricted to on-line microbial measurements in which time increments are experimentally or industrially targeted. Performing predictions based on information obtained from off-line techniques, like colony counts, requires the development of autonomous modelling structures like the one developed in the current study.

References

Adams, M. R. & Moss, M.O. (2000). *Food microbiology*, 2nd ed. Royal Society of Chemistry, Cambridge.

Alho, J. & Spencer, B.D. (2005). *Statistical demography and forecasting*. Springer, New York.

Alvarez, I., Virto, R., Raso, J., Condon, S. (2003). Comparing predicting models for the *Escherichia coli* inactivation by pulsed electric fields. *Innovative Food Science and Emerging Technologies,* 4, 195-202.

Arfken, G. (1985). The incomplete gamma function and related functions. *In Mathematical methods for physicists*, 3rd ed. Academic Press, Orlando, FL.

Arsene, F., Tomoyasu, T. and Bukau, B. (2000). The heat shock response of *Escherichia coli. International Journal of Food Microbiology,* 55, 3–9.

Augustin, J.C., Carlier, V., Rozier, J. (1998). Mathematical modeling of the heat resistance of *Listeria monocytogenes. Journal of Applied Microbiology,* 84, 185-191.

Balakrishnan, N. & Basu, A.P. (1996). *The exponential distribution: theory, methods, and applications*. Gordon and Breach, New York.

Bishop, C. (1995). *Neural networks for pattern recognition*, Oxford University Press

Bunning, V. K., Crawford, R.G, Tierney, J.T., Peeler, J.T. (1990). Thermotolerance of *Listeria monocytogenes* and *Salmonella typhimurium* after sublethal heat shock. *Applied Environmental Microbiology,* 56, 3216-3219.

Buzrul, S., Alps, H., Bozoglu, F. (2005). Use of Weibull frequency distribution model to describe the inactivation of *Alicyclobacillus acidoterrestris* by high pressure at different temperatures. *Food Research International,* 38, 151-157.

Coote, P. J., Holyoak, C.D., Cole, M.B. (1991). Thermal inactivation of *Listeria monocytogenes* during a process simulating temperatures achieved during microwave heating. *Applied Microbiology,* 70, 489–494.

Cheroutre-Vialette, M. & Lebert, A. (2000) Modelling the growth of Listeria monocytogenes in dynamic conditions. *International Journal of Food Microbiology* 55, 201-207.

Corradini, M. G. & Peleg, M. (2009) Dynamic Model of Heat Inactivation Kinetics for Bacterial Adaptation. *Applied and Environmental Microbiology* 75, 2590-2597.

Corradini, M.G., Normand, M.D., Peleg, M. (2010). Stochastic and Deterministic Model of Microbial Heat Inactivation. *Journal of Food Science,* 75(2), R59-R70.

Corvalan, C., Campanella, O.H., Haley, T.A. (2003). A new model to predict the inactivation and growth of microorganisms taking into account previous processing history. Presented at the Annual Meeting of the Institute of Food Technologists, Chicago, IL, 12 to 16 July 2003.

Corvalan, C., Chen, G., Campanella, O.H. (2004). A new approach using a fractional kinetic model to describe microbial inactivation. Presented at the Annual Meeting of the Institute of Food Technologists, Las Vegas, NV, 12 to 16 July 2004.

Craig, E.A. (1985). The heat shock response. *Critical Reviews in Biochemistry,* 18, 239-280.

de Marañón, I. M., Chaudanson, N., Joly, N., Gervais, P. (1999). Slow heat rate increase yeast thermotolerance by maintaining plasma membrane integrity. *Biotechnology and Bioengineering*, 65(2), 176−181.

El-Samad, H., Kurata, H., Doyle, J.C., Gross, C.A., Khammash, M. (2005). Surviving heat shock: Control strategies for robustness and performance. *Proceedings of the National Academy of Sciences,* 102(8), 2736-2741.

Enelund, M. & Olsson, P. (1999). Damping described by fading memory: analysis and application to fractional derivative models. *International journal of Solids and Structures,* 36, 939–970.

Fabisiewicz, A. & Piechowska, M. (1988). Heat-shock proteins in membrane vesicles of *Bacillus subtilis. Acta Biochimica Polonica,* 35, 367-376.

Farber, J. M. & Brown, B.E. (1990). Effect of prior heat shock on heat resistance of *Listeria monocytogenes* in meat. *Applied and Environmental Microbiology,* 6, 1584–1587.

Feller, W. (1967). An introduction to probability theory and its applications, 1, 3rd ed., Wiley, New York.

Fernandez, A., Salmeron, C., Fernandez, P.S., Martinez, A. (1999). Application of a frequency distribution model to describe the thermal inactivation of two strains of *Bacillus cereus. Trends in Food Science & Technology,* 10, 158–162.

Frazier, W. C. & Westhoff, D.C. (1988). *Food microbiology*, 4th ed. McGraw-Hill, New York.

Garcia-Gimeno, R. M., Hervas-Martinez, C., Rodriguez-Perez, R. & Zurera-Cosano, G. (2005) Modelling the growth of *Leuconostoc mesenteroides* by artificial neural networks. *International Journal of Food Microbiology* 105, 317-332.

Geeraerd, A. H., Herremans, C. H., Cenens, C. & Van Impe, J. F. (1998), Application of artificial neural networks as a non-linear modular modeling technique to describe bacterial growth in chilled food products. *International Journal of Food Microbiology* 44, 49-68.

Geeraerd, A. H., Herremans, C. H. & Van Impe, J. F. (2000). Structural model requirements to describe microbial inactivation during a mild heat treatment. *International Journal of Food Microbiology,* 59, 185-209

Guyot, S. Pottier, L. Ferret, E., Gal, L. & Gervais, P. (2010). Physiological responses of *Escherichia coli* exposed to different heat-stress kinetics. *Archives of Microbiology,* 192, 651-661.

Hahn, G. M. & Li, G.C. (1982). Thermotolerance and heat shock proteins in mammalian cells. *Radiation Research,* 92, 452-457.

Hajmeer, M. N., Basheer, I. A. & Najjar, Y. M. (1997). Computational neural networks for predictive microbiology .2. Application to microbial growth. *International Journal of Food Microbiology* 34, 51-66.

Hajmeer, M., Basheer, I. & Cliver, D. O. (2006). Survival curves of *Listeria monocytogenes* in chorizos modeled with artificial neural networks. *Food Microbiology* 23, 561-570.

Hassani, M., Manas, P., Raso, J., Condon, S. & Pagan, R. (2005). Predicting heat inactivation of *Listeria monocytogenes* under nonisothermal treatments. *Journal of Food Protection,* 68, 736–743.

Hassani, M., Condon, S. & Pagan, R. (2007) Predicting microbial heat inactivation under nonisothermal treatments. *Journal of Food Protection* 70, 1457-1467.

Haykin, S. (1994). *Neural Networks: a comprehensive foundation*, Maxwell Macmillan International.

Huang, J.S. (1981). On a "lack of memory" property. *Annals of the Institute of Statistical Mathematics, 33, Part A,* 131-134.

Humphrey, T. (2004). *Salmonella,* stress responses and food safety. *Nature Reviews Microbiology,* 2, 504-509.

Juneja, V. K., Klein, P.G., Marmer, B.S. (1998). Heat shock and thermotolerance of *Escherichia coli* O157:H7 in a model beef gravy system and ground beef. *Journal of Applied Microbiology,* 84, 84:677–684.

Juneja, V.K. & Novak, J.S. (2001). Heat resistance of *Clostridium perfringens* vegetative cells as affected by prior heat shock. *Journal of Food Safety,* 21, 127-139.

Kaur, A., Takhar, P.S., Smith, D.M., Mann, J.E., Brashears, M.M. (2008). Fractional Differential Equations Based Modeling of Microbial Survival and Growth Curves:

Model Development and Experimental Validation. *Journal of Food Science,* 73(8), E403-E414.

Lema, M.W., Brown, A., Butler, C.A., Hoffman, P.S. (1988). Heat-shock response in *Legionella pneumophila. Canadian Journal of Microbiology,* 34, 1148-1153.

Lewis, M. & Heppell, N.J. (2000). *Continuous thermal processing of foods.* Aspen Publishers, Inc., Gaithersburg, MD.

Lindquist, S. & Kim, G. (1996). Heat-shock protein 104 expression is sufficient for thermotolerance in yeast. *Proceedings of the National Academy of Sciences, USA.* 93, 5301-5306.

Lindquist, S. (1986). The heat-shock response. *Annual Review of Biochemistry,* 55, 1151-1191.

Linton, R. H., Webster, J.B., Pierson, M.D., Bishop, J.R., Hackney, C.R. (1992). The effect of sublethal heat shock and growth atmosphere on the heat resistance of *Listeria monocytogenes* Scott A. *Journal of Food Protection,* 55, 84-87.

Linton, R.H., Pierson, M.D., Bishop, J.R. (1990). Increase in heat resistance of *Listeria monocytogenes* Scott A by sublethal heat shock. *Journal of Food Protection* 53, 924-927.

Lou, Y. & Yousef, A.E. (1996). Resistance of *Listeria monocytogenes* to heat after adaptation to environmental stresses. *Journal of Food Protection,* 59(5), 465-471.

Mackey, B.M. & Derrick, C.M. (1986). Elevation of the heat resistance of *Salmonella typhimurium* by sublethal heat shock. *Journal of Applied Bacteriology,* 61, 389-393.

Mackey, B.M. & Derrick, C.M. (1987a). Changes in the heat resistance of *Salmonella typhimurium* during heating at rising temperatures. *Letters in Applied Microbiology,* 4, 13-16.

Mackey, B.M. & Derrick, C.M. (1987b). The effect of prior heat shock on the thermoresistance of *Salmonella Thompson* in foods. *Letters in Applied Microbiology,* 5, 115-118.

Mafart, P., Couvert, O., Gaillard, S. & Leguerinel, I. (2002). On calculating sterility in thermal preservation methods: application of the Weibull frequency distribution model. *International Journal of Food Microbiology,* 72, 107-113.

Mattick, K. L., Legan, J.D., Humphrey, T.J., Peleg, M. (2001). Calculating *Salmonella* inactivation in nonisothermal heat treatments from isothermal nonlinear survival curves. *Journal of Food Protection,* 64, 606–613.

McAlister, L. & Finkelstein, D.B. (1980). Heat shock proteins and thermal resistance in yeast. *Biochemical and Biophysical Research Communications,* 93, 819-824.

Michalski, C. R., Brackett, R.E., Hung, Y.C., Ezeike, G.O.I. (1999). Use of capillary tubes and plate heat exchanger to validate U.S. Department of Agriculture pasteurization protocols for elimination of *Salmonella* Enteritidis from liquid egg products. *Journal of Food Protection,* 62, 112–117.

Mizrahi & Karel. (1978). Evaluation of kinetic model for reactions in moisture-sensitive products using dynamic storage conditions. *Journal of Food Science,* 43(3), 750-753.

Murano, E. A. & Pierson, M.D. (1992). Effect of heat shock and growth atmosphere on the heat resistance of *Escherichia coli 0157:*H7. *Journal of Food Protection,* 55, 171-175.

Murphy, R.Y., Marks, B.P., Johnson, E.R., Johnson, M.G. (2000). Thermal inactivation kinetics of *Salmonella* and *Listeria* in ground chicken breast meat and liquid medium. *Journal of Food Science,* 65, 706–710.

Neidhardt, F.C. & VanBogelen, R.A. (1982). *Heat shock response*, p. 1334-1345. *In* M.J. Schlesinger, M. Ashburner, and A. Tisieres (ed.), Heat shock from bacteria to man. Cold Spring Harbor Laboratory, Cold Spring Harbor, N.Y.

Panagou, E. Z. (2008). A radial basis function neural network approach to determine the survival of *Listeria monocytogenes* in Katiki, a traditional Greek soft cheese. *Journal of Food Protection* 71, 750-759.

Peleg, M. & Cole, M.B. (1998). Reinterpretation of Microbial Survival Curves. 1998. *Critical Reviews in Food Science,* 38(5), 353-380.

Peleg, M. & Penchina, C.M. (2000). Modeling microbial survival during exposure to a lethal agent with varying intensity. *Critical Reviews in Food Science and Nutrition,* 40(2), 159-172.

Peleg, M. (2000). Microbial survival curves – the reality of flat "shoulders" and absolute thermal death times. *Food Research International,* 33, 531-538.

Peleg, M. (2006). *Advanced quantitative microbiology for foods and biosystems*. CRC Press, Boca Raton, FL.

Peleg, M., Normand, M.D., Campanella, O.H. (2003). Estimating microbial inactivation parameters from survival curves obtained under varying conditions—the linear case. *Bulletin of Mathematical Biology,* 65, 219–234.

Peleg, M., Normand, M.D., Corradini, M.G. (2005). Generating microbial survival curves during thermal processing in real time. *Journal of Applied Microbiology,* 98, 406-417.

Peleg, M., Penchina, C.M., Cole, M.B. (2001). Estimation of the survival curve of *Listeria monocytogenes* during non-isothermal heat treatments. *Food Research International,* 34, 383-388.

Periago, P.M., van Zuijlen, A., Fernandez, P.S., Klapwijk, P.M., ter Steeg, P.F., Corradini, M.G., Peleg, M. (2004). Estimation of the non-isothermal inactivation patterns of *Bacillus sporothermodurans* IC4 spores in soups from their isothermal survival data. *International Journal of Food Microbiology,* 95, 205-218.

Podlubny, I. (1999). *Fractional differential equations*: an introduction to fractional derivatives, fractional differential equations, to methods of their solution and some of their applications. Academic Press, San Diego, CA.

Powers, J. J., Lukaszewicz,W,. Wheeler, R., Dornseifer, T.P. (1965). Chemical and microbial activity rates under square wave and sinusoidal temperature fluctuations. *Journal of Food Science,* 31, 187–197.

Saxena, R.K., Mathai, A.M., Haubold, H.J. (2002). On fractional kinetic equations. *Astrophysics and Space Science,* 282(1), 281-287.

Saxena, R.K., Mathai, A.M., Haubold, H.J. (2004). On generalized fractional kinetic equations. *Physica A: Statistical Mechanics and its Applications,* 344, 657-664.

Schlesinger, M.J. (1986). Heat shock proteins: the search for functions. *Journal of Cell Biology,* 103, 321-325.

Shimoni, E. & Labuza, T.P. (2000). Modeling pathogen growth in meat products: future challenges. *Trends in Food Science and Technology,* 11(11), 394-402.

Stephens, P.J., Cole, M.B., Jones, M.V. (1994). Effect of heating rate on the thermal inactivation of *Listeria. Journal of Applied Microbiology,* 77, 702-710.

Streips, U.N. & Polio, F.W. (1985). Heat-shock proteins in bacilli. *Journal of Bacteriology,* 162, 434-437.

Tanguay, R.M. (1983). Genetic regulation during heat shock and function of heat-shock proteins: a review. *Canadian Journal of Biochemistry and Cell Biolog,* 61, 387-394.

Tjakko, A. & Wouters, J.A. (1999). Microbial stress response in minimal processing. *International Journal of Food Microbiology,* 50, 65-91.

Vaidya, N. & Corvalan, C.M. (2007). A closed analytical solution for non-isothermal inactivation of *Listeria monocytogenes* and *Salmonella* spp. based on non-linear isothermal survival data. Presented at the Annual Meeting of the Institute of Food Technologists, Chicago, IL, 28 July to 1 August 2007.

Vaidya, N. & Corvalan, C.M. (2009). An Integral Model of Microbial Inactivation Taking into Account Memory Effects: Power-Law Memory Kernel. *Journal of Food Protection,* 72(4), 837-842.

Vaidya, N. & Corvalan, C.M. (2011). Thermal inactivation and memory effects. *Journal of Food Protection* (submitted).

Valdramidis, V. P., Geeraerd, A. H., Bernaerts, K. & Van Impe, J. F. (2006) Microbial dynamics versus mathematical model dynamics: The case of microbial heat resistance induction. *Innovative Food Science & Emerging Technologies,* 7, 80-87.

Valdramidis, V. P., Geeraerd, A. H. & Van Impe, J. F. (2007) Stress-adaptive responses by heat under the microscope of predictive microbiology. *Journal Applied Microbiology,* 103, 1922-1930.

Valdramidis V.P., Yannakakis G., Cullen, P.J., Geeraerd, A.H. & Van Impe, J.F. (2009). Memory embedded structures of Artificial Neural Networks: limitations and constraints in Predictive Modelling in Foods. *Predictive Modelling in Foods – Conference Proceedings*, CDROM p. 124-127 [Sixth International Conference on Predictive Modelling in Foods,Washington D.C. (USA), 8-12 September, 2009].

van Boekel, M. A. J. S. (2001). On the use of the Weibull model to describe thermal inactivation of microbial vegetative cells. *International Journal of Food Microbiology,* 74, 139-159.

Watson, K. (1990). Microbial stress proteins. *Advances in Microbial Physiology,* 31, 184-223.

Yanushkevichius, R. (2005). Stability estimation of characterization by the lack of memory property. *Journal of Mathematical Sciences,* 131, 5697-5699.

In: Progress on Quantitative Approaches in Thermal … ISBN: 978-1-62100-842-2
Editors: V. P. Valdramidis and J. F. M. Van Impe © 2012 Nova Science Publishers, Inc.

Chapter 10

STERILISATION-PASTEURISATION VALUES TAKING INTO ACCOUNT MICROBIAL NON-LINEARITIES

Pierre Mafart and Louis Coroller
Université Européenne de Bretagne, France - Université de Brest,
Laboratoire Universitaire de Biodiversité et Ecologie Microbienne,
Quimper, France

Abstract

Traditional calculations of heat processes are based on the conventional and popular *F*-value concept which is based on the log-linearity of survival curves. This standard indicator is independent of any target strain and is a useful tool for the assessment and comparison of heat treatments. While classical calculations are based on the property of the *F*-value additive character, when the survival curves are non-log-linear, this indicator loses this character and cannot be directly assessed without the intermediate calculation of the log-reduction of a target strain. Consequently, to retain the traditional definition of the sterilisation (pasteurisation) value (time needed to achieve a required or observed reduction in microbial number at a reference temperature, or any equivalent time-temperature combination yielding the same reduction), the calculation of any *F*-value linked to non-linear curve requires both the knowledge of the resistance parameters of a target strain and the implementation of a non-linear inactivation model.

Two examples of non-linear models are given in this chapter and implemented to calculate the required *F*-value (which depends on the required level of safety) and the actual *F*-value (which is actually observed during a thermal cycle) in isothermal and dynamic conditions.

1. Introduction

Traditional calculations of heat processes (sterilisation, pasteurisation and cooking) are based on the popular *F*-value concept which was introduced in 1923 by Ball. Ball can be regarded as the first author to integrate both microbiological and engineering aspects in the assessment and optimisation of heat treatments. The *F*-value is defined as the time needed to achieve a required or observed reduction in microbial numbers at a reference temperature, or as any equivalent time-temperature combination yielding the same reduction. This parameter then reflects the efficiency of a time-temperature combination and is a relevant indicator for comparing the intensity of two heat treatments. If one considers the *actual F-value* which is measured after a thermal cycle, this indicator should be regarded as an *assessment tool* whereas the *required F-value* which is determined from a necessary safety level acts more as an *optimisation tool.*

In order to understand the implications and limitations of this concept, it is worth recalling the context of this early period which is considered as the beginning of what is presently called predictive microbiology.

- Despite the previous works of Madsen and Nyman (1907) and Chick (1908) who pointed out the first order nature of spore survival kinetics and then introduced the specific inactivation rate as the relevant heat sensitivity parameter, the heat resistance of spores was still characterised by the simplistic Bigelow *TDT* concept (1921) which corresponded to the death time of a global spore population at a given heating temperature.
- The famous Arrhenius equation (1889) was successfully applied to quantify the effect of temperature on the specific inactivation rate, but Bigelow (1921) introduced his well known *z*-concept for describing the effect of heating temperature on *TDT*. This *z*-value is still extensively used in the area of food industries.

1.1. The Two Basic Characteristics of the Conventional *F*-value

- Additive parameter

Let us consider two following heat treatments yielding two decimal reduction ratios n_1 and n_2 respectively:

$$n_1 = \log\frac{N_0}{N_1} \tag{1}$$

$$n_2 = \log\frac{N_1}{N_2} \tag{2}$$

Where N_0, N_1 and N_2 represent the size of the initial alive population, the number of surviving cells after the first heating step and the number of surviving cells after the second heating step, respectively. The overall decimal reduction ratio after the two heat treatments is then:

$$n = \log\frac{N_0}{N_2} = \log\frac{N_0}{N_1}\frac{N_1}{N_2} = \log\frac{N_0}{N_1} + \log\frac{N_1}{N_2} = n_1 + n_2 \tag{3}$$

Note that this additive relationship always remains valid, whatever the pattern of the survival curve of heated spores or vegetative cells. Now, *if and only if the survival curves are log-linear* (first order kinetic), the *F*-values obtained after each heating phase and the overall *F*-value can be written as follows:

$$F_1 = n_1 D^* \tag{4}$$

$$F_2 = n_2 D^* \tag{5}$$

$$F = nD^* \tag{6}$$

where D^* is the decimal reduction time at the reference temperature T^*. From the combination of equations (4)-(6), the following can be written:

$$F = F_1 + F_2 \tag{7}$$

This additive relationship is the basis of traditional calculation of F-values, whatever the numerical or analytical methods implemented, provided the studied survival curves at isothermal conditions are log-linear. In the frequent cases of non-linearities, this relationship is no longer valid and every calculation must consider the decimal reduction ratio instead of the F-value directly.

- ***Standard indicator independent of the target organisms***

Considering an isothermal time-temperature combination (t, T), the corresponding F-value is given by the following equation:

$$F = 10^{\frac{T-T^*}{z}} t \tag{8}$$

where T^* is the reference temperature. From this last equation, it can be seen that the determination of F does not require any D-value, but only the z-value which is conventionally fixed at 10°C for sterilisation or 7°C for some pasteurisation applications. The determination of cooking-values is more complex because cooking leads to a large number of modifications (change of texture, colour, taste etc.) with different corresponding kinetic parameter values. For this reason, the corresponding z-values are more or less arbitrarily fixed (most often at 16°C, 25°C or 33°C).

1.2. Consequences of Microbial Non-linearities on the *F*-concept

When the survival curve linked to the process is not log-linear, the F-value loses its additive character and conventional calculations can no longer be applied. Consequently, any assessment or optimisation calculation requires a suitable primary model for determining log-decrease values (n) which retain their additive character. From this intermediate step, it is possible to derive a sterilisation value which may be close to the F- concept, but this requires the knowledge of resistance parameters of a target strain. For example, Valdramidis et al. (2011) introduced the concept of *MH*- value (Mild Heat Value) which is the time needed to achieve a predefined microbial reduction at a reference temperature with a known thermal resistant constant and a known z-value. Note that this modification of the standard F-concept also changes its meaning: while the conventional F-value is a tool for the standard comparison of processes regardless of any target strain, the *MH*-value rather reflects the efficiency of a process on the reduction of a specific target strain.

Over the last 30 years, a number of inactivation models have been developed with the aim of describing non-log-linear microbial survival curves. An overview of this category of primary models was given by Geeraerd et al. (2005). Among these models, two examples will be taken here as case studies. The Geeraerd model (2000) is a set of differential equations which describes survival curves linked to mild heat treatments by incorporating physiological adjustments during microbial inactivation experiments (for describing the shoulder) and a braking function (for describing the tail). The Mafart equation (2002) is based on a Weibull frequency distribution of individual thermal resistances of cells and is a reparameterisation of the equation previously proposed by Peleg and Cole (1998).

- ***Geeraerd model***

This model includes the conventional specific inactivation k_{max} but adds two complementary parameters: C included in the adjustment function is related to the physiological state of cells whereas N_{res} included in the braking function corresponds to a fraction of resistant cells which are not killed during the heat treatment. The first differential equation is as follows:

$$\frac{dN}{dt} = -k_{\max}\left(\frac{1}{1+C(t)}\right)\left(1-\frac{N_{res}}{N}\right)N \tag{9}$$

The second differential equation describes the evolution of C as a function of time:

$$\frac{dC}{dt} = -k_{\max}C \tag{10}$$

- ***Mafart equation***

This model which is suitable for non-linear survival curves showing a downward or an upward concavity includes a resistance parameter δ corresponding to the time of first decimal reduction and a shape parameter p . In the particular case of a log-linear kinetic, $p = 1$ whereas for $p < 1$, the curve presents an upward concavity and for $p > 1$, a downward concavity.

$$N = N_0 10^{-\left(\frac{t}{\delta}\right)^p} \tag{11}$$

The corresponding decimal reduction ratio as defined in equations (1) and (2) can then be written as follows:

$$n = \left(\frac{t}{\delta}\right)^p \tag{12}$$

2. Required *F*-value

In the case of log-linear kinetics, the required *F*-value is the product of two factors (see equation (6)): the decimal reduction ratio (or log-decrease) which represents the required level of safety and the *D*-value (decimal reduction time) at the reference temperature which represents the thermal resistance of the target strain. Applying the conventional definition of the sterilization value, the required F-value can be determined from non-log-linear models.

2.1. Determination from the Geeraerd Model

At isothermal conditions, the explicit version of equations (9) and (10) is as follows:

$$N = (N_0 - N_{res})e^{-k_{max}t}\frac{1+C_0}{1+C_0e^{-k_{max}t}} + N_{res} \tag{13}$$

Let us consider the initial ratio of resistant cells: $\alpha = N_{res}/N_0$. The equation (13) can be rearranged into:

$$n = \log\frac{N_0}{N} = -\log\left[(1-\alpha)e^{-k_{max}t}\frac{1+C_0}{1+C_0e^{-k_{max}t}} + \alpha\right] \tag{14}$$

At the reference temperature, the maximum reduction ratio takes the particular value k^*_{max}. By definition, the time needed to reach the required log-decease is the F-value which can be determined by solving the following equation:

$$n = -\log\left[(1-\alpha)e^{-k^*_{max}F}\frac{1+C_0}{1+C_0e^{-k^*_{max}F}} + \alpha\right] \tag{15}$$

The solution of this last equation is:

$$F = \frac{1}{k^*_{max}}Ln\left[\frac{(1-10^{-n})C_0 + 1 - \alpha}{10^{-n} - \alpha}\right] \tag{16}$$

It is clear that the quantities 10^{-n} and α can be neglected with respect to the unit, so that a good approximation of the equation (16) is given by the following expression:

$$F = \frac{1}{k^*_{max}}Ln\left[\frac{C_0 + 1}{10^{-n} - \alpha}\right] \tag{17}$$

Note that in the absence of resistant cells (α =0), the equation (16) can be reduced into the following form:

$$F = \frac{Ln10}{k^*_{max}}\left[n + \log(1 + C_0)\right] \tag{18}$$

Similarly, in the absence of shoulder ($C_0 = 0$), the equation (16) becomes:

$$F = \frac{1}{k^*_{max}}Ln\frac{1-\alpha}{10^{-n} - \alpha} \tag{19}$$

Not surprisingly, in the particular case of first order inactivation kinetics, ($\alpha = 0$ and $C_0 = 0$), the equation (17) can be reduced into:

$$F = n\frac{Ln10}{k^*_{\max}} = nD^* \qquad (20)$$

Let us compare two strains of *Listeria innocua*. The first strain will be assumed to be inactivated according to first order kinetics with $D^* = 0.53$ min (reference temperature: 60°C). The curve of the second strain will be assumed to present a shoulder and a tail and to be fitted according to the Geeraerd model with $AsymD^* = 0.53$ min (which corresponds to $k^*_{max} = 4.34\ \text{min}^{-1}$), $C_0 = 13.8$ and $\alpha = 10^{-6}$. Because the shoulder and the tail portions of the curve correspond to lower thermal resistances than the maximum resistance related to the asymptotic line of the middle part of the curve, the required pasteurization value of the non-linear strain (P_{NL}) will be systematically higher than the pasteurization value of the linear strain (P_L). As shown in the following table, the ratio between both required P-values depends on the required log-decrease n and will obviously be minimal for the n-values corresponding to the central linear portion of the survival curve.

Table 1. Pasteurization values assuming log-linear kinetics (P_L) or non-log-linear kinetics (P_{NL}) for given required decimal reduction values

n	P_L (Eqn 6)	P_{NL} (Eqn 17)	P_{NL}/P_L
1	0.53	1.15	**2.17**
2	1.06	1.68	1.58
3	1.59	2.21	1.39
4	2.12	2.75	1.30
5	2.65	3.30	**1.25**
5.9	3.13	4.12	1.32
5.99	3.17	4.67	1.47
5.999	3.18	5.20	**1.64**

2.2. Determination from the Mafart Equation

By defining the F-value the following expression of the required sterilization value can be deduced from the equation (12)

$$F = n^{\frac{1}{p}}\delta^* \qquad (21)$$

where δ^* represents the resistance parameter at the reference temperature. In the case of log-linear inactivation kinetics ($p = 1$), the equation (21) will come down to the conventional equation (6).

Comparing two survival curves, one being log-linear, the other presenting a concavity, beyond a certain reduction ratio, both curves generally intersect. Consequently, if $p < 1$ (upward concavity) the required F-value corresponding to the non-linear curve (F_{NL}) will be lower than the value related to the log-linear curve (F_L) for low n-values required. Contrarily, for high reduction ratios required, the opposite will be observed.

Let us compare two strains of *Clostridium botulinum* . The first strain will be assumed to be inactivated according to a first order kinetic with $D^* = 0.2$ min (reference temperature: 121.1°C). The curve of the second strain will be assumed to present an upward concavity with $\delta^* = 5.27.10^{-3}$ min and $p = 0.346$. The figure 1 presents for both strains the required F-value as a function of the required decimal reduction ratio n.

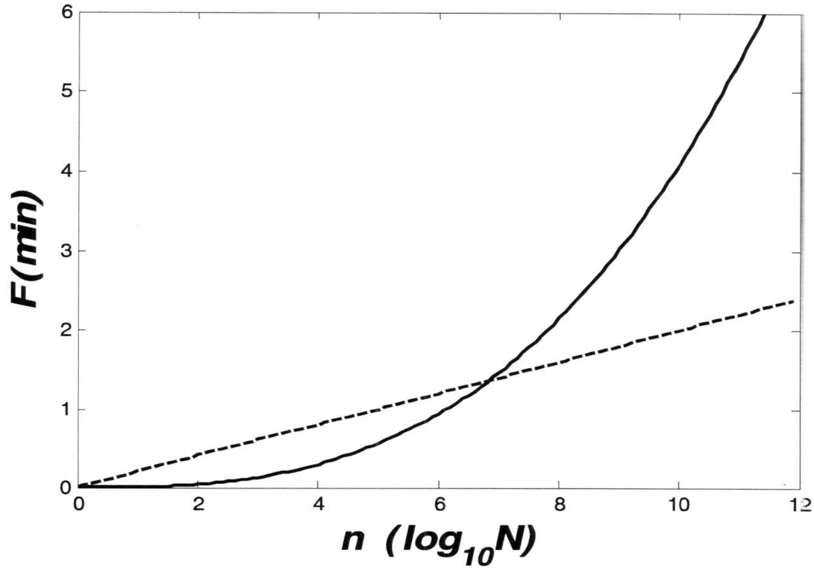

Figure 1. Sterilization value (F) as a function of the required decimal reduction, assuming the loglinear inactivation (dotted line) or the non-loglinear inactivation (straight line) of the bacterial population.

Obviously, the opposite observation could be achieved with regard to the survival curve of a downward concavity strain: $F_{NL} > F_L$ for low n-values and $F_{NL} < F_L$ for high n-values.

3. Actual Sterilization Value

In this paragraph, we will not consider the required sterilization value which would allow a target level of safety, but the sterilization (pasteurization, cooking) value which is actually reached after a given time-temperature scenario of a heating process.

In conventional calculations, based on first order inactivation kinetics and at dynamic conditions (changing temperatures), the equation (8) has to be transformed into:

$$F = \int_0^t 10^{\frac{T-T^*}{z}} dt \qquad (22)$$

This equation was the base of the early classical calculation procedures, i.e., (i) the reference Bigelow empirical method which is more a measurement tool and (ii) the theoretical Ball method which is more suitable for simulation and optimization applications.

In the case of non-linear kinetics, calculations are more complex and, even from the simplest models, require numerical solutions of differential equations.

3.1. Determination from the Geeraerd Model

The most convenient criterion for characterizing a sterilization value in application of this model is the *MH*-value defined by Valdramidis et al. (2010), i.e., time needed to achieve a predefined reduction *among the non-resistant cells* at a reference temperature *T** and a known thermal resistant constant which is assumed to follow the Bigelow relationship (*z*-concept). The part of maximum rate of inactivation of the curve is regarded as an asymptotic linear and the resistance parameter is defined as follows:

$$AsymD = \frac{Ln10}{k_{max}} \tag{23}$$

Then, the achieved microbial reduction is given as follows:

$$\int_{N_0}^{N} \frac{dN}{N} = -\int_0^t k_{max}\left(\frac{1}{1+C}\right)\left(1-\frac{N_{res}}{N}\right)dt \text{ or,}$$

$$Ln\frac{N-N_{res}}{N_0-N_{res}} = -\frac{Ln10}{AsymD^*}\int_0^t \exp\left[\frac{Ln10}{z}(T-T^*)\right].\left(\frac{1}{1+C}\right)dt \tag{24}$$

In the asymptotic linear portion of the curve, by analogy with the classical equation (6) the *MH*-value can be written as follows:

$$MH = AsymD^* \, Ln\left(\frac{N_0-N_{res}}{N-N_{res}}\right) \tag{25}$$

The combination of equations (24) and (25) then yields:

$$MH = Ln10\int_0^t \exp\left[\frac{Ln10}{z}(T-T^*)\right].\left(\frac{1}{1+C}\right)dt \tag{26}$$

With
$$\frac{dC}{dt} = -k_{max}(T(t)).C(t) \tag{27}$$

3.2. Determination from the Mafart Equation

The differential version of the equation (12) is as follows:

$$dn = p\delta^{-p}t^{p-1}dt \tag{28}$$

where
$$\delta = \delta^* 10^{-\frac{T-T^*}{z}} \tag{29}$$

The combination of equations (28) and (29) then yields:

$$dn = p\left(\delta * 10^{-\frac{T-T*}{z}}\right)^{-p} t^{p-1} dt \tag{30}$$

$$n = p\delta *^{-p} \int_0^t \left[\exp\frac{Ln10}{z}(T - T*)\right]^p t^{p-1} dt \tag{31}$$

By substituting the n-value obtained into equation (21), the obtained F-value is as follows:

$$F = \left[p\int_0^t \left[\exp\frac{Ln10}{z}(T - T*)\right]^p t^{p-1} dt\right]^{\frac{1}{p}} \tag{32}$$

Interestingly, the determination of the sterilization value does not require the knowledge of the scale parameter $\delta*$, but only that of the shape parameter p and that of z.

The main drawback of the conventional F-value when applied to non log-linear kinetics is the absence of the additive character. Let F be the overall sterilization value resulting from two successive heat treatments whose sterilization values would be F_1 and F_2 respectively. The following relationship can be derived from the Mafart equation (see Mafart et al., 2002):

$$F^p = F_1^p + F_2^p \tag{33}$$

It can then be convenient to replace the classical F-value by a close parameter $\varphi = F^p$. Accordingly, the equation (32) has to be transformed into:

$$\varphi = p\int_0^t \left[\exp\frac{Ln10}{z}(T - T*)\right]^p t^{p-1} dt = p\int_0^t [L(T)]^p t^{p-1} dt \tag{34}$$

where the function $L(T)$ is the so called lethality factor.

It is then possible to apply the classical methods for calculating the φ-value and to deduce the F-value from:

$$F = \varphi^{\frac{1}{p}} \tag{35}$$

At isothermal conditions, the solution of the equation (34) is as follows:

$$\varphi = [L(T)t]^p \tag{36}$$

4. Determination of the *F*-value at Dynamic Conditions

4.1. Empirical Approach

The function *T(t)* is rarely known and in most cases, managers have a recording of the retort temperature and of the core temperature of the heated product. Each time interval Δt, corresponds to the following variation of the φ function:

$$\Delta\varphi = p[L(T)]^p t^{p-1}\Delta t \quad (37)$$

In most cases the p shape parameter can be assumed to be independent of temperature (Couvert et al., 2005; Mafart et al., 2002; Van Bockel et al., 2002). In the rare cases where a clear effect of temperature on the p-value can be observed, a secondary model would have to be derived and to be combined with the equation (37).

At the end of the thermal cycle, the cumulated φ-value is then:

$$\varphi = p\sum_1^n [L(T)]^p t^{p-1}\Delta t_i \quad (38)$$

Because the survival curves are not log-linear, in calculations the origin of time cannot be taken at the beginning of the thermal cycle, but has to be taken at the time when cells start to die. The application of the equation (38), like that of any non-linear dynamic model, then requires the knowledge of the corresponding minimum lethal temperature. For example, Zanoni et al. (1997) observed that during the cooking of Bologna sausage, *Enterococcus faecium* cells started to die at 55°C.

The equation (8) allows the achieved φ-value of a heating process to be obtained from a Bigelow-like method. Similarly, the log-reduction will be given by the following equation:

$$n = p\delta^{*-p}\sum_1^n [L(T)]^p t^{p-1}\Delta t_i \quad (39)$$

Let us consider the sterilization cycle represented by Figure 2 and its efficiency on two strains of *Clostridium botulinum:*

- A strain presenting a log-linear survival curve (D^* = 0.2 min; z = 10°C)
- A Weibullian strain ($\delta^* = 5.27.10^{-3}$; p = 0.346; z = 10°C)

Table 2 compares the achieved F and n values related to both strains (FL, nL) and (FNL, nNL) respectively. The increments Δt of equations (38) and (39) were fixed at 1 min (as in the traditional Bigelow method). It was also assumed that the minimum lethal temperature at which the death of spores started was 100°C.

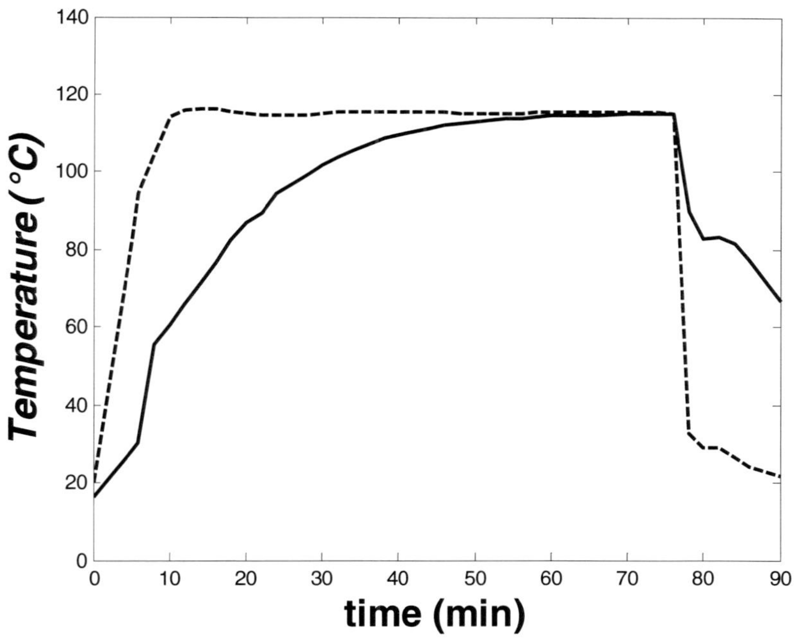

Figure 2. Example of a temperature follow up during a heat treatment. The ambient temperature of the can in the autoclave is represented by the dotted line and the temperature at the cold point of the can is represented by the straight line.

Table 2. Sterilization values (F) and decimal reduction values (n) during the heating process (see figure 3). These values were computed following the loglinear (L) or the non-log-linear (NL) assumptions

t	F_L	n_L	F_{NL}	n_{NL}
36 min	0.18	**0.9**	0.02	**1.5**
45 min	0.91	**4.5**	0.13	**3.0**
60 min	3.49	**17.4**	0.60	**5.2**
76 min	6.32	**32.6**	1.20	**6.6**

It can be seen that the log-reduction regarding the Weibullian strain exceeds that of the linear strain only at the beginning of sterilization whereas, at the end of the cycle the achieved log-reduction of the first strain is clearly lower than that which would be expected from the conventional calculations based on first order kinetics. This observation can easily be explained by considering the pattern of the non-linear survival curve which presents an upward concavity. At the beginning of the cycle, the thermal resistance of the concerned strain is relatively low, whereas it becomes quite high at the end of the sterilization. Of course, for a strain showing a downward concavity ($p > 1$) the opposite trend would be observed and conventional calculations would underestimate the efficiency of the heating process.

4.2. Theoretical Approach

When the heat transfer equation regarding the heated product is known, the traditional Ball approach can be adapted to the cases of non log-linear survival curves from the following equation:

$$n = p\delta^{*-p} \int_0^t 10^{p\frac{T-T^*}{z}} t^{p-1} dt \tag{40}$$

This can be combined with the following classical transfer equation (valid only after a certain heating time):

$$T = T_\infty - j(T_\infty - T_0)e^{-kt} \tag{41}$$

where T_0 and T_∞ represent the initial core temperature of the product and the retort temperature respectively, whereas j and k are the thermal empirical parameters of the product.

Similarly, during the chilling step of the thermal cycle and after a certain chilling time, the corresponding transfer equation can be written as follows:

$$T = T_c + (T_g - T_c)e^{-kt} \tag{42}$$

Let us once again take the previous example of sterilization regarding the same strains of *Clostridium botulinum.* From the temperature recording represented in Figure 2, the following thermal parameters could be estimated:

$$k = 0.088 \text{ min}^{-1}$$

$$j = 1.92$$

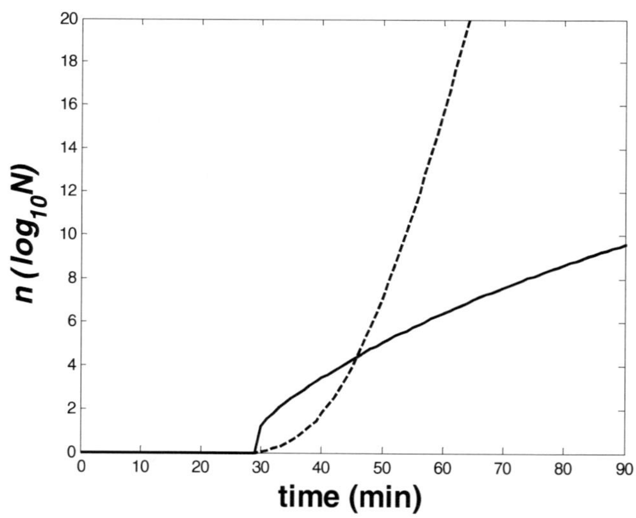

Figure 3. Decimal reduction number during the sterilization process, assuming the loglinear inactivation (dotted line) or the non-loglinear inactivation (straight line).

The equation resulting from the combination of equations (40) and (41) could be solved numerically (Matlab R2010a, Mathworks, MA, USA). Assuming that the death of spores starts at 100°C, Figure 3 illustrates the compared simulations of the evolution of the *n* value of both strains (the linear and the Weibullian strains) during the first 76 minutes of the heating phase.

The following table 3 shows the compared *n* and *F*-values at different phases of heating.

Table 3. Sterilization values (F) and decimal reduction values (n) during the heating process (see figure 3). These values were computed following the log-linear (L) or the non-log-linear (NL) assumptions

t	F_L	n_L	F_{NL}	n_{NL}
36 min	0.15	**0.7**	0.05	**2.1**
45 min	0.79	**4.0**	0.35	**4.3**
60 min	3.14	**15.7**	1.13	**6.4**
76 min	6.63	**33.2**	2.34	**8.2**

Conclusion

The occurrence of non-log-linear survival curves does not make the *F*-concept out of date, but highly complicates its determination and modifies its meaning. While the conventional sterilization value, based on first order survival kinetics, characterizes only a given heat process regardless both of any target strain and any property of the treated products, the modified sterilization value is necessarily linked to a target organism. Consequently, its determination requires not only a *z*-value, but generally also the two (or more) resistance parameters of the target strain.

Another difficulty is the variability not only of the parameter resistance, but also of the shapes of curves regarding a given strain. A considerable amount of works would be needed in order to elucidate the effects of some factors such as the age of spores, the physiological state of vegetative cells, the main environmental factors etc., on the shape parameters which determine the patterns of the survival curves. In particular, most survival curves were studied in liquid media and laboratory conditions, but only few data are available on the shapes of curves in canned foods.

The fact that the heat resistance of cells is variable with heating time adds a further difficulty because it requires that the temperature at which the cells start to die is known. For this reason, the origin of time has to be taken, not at the beginning of the heating, but at the moment when the lethal core temperature is reached, otherwise at this moment, the heat resistance would be either overestimated (curves presenting an upward concavity) or underestimated (curves presenting a shoulder or a downward concavity).

Because the modified sterilization value is dependent on a target strain, it loses its interesting function of a standard indicator allowing an intrinsic comparison of heat processes. It would then be worth selecting a limited number of (real or fictive) reference pathogenic and spoilage strains to which standard parameter values would be allocated. A relevant choice of such reference strains would justify the implementation of a large international research program.

The modification of the *F*-concept taking into account non-linearities can be regarded as a "primary extension" because it involves new primary survival models. Further improvement of this concept would consist in adding a "secondary" extension through the implementation of secondary models taking into account the environmental property of the product. To this effect, Mafart (2000) proposed the extension of the conventional lethal factor which became a function not only of the heating temperature, but also, of the pH and the water activity. While the traditional sterilization value characterizes only a given heat process, the primary extension of this concept takes the resistance of the target strain into account. Furthermore, a secondary extension of this concept also adds the consideration of some properties of the treated foodstuff.

References

Arrhenius, S. (1889). Über die Reaktiongeschwindigkeit bei der Inversion von Rohmzucker durch Saüren. *Z. Phys. Chem.*, 4 ,226-248.

Ball, C. (1923). Thermal process time for canned food. *Nat. Res. Council*, DC. Bull. 7-1.

Bigelow, W. (1921). The logarithmic nature of thermal death curves. *J. Infec. Des.*, 29,528-536.

Chick, H. (1908). An investigation of the laws of disinfection. *J. Hyg.*, 8, 92.

Couvert, O., Gaillard, S., Savy, N., Mafart, P. & Leguérinel, I. (2005). Survival curves of heatred bacterial spores: effect of environmental factors on Weibull parameters. *Int. J. Food Microbiol.*, 101,73-81.

Geeraerd, A., Herremans, C. & Van Impe, J. (2000). Structural requirements to describe inactivation during a mild treatment. *Int. J. Food Microbiol.*, 59,185-209.

Geeraerd, A., Valdramidis, V. & Van Impe, J. (2005). GinaFIT, a freeware tool to assess non-log-linear microbial survivor curves. *Int. J. Food Microbiol.*, 102, 95-105.

Madsen, T. & Nyman, M. (1907). Zur Theorie der Desinfektion. *Zeitschr. Hyg.*, 17, 388-404.

Mafart, P. (2000). Taking injuries of surviving bacteria into account for optimizing heat treatments. *Int. J. Food Microbiol.*, 55, 175-179.

Mafart, P., Couvert, O., Gaillard, S. & Leguérinel, I. (2002). On calculating sterility in thermal preservation methods : application of the Weibull frequency distribution model. *Int. J. Food Microbiol.*, 72, 107-113.

Peleg, M. & Cole, (1998). M. Reinterpretation of microbial survival curves. *Critic. Rev. Food Sci.*, 38, 353-380.

Valdramidis, V., Geeraerd, A., Tiwari, B., Cullen, P., Kondjoyan, A. & Van Impe, J. (2011). Estimating the efficacy of mild heating processes taking into account microbial non-linearities: a case study on the thermisation of food simulant. *Food control*, 22, 137-142.

Van Boekel, M. (2002). On the use of the Weibull model to describe thermal inactivation of microbial vegetative cells. *Int. J. Food Microbiol.*, 74, 139-159.

Zanoni, B., Peri, C., Garzaroli, C. & Pierucci, S. (1997). A dynamic mathematical model of the thermal inactivation of *Enterococcus faecium* during Bologna Sausage cooking. *Lebens.-Wiss.- Technol.*, 97, 727-734.

In: Progress on Quantitative Approaches in Thermal … ISBN: 978-1-62100-842-2
Editors: V. P. Valdramidis and J. F. M. Van Impe © 2012 Nova Science Publishers, Inc.

Chapter 11

TIME-TEMPERATURE INTEGRATORS IN THERMAL PROCESSING

S. Bakalis, M.J.H. Simmons, K. Meahauden and P.J. Fryer
Centre for Formulation Engineering, Chemical Engineering,
University of Birmingham, Birmingham, UK

Abstract

Thermal treatment is the most common method used by industry to ensure food is safe for consumption and to increase its storage life. To ensure safety, food is often overprocessed which can significantly affect its nutritional value as well as taste and flavour attributes. However accurate and reliable characterisation of thermal processes remains a challenge as complicated and moving processing equipment prohibits the use standard temperature measurement techniques such as thermocouples. Time Temperature Integrators (TTIs) has been suggested as an alternative that would provide information for process validation and design. TTIs are small unattached measurement devices which contain a thermally labile enzyme: determination of the degree of degradation of the enzyme at the end of the thermal process enables the integrated temperature history to be obtained. Enzymatic Time Temperature Integrators (TTIs) are commonly used to determine the heat treatment efficiency. Experiments permormed in pilot scale equipment demonstrate limitations in using thermocouples to validate thermal processes and demonstrate the potential of TTIs as measurement tools to validate both the effect but also the uncertainty of thermal processes in conditions.

1. Introduction

The main objective of food manufacturers is to offer products that are safe while have the highest possible nutritional value. Over the last decades pressure on food manufacturers is high as they are responsible for the safety of their products. In addition processing is in practice determined to a large extent by a regulatory framework (e.g. hygienic design, thermal processing).

In the area of food microbiology a range of fundamental questions remains unanswered, e.g. determination of accurate inactivation kinetics. In addition the inactivation (thermal or non-thermal) that can be achieved in large scale depends on various processing conditions, such as convective heat transfer, that are often not very well understood. Despite the lack of clear understanding food manufacturers still have the challenge of producing large quantities of safe foods. This in practice is typically achieved by over processing resulting in a reduction of both nutritional and sensorial aspects of foods. With a current pressure in fresher, higher quality products over processing is a less attractive option and a better understanding of the significance of the processing variables should be obtained.

Traditionally to guarantee safety of food products, processors use a range of methods to reduce the amount of micro organisms and/or to eliminate the pathogenic micro organisms present in the food.

Improved Hygiene. The micro organism growth can be delayed or prevented by observing basic hygiene rules during the production of the food (Frazier and Westhoff, 1978). Hazard Analysis and Critical Control Points (HACCP) is an example of a food safety methodology which can be applied in order to reduce food contamination. It relies on the identification of Critical Control Points (CCP's) in food production and preparation processes, which are then closely monitored in order to ensure that food is safe for consumption (Mortimer and Wallace, 1998).

Chemical preservation. Chemicals such as salt or acid can be added to the food in order to decrease the pH of the food or to reduce the water activity, limiting the growth of the micro organisms (Goff, 1999).

Biological fermentation. The aim of food fermentation is to increase the number of desired micro organisms. Desirable micro organisms are added to the food and cause the competitive disappearance of the non desirable micro organisms. This is one of the oldest methods of food preservation (Goff, 1999).

Active packaging. Smart or active packaging are used to extend the shelf life of the food product by acting as oxygen scavengers, carbon dioxide scavengers, moisture control agents, or anti microbial agents (Sacharow, 2006; Kerry et al., 2006; Nerin et al., 2006).

Removal of micro organism by micro filtration. This technique is considered as a cold pasteurisation and it is only used for liquid products such as milk or beer (Goff, 1999).

Irradiation. In this technique, electro magnetic waves or electrons are applied to the food. Ionizing and UV radiations damage the microbial DNA and cause potentially lethal DNA lesions. However, the application of this technique is limited due to current public concern.

High pressure. Pressures ranging from 100 to 1000 MPa are applied to the food causing damage to the micro organism cell membrane and denaturation of its proteins. This technique is already being used by some industries and is starting to attract a number of industries (Rastogi et al. 2007) Manas and Pagan, 2005; Stewart, 2004; Devlieghere et al., 2004; Lado and Yousef, 2002).

Pulsed electric field. Short duration (1-100μs) high electric field pulses (10-50kV cm^{-1}) are applied to food between two electrodes. The structure of the membrane of the micro organism is generally damaged during the exposure at high voltage field. This technique can only be applied to liquid foods such as orange juice and liquid eggs (Hulsheger et al. 1981; Guerrero-Beltran et al. 2010) ; Manas and Pagan, 2005; Stewart, 2004; Devlieghere et al., 2004; De Haan et al., 2002; Lado and Yousef, 2002).

Thermal treatment. This is one of the most common methods used to produce microbiologically safe products. Typically high temperatures (>60°C) are applied to foods to reduce the quantity of micro organisms and spores present is a commonly used thermal treatment (Fryer et al., 1997) and will be described more in detail in the next section.

1.1. Thermal Treatment

Heat treatment is a commonly used unit operation aiming to cook the product and at the same time to reduce the quantity of micro organisms and spores by applying high temperatures to the food (Fellows, 2000; Fryer et al., 1997). Heat treatment can be classified according to the temperatures applied to:

Pasteurisation is the process of relatively mild heating (generally does not exceed 100°C) of food for the purpose of reducing the microbial load (Lloyd et al., 1999). The pasteurised product is typically refrigerated and consumed before its expiration date.

Sterilisation is a high level of treatment and the commonly used temperature range is 115°C – 125°C. The aim of this treatment is to ensure the destruction of all the pathogenic micro organisms, their toxins and their spores. Sterilisation treatment is based on the destruction of *Clostridium botulinum*, which can prove fatal if ingested.

The US Food and Drug Administration define the 'minimal thermal process' as 'the application of heat to food, either before or after sealing in a hermetically sealed container, for a period of time and at a temperature scientifically determined to be adequate to ensure the destruction of micro organisms of public health concern' (Lopez, 1987). As process engineers determining the optimal process conditions (i.e. process temperature and time) is important. Theoretically, the duration of the thermal treatment chosen depends on several parameters (Fellows, 2000):

- the thermal resistance of the micro organisms and spore potentially present into the food
- the heating characteristics
- the pH of the food and other nutrient characteristics
- the shape and size of the container
- the physical conditions of the food (liquid, solid, mixture).

If one assumes that there is no heat transfer limitation, i.e. uniform temperature in the product, the process time would depend on kinetics of thermal inactivation of the targeted microorganism – spores (Holdsworth 1985). The calculation of the thermal process is typically based on the destruction of the heat resistant spore of Clostridium botulinum (Fellows, 2000).

At constant temperature T, the death of micro organisms is given by (Ball and Olson 1957):

$$\frac{N_{final}}{N_{initial}} = 10^{-t/D_T} \tag{1}$$

where $N_{initial}$ and N_{final} are the initial and final number of micro organisms, t is the duration of the heat treatment and D_T is the Decimal reduction time required to reduce the number of micro organisms by 90% at temperature T, and which varies with the type of organism (Valentas et al., 1997).

Traditionally, in canning industry, the concept of 12 D_T cook (botulinum cook) is used. This means processes are designed for a 12 log reduction in the micro organism population (Richardson, 2004).

Equation 2.1 can be rearranged to give the relationship in a form that can be used for the determination of the thermal processing time, *t*:

$$t = D_T \cdot log\left(\frac{N_{initial}}{N_{final}}\right) \tag{2}$$

Thermal processes are commonly compared using parameters such as the F or P value defined as follows:

$$F\ (or\ P) = D_{T(ref)} \cdot log\left(\frac{N_{initial}}{N_{final}}\right) \tag{3}$$

Where *F* or *P* are the process values (*F* for sterilisation and *P* for pasteurisation). If the process temperature is not constant with respect to time F (or P) values represent an equivalent time at a chosen reference temperature (Ball and Olson, 1957; Bhowmik and Tandon, 1987; Holdsworth, 1985; Ohlsson, 1980) and can be calculated as follows:

$$P\ or\ F = \int_0^t 10^{\frac{T(t) - Tref}{z}} \cdot dt \tag{4}$$

where *T(t)* is the product temperature (°C), T_{ref} is the reference temperature for the D_T value (°C), *t* is the process time (min) and *z* is the number of degrees Celsius needed to bring about a ten-fold change in decimal reduction time. The *z* value represents the temperature sensitivity of the system and should ideally be the same as that for the targeted micro organism. In case of estimation the effect of processes where temperature is non uniform the temperature at the coldest point is commonly used.

Clostridium botulinum has a *z* value of 10°C (Hendrickx et al., 1995; Richardson, 2001). Most of the micro organisms vegetative cells and spores have a *z* value ranging from 5°C to 12°C while the *z* value for the destruction of thermal labile chemical substances start from 25°C to 30°C. Thus the destruction of nutrients is less temperature sensitive than the destruction of micro organisms.

The *P* or *F* value represents the thermal treatment received by the food. These values have to be in practice evaluated for the worst case scenario for thermal treatment, i.e. at the coldest location. As such significant effort has been devoted over the years towards understanding heat transfer in food processes.

1.2. Heat Transfer during Thermal Treatment

When conduction is the controlling heat transfer mechanism the temperature field can be estimated from the solution of the partial differential equation:

$$\rho c_p \frac{\partial T}{\partial t} = \nabla(\lambda \nabla T) \tag{5}$$

The solution of which requires knowledge of the spatial variation of the thermal conductivity, λ, the density, ρ and the specific heat, c_p. Initial and boundary conditions are also required. The simplest is constant temperature; however, a heat transfer boundary condition is often necessary. The overall rate of heating of a solid will depend on consecutive processes; heat must move to the product and then within it. Equation (5) has been used in the literature (Bakalis et al. 2001; Miri et al. 2008) to predict and optimise food processes that heat transfer is dominated by conduction. These would include heating of solids and very viscous liquids.

In many food processes though convective heat transfer mechanisms occur. In those cases the heat transfer equation is coupled with the momentum transport and as such both equations need to be solved simultaneously. In order to obtain the temperature field and the thermal treatment the partial differential equations describing the heat and momentum transport in vector notations required to be solved are (Bird et al. 2007):

Equation of continuity:

$$\nabla \cdot \vec{u} = 0 \tag{6}$$

Equations of motion:

$$(\vec{v} \cdot \nabla)\vec{v} = -\nabla P + \mu \nabla^2 \vec{v} \tag{7}$$

Equation of energy:

$$\rho C_p \vec{v} \cdot \nabla T = \lambda \nabla^2 T \tag{8}$$

In the above equation physical properties of the fluid, i.e. density, specific heat and thermal conductivity, were assumed constant. Further to that heat generation by viscous dissipation was assumed negligible.

In order to overcome problems typically associated with batch processes, i.e. large processing time and relatively low heat transfer coefficient, effort has been devoted towards developing continuous processes. Typically these would be High Temperature Short Time process (HSTS) that would ensure microbial inactivation while maximising nutrient content (Fryer et al., 1997). In a typical sterilisation process the product passes through a heater and is then cooled. In many cases sterility is achieved in a holding tube placed between the heater and the cooler. Tubular heat exchangers are commonly used in the food industry, providing a relatively simple geometry for solving the above system of equations.

The concept of F value, described for batch processes using Equation 3 can be extended to for the non-isothermal case of continuous flow the equation as follows:

$$F(r,L) = \int_0^t 10^{(T(r,t)-T_o(r,t))Z}\, dt = \int_0^L 10^{(T(r,t)-T_o(r,t))Z} \left(\frac{1}{dz/dt} \right) dz = \int_0^L 10^{(T(r,z)-T_o(r,z))Z} \left(\frac{1}{u(r,z)} \right) dz \tag{9}$$

This equation assumes that microorganisms remain on the streamlines throughout the process. The lethality given by equation (4) is used for the slowest heating zone, i.e. the centre line. A volume average lethality is often estimated as follows:

$$\hat{F}(z) = \frac{\int_0^R F(r,z)u(r,z)2\pi r dr}{\int_0^R u(r,z)2\pi r dr} = \frac{\int_0^R F(r,z)u(r,z)2\pi r dr}{\dot{Q}} \tag{10}$$

An analytical method for aseptic processes has been developed, albeit neglecting any temperature effects on viscosity (Simpson and Williams 1974). The authors suggested a total dimensionless length ($\zeta = z\alpha/(R^2 U_{avg}$) of 1.2, with 0.8 for the heating section and 0.4 for the cooling (a holding section was not considered).

The above systems of equations assume a continuous – homogeneous fluid. As a large number of food products contain up to 60% particles, two phase flow systems has been an area of active research over the last decades (Lareo et al. 1997). This presents a challenge as particle – particle interactions can result in non-uniform and often unpredictable flows. In order to design such processes, in practice the thermal inactivation of the fastest moving particles is required. Locating this particle though is far from trivial (Lareo et al. 1997). In addition the heat transfer coefficients between the fluid and the particle are needed. Although a number of correlations for the heat transfer coefficient have been published in the literature (Barigou et al. 1998) the accuracy of these models for food systems is questionable.

In order to model particulate flows a set of equations describing the motion of the fluid and the particles has to be solved simultaneously. The motion of fluid is described from the continuity and momentum equations (equations (6) and (7)), while the motion of particles is given from a force balance on each particle (Sastry et al. 1989). The energy equations for the particles and the fluid have to be solved under transient conditions to estimate temperature at various locations. From the above it is clear that solving the full problem for realistic process conditions pauses a tremendous computational challenge.

From the above it is clear that designing and optimising safe food processes from a first principles is a challenge and a range of experimental techniques would have to be developed to aid process design and evaluation.

1.2.1. Evaluating Velocity Fields

From the previous section it is clear that flow field has a profound effect in cases where heat convection is significant. A number of investigators have used various methods in order to measure velocity distributions. Laser Doppler Anemometry (LDA) and Particle Imaging Velocimetry (PIV) are commonly used techniques to characterise fluid flow (Sastry, Heskitt et al. 1989; Bakalis and Karwe 1998; Gabriele et al. 2009). However, as both LDA and PIV are methods that require optical access to the flow, not only the materials used but also the equipment has to be transparent to allow flow materials to be quantified. Magnetic Resonance Imaging (MRI) has also been used to measure velocity distributions of complex rheological properties in opaque systems (Wang et al. 2000). Although velocity measurements can be performed in opaque materials, a significant limitation is imposed from the fact that it is not possible to use metal in the experimental equipment not allowing measurements under in realistic heat transfer conditions. Ultrasound velocity measurements have been also used in food materials (Fischer et al. 2009) but there has not been any application relevant to thermal processing of foods.

The School of Physics at Birmingham has developed a unique way of following particles and flows in opaque fluids, using Positron Emission Particle Tracking (PEPT). In this technique a radioactive tracer particle (about 100 μm in diameter) is introduced in the flow. The position of the tracer can be then located using sub mm accuracy in pilot scale equipment (Parker et al. 1993). In relevance to food thermal processes the technique was used to measure velocity distributions inside axially rotating horizontal cans (Cox et al. 2003). It is a common industrial practice to add air in the cans, the effect of the headspace was investigated. The authors concluded that the addition of the headspace alters the character of the flow resulting in a fully three dimensional flow. This is believed to result in a higher heat transfer coefficient.

1.2.2. Reaction Rates in Heating

It is important to keep in mind that during thermal processing not only microorganisms are getting deactivated but also chemical reactions responsible for flavour generation are taking place. Flavour generation involves a system of complex reactions with Maillard reactions being common where reducing sugars react with protein to produce a wide range of characteristic flavour compounds (Fennema 1996). The kinetics of the reactions depend on a range of parameters such as water activity, pH and most importantly temperature. Reaction schemes are typically simplified to a first order reaction and an Arrhenius equation is used to describe the dependence of reaction rates on temperature.

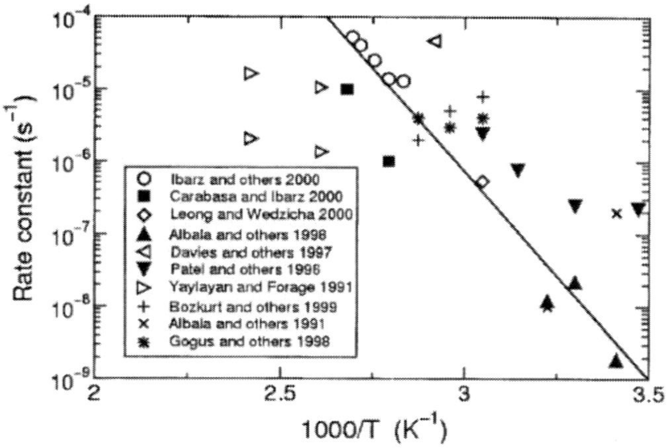

Figure 1. First order rate constants for the formation of flavour compounds compounds (hydroxymethylfurfural or other O containing heterocyclic compounds) (with permission from Jousse et al. 2002).

Figure 1 shows rate constant data where a simplified reaction scheme was used (Jousse, Jongen et al. 2002) and data from literature show a surprisingly good agreement with an Arrhenius relationship. Alongside the development of desirable flavour compounds, thermal treatment results in loss of nutrients as well as loss of textural quality (De Roeck et al. 2010). Van Boekel (Van Boekel 2008) gives a detailed review of kinetic models in foods.

As a result of these reactions excessive thermal treatment will produce low quality food products, both in taste and nutrition. The aim of thermal processing is thus to provide the minimum thermal treatment which gives the required level of microbial kill and thus safety and shelf life.

1.3. Traditional Techniques for Monitoring the Effectiveness of Heat Transfer

The effectiveness of the thermal treatment has to be validated in order to ensure that the food product is safe for consumption. It is a common industrial practice to monitor temperature and temperature history of products.

Typically Thermocouples and data loggers are used as measurement tools. Temperature is measured as a function of time inside the food container. This time temperature history can therefore be translated into an equivalent time at a reference temperature. This validation technique is straightforward to apply and the data analysis is relatively fast and easy. However, thermocouples and data logger are not convenient for all types of thermal processes. In some cases, they can interfere on the movement of the fluid and lead to incorrect time temperature history and sometimes it is impossible to put them at the coldest location of the container. Moreover, the size of the wireless data logger (caused by the battery), is an issue since it has been not possible so far to miniaturize them sufficiently (Hendrickx et al. 1995; Marra and Romano 2003).

The efficiency of the thermal treatment can be also determined by measuring the change of a quality or safety attribute (e.g. micro organism counts) before and after thermal treatment. A sample of the food is taken before and after thermal treatment and the micro organisms are counted. The *F* or *P* value is calculated using Equation 4 (Richardson, 2001). However this technique is laborious, lengthy (incubation takes days), expensive (the media used to grow the micro organism can be expensive) and sometimes the analysis is problematic due to detection limits (Van Loey et al. 1996). The disadvantages encountered by these two techniques have led to the development of a new monitoring technique, the Time Temperature Integrators (TTIs).

2. Time Temperature Integrators (TTIs)

As described previously, various techniques are available to determine the efficiency of thermal processes. Nonetheless, these techniques present various limitations. The evaluation of the level of a food attribute before and after heat treatment is an efficient and widely used technique. However, when the parameter of interest is the determination of the presence of some micro organisms, the analysis can take several days while the shelf life of a product can be limited (Hendrickx et al., 1995).

The analysis of thermocouple data is relatively fast and easy. However, the development of new technologies makes the access for probes sometimes impossible. This is the case for agitated vessels where surface scraped heat exchangers are used (Weng et al. 1991; Hendrickx, Maesmans et al. 1995). Another problem emerges when industry needs to know the thermal process that a food particulate floating in liquid (e.g. soup with pieces of meat) receives. If a thermocouple is attached to a food particulate in liquid, the thermocouple will interact with the fluid motion and will modify the process. Therefore, it will not be possible to know what the food particulate undergoes during the process.

Time Temperature Integrators (TTIs) are alternatives to conventional temperature probe systems. TTIs are a relatively new technology which allow determination of the impact of a process on a product attribute (Van Loey, Hendrickx et al. 1996). TTIs are devices which contain a thermally labile substance, which under a heat treatment undergoes irreversible changes which can be quantified as a *F* or *P* value (Taoukis and Labuza 1989; Hendrickx et al. 1992; Hendrickx, Maesmans et al. 1995; Van Loey, Hendrickx et al. 1996; Tucker 1999; Guiavarc'h et al. 2005).

TTIs are relatively small (a few cm in size) (Hendrickx, Weng et al. 1992), shock resistant monitoring devices which can be made neutrally buoyant from materials with the same thermal conductivity as the food. Their shape can be modified to simulate food particles and they can be put inside a particle without interfering with the fluid motion. Their analysis is fast and relatively simple (VanLoey et al. 1996; Guiavarc'h, Zuber et al. 2005) (Guiavarc'h et al., 2005b; Van Loey et al., 1996b). Conversely to the thermocouples, the time temperature history of the product is not needed to determine the impact of thermal treatments (Hendrickx, Maesmans et al. 1995; Van Loey, Hendrickx et al. 1996; Guiavarc'h, Zuber et al. 2005).

2.1. Types of the TTI

TTIs can be classified according to the substance that they contain, which can be microbiological, enzymatic, chemical or more rarely physical. Alternatively, TTIs can be classified according to their origin and application within the food.

Intrinsic TTI. They are naturally present in the food and, the efficiency of the thermal treatment can be evaluated by quantifying it before and after treatment. Intrinsic TTI presents some advantages such as being homogeneously dispersed inside the food. Researches have been undertaken to assess the heat treatment received by the milk (Claeys et al., 2002). In this study, some components naturally present in the milk are used as TTIs.

Extrinsic TTI. Extrinsic TTIs are artificially added to the food (Claeys et al. 2002) . Extrinsic TTIs can be subdivided into 3 groups (Hendrickx, Maesmans et al. 1995; Van Loey, Hendrickx et al. 1996):

- dispersed TTIs are added and mixed to the food and are in direct contact with the food;
- permeable TTIs are located in separate units with a permeable barrier allowing exchange between the food and the TTIs;
- isolated TTIs are located in separate units and their barrier is not permeable and therefore does not allow exchange with the food material.

A summary of different TTIs is given in

Table 1. Classification of TTIs

	Microbiological TTI	Enzymatic TTI	Chemical TTI	Physical TTI
Principle	Microbiological TTIs can be divided into 2 types of techniques. In the first one, the impact of the process is quantified from the number of micro organisms that survives. The second one, just show if there is growth or no growth of micro organisms	Quantification of the activity of a thermally labile enzyme that remains after the heat treatment. The amount of remaining activity reveals the impact of the temperature history	Detection of concentration change of a chemical substance voluntarily added to the food.	Diffusion of a coloured chemical compound in a paper wick
Popularity	Most commonly used type of TTIs	Relatively new technique. Food companies starting to use them	Rare because Chemical TTIs still cannot be used for food monitoring since no chemical reactions have been identified so far with an applicable *z* value	Development limited - it is not currently used in food safety determination
Application	Monitoring of the sterilisation of food or pharmaceutical products	Their high thermostability allows them to be used at very wide range of temperatures (pasteurisation, sterilisation) for food and pharmaceutical products	Used for the monitoring of the deterioration of other quality attributes	Limited application
Advantage	Widely spread	Small, wireless and shock resistant	Fast and analytically accurate	Easy to prepare to calibrate, user-friendly, fast to obtain the results and easy to recover
Inconvenient	Long and analytically not precise enough	Not suitable for on-line monitoring. Retrieval from the food product can be an issue	Can not be used for food safety monitoring	This system is activated by steam therefore no other heating media can be monitored
Reference	(Hendrickx, Maesmans et al. 1995; Van Loey, Hendrickx et al. 1996)	(Hendrickx, Maesmans et al. 1995; Van Loey, Hendrickx et al. 1996)	(Adams and Langley, 1998)	(Hendrickx, Maesmans et al. 1995; Van Loey, Hendrickx et al. 1996)
Origin	Intrinsic or extrinsic	Intrinsic or extrinsic	Extrinsic	Extrinsic

2.1.1. Enzymatic TTI

The usage of enzymatic TTIs is currently increasing in the determination of heat treatment effectiveness since they do not cause any food safety issues, unlike microbiological TTIs (Maesmans et al. 1994; Tucker 1999; Guiavarc'h et al. 2002; Guiavarc'h et al. 2004; Guiavarc'h et al. 2004). However, the selection of the enzyme can be difficult since it has to fulfil several requirements:

- the enzyme isothermal inactivation must follow a known order of kinetics (Welt et al. 2003);
- the enzyme must be thermally resistant;
- the enzyme must have a *z* value that matches organisms of interest typically $z = 8 - 10$°C in order to be used as a food safety tool (Van Loey, Hendrickx et al. 1996).;

Studies have been undertaken on various enzymes in order to determine the most suitable to use for heat treatment effectiveness. Tomato and cucumber Pectinmethylesterase (PME) has been studied as a potential TTI (Guiavarc'h et al. 2003), who showed that the purified tomato PME could follow a first order kinetics. D_{Tref} and *z* could be changed by modifying a stabilizer agent (buffer pH, presence of glycerol). However, the cucumber PME exhibits a biphasic inactivation curve which made it not useable as a TTI. Peroxidase enzyme has been used to determine the efficiency of pasteurisation and gave excellent correlation with the measured time-temperature values (Hendrickx, Weng et al. 1992). However, most studies have focused on use of α-amylase from *Bacillus spp.*, which possesses good thermal resistance, has a *z* value more or less equal to that of the targeted micro organism (*Clostridium botulinum*) and its denaturation follows first order (Tucker et al. 2002). Amylases which are amylolytic enzymes catalyse the breakdown of starch, glycogen and some polysaccharides. The α-amylase hydrolyses the alpha 1,4 glycosidic bonds in the inner region of the starch polymer. This enzyme has important industrial applications in starch liquefaction, brewing, sizing in textile industries, and paper and detergent manufacturing processes. α-amylase is produced on the industrial scale using *Bacillus spp.*

The α-amylase directly extracted from the *Bacillus spp* is stable for pH values ranging from 5.5 to 8. By adding calcium this stability can be changed. With the optimum amount of calcium, the α-amylase from *Bacillus spp* can resist high temperatures for relevant times (90°C). Several studies have been undertaken to determine the temperature ranges in which amylases from various micro organisms could be used (Haentjens et al. 1998; Guiavarc'h, Van Loey et al. 2004; Guiavarc'h, Van Loey et al. 2004). However, when the enzyme does not fulfil these requirements, it is still possible to change the thermal sensitivity of the enzyme by changing the conditions of the enzyme (immobilised or not) and its environment (e.g. pH, concentration in Calcium, NaCl).

2.1.2. Application of Enzymatic TTIs

Enzymatic TTIs have been used extensively from the food industry to quantify the effect of thermal processing. Some of the applications are as follows:

Pasteurisation: It is relatively easy to find enzymes which are thermostable at pasteurisation temperatures. No specific system is needed to protect the enzyme thus, these experiments are easily undertaken. As an example, some TTIs have been added to batches of fruit preparations in order to evaluate the pasteurisation achieved by two different ways of heating, namely by using an ohmic column or a tubular heat exchanger (Tucker, Lambourne et al. 2002). The successful results gave a *P* value in various cases where the thermocouples could not be used. The use of α-amylase to evaluate a pack pasteurisation process has also given promising results. In the pasteurisation domain, the comparison of the results from the read out of the TTI and the results from the integration of the processing value shows that the TTI could be used as an alternative process evaluation tool when others more accurate methods can not be applied (Van Loey, Hendrickx et al. 1996). TTIs were used to measure the pasteurisation in baked mushroom quiches. Usually the temperature measurements of product which undergo a change of phase during processing are not easy. The use of temperature probes to measure the temperature in the centre of the quiches was not possible since during the process the quiches travel along a belt of a continuous oven. However, TTI could be used since they would stay at the correct location during processing. TTI were made from the α-amylase from *Bacillus amyloliquefaciens*, their D_{70} value was 8.4 min with a *z* value of 8.9°C. For 10 quiches, one TTI was put in the centre and one on the edge. The quiches were then put on the conveyor belt of a continuous oven set at a 210°C for 25 minutes. In food safety point of view, the target thermal treatment was 70°C for 2 minutes in order to kill the strains of *Salmonella* and *Listeria monocytogenes* and all 10 quiches received the targeted thermal treatment. However, the centre of the quiches received a lower thermal treatment than the edge. The average *P* value at 70°C of the 10 quiches for the centre was 12.3 min and for the edge was 14.7 min. This result was not surprising since the centre was the expected cold spot. This study shows a successful application of the Enzymatic TTI to monitor thermal processes undergone at pasteurisation temperature.

Sterilisation: Finding an enzyme with the adequate thermostability in these ranges of temperatures (around 121°C) is challenging to say the least. Some studies have been undertaken with some enzymes under reduced water content (Haentjens, Van Loey et al. 1998; Guiavarc'h, Van Loey et al. 2004; Guiavarc'h, Van Loey et al. 2004). This is easily realised in laboratories; however this is not the case in industrial processes. The hyperthermophile *Pyrococcus furiosus* presents some good characteristics in terms of heat resistance. Indeed, this micro organism has been isolated from a shallow marine solfatara at Vulcano Island off southern Italy where the temperature, pH and oxygen availability are not favourable for common micro organism development (Figure 2).

The maximal temperature at which the organism will grow is 103°C with an optimum growth rate at 100°C (Tucker et al. 2007). The enzyme of these micro organisms is potentially interesting to use, however, it is difficult to culture or to obtain as it is produced only by few laboratories. Recent work with the amylase of the *Pyrococcus furiosus* shows that the *z* value of the amylase of the *Pyrococcus* is 10°C and the industrial trials demonstrate promising results. However, this enzyme is difficult to obtain and more tests are necessary to validate its effectiveness for sterilisation processes (Tucker, Brown et al. 2007).

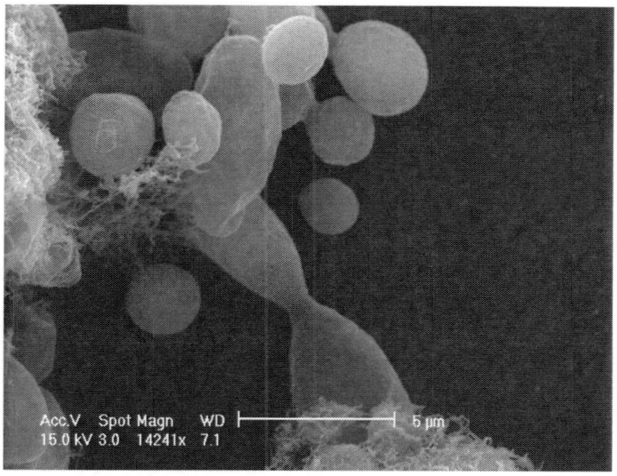

Figure 2. SEM images of *Pyrococcus furiosus* cells.

2.2. Types of TTI Responses

The impact of the process on the TTI is determined by the quantification of the changes occurring within the TTI. In the case of enzymatic TTIs, the enzyme is denatured by heat treatment and the enzyme activity decreases. The quantification of the change of activity before and after heat treatment shows its impact on the TTI. However, there are two cases to be considered:

When the z value of the TTIs is equal to the z value of the targeted micro organism (*Clostridium botulinum* for food safety), the modification in enzyme activity will correlate well with the decrease of the number of targeted micro organisms. In this case, the time temperature history of the product is not needed in order to determine the impact of the thermal treatment on the micro organism population present in the food and can be directly determined from the reading of the TTI.

However, when the z value of the TTIs is not equal the targeted micro organism z value, the correlation between the change of enzyme activity and the reduction of micro organism number can not be done. In this case, the time temperature history of the product is needed. However, it is possible to use multi component TTIs (composed of several units containing enzymes with different z values) in order to determine the impact of the thermal treatment when the z values are different and the time temperature history of the product is not available (Vanloey et al. 1995; Van Loey, Hendrickx et al. 1996).

Until recently, the use of multi components TTIs as a tool to evaluate the impact of thermal treatment on the targeted micro organism was limited since no reliable extrapolation technique was available. The multi component TTIs could not be used without information about the time temperature history (Maesmans et al. 1994; Maesmans et al. 1995). However, a recent study showed that multi component TTIs having a z value relatively close to the targeted z value can now be used instead of the single component TTIs when these are not available. It was shown that two components TTIs with z values respectively below and above the targeted z value could be potentially used to evaluate thermal efficiency of food for sterilisation processes (Hendrickx, Maesmans et al. 1995; Guiavarc'h et al. 2005).

2.3. Other TTI Applications: Monitoring the Cold Chain

TTIs are already commercially available as tools for quality monitoring and control of frozen products. Several studies were undertaken on the reliability of these commercially available TTIs. The A12 Fresh Check TTIs label (Lifetimes Technologies, NJ, USA) were used to monitor the distribution chain in real world applications. These TTIs have the shape of a label and can be stuck on the food packaging. A chromatic variation appears on the label depending on the time temperature that they have been exposed to, which can be related to a quality loss. This study showed that the temperature effects recorded by the TTIs were well correlated to the decreasing quality of the food. In this study, it was found out that the TTIs could be used as an alternative technique to the direct temperature recording (Riva et al. 2001). Other TTIs are available for commercial uses such as: the MonitoMark TTI (3M, St Paul, USA) which was used in research on the reliability of TTIs when temperature abuse occurs during the distribution and the storage of frozen food (Shimoni et al. 2001; Giannakourou et al. 2005).

Other studies showed that TTIs can be used to monitor changes in quality attributes of perishable and semi perishable foods (e.g. tomatoes, naked and wrapped lettuce, canned fruitcake and UHT sterilized milk) and that they can be used as effective tools to monitor the cold chain (Wells and Singh 1988; Giannakourou, Koutsoumanis et al. 2005).

Time-temperature indicators (TTIs) provide a visual summary of a product's accumulated chill-chain history, recording the effects of both time and temperature. Attempts have been made to develop a prototype of a time-temperature indicator based on the diffusion of lactic acid, which is temperature-dependent. Four lactic acid-based TTIs were made in different substrate concentrations. Color changes associated with the diffusion of lactic acid were monitored. In the vapor diffusion of lactic acid, an irreversible color change of a chemical chromatic indicator (from green to red) clearly and progressively occurred due to the pH reduction (Wanihsuksombat et al. 2010).

2.3.1. Advantages and Disadvantages of TTIs

TTIs are small, wireless and shock resistant which made them suitable for processes where the thermocouples and data loggers can not be used. TTIs are relatively fast to analyse compared with the microbiological method where incubation times are long (3 days), inexpensive and minimal training is needed for their analysis. The major advantage of the TTIs is that there is no need to know the time temperature history of the product in order to determine the impact of the thermal treatment on the food. TTIs can be used as a measuring tool in process validation or optimisation when the use of thermocouples or other methods is not possible. However, TTIs are not suitable for on-line monitoring of the thermal treatment efficiency since their analysis is not immediate. In addition, since they are very small, in practice their retrieval from the food product can be an issue. TTIs can be used for determining the efficiency of thermal treatments in order to improve the design of processes or to improve the quality of a product by validating processes routinely used. Another limitation of the TTIs is the need to know their thermal inactivation characteristics (D_T and z values) in order to process the data. Preliminary experiments are needed before being able to use them and each batch of enzyme has to be calibrated (Maesmans et al., 2005; Van Loey et al., 1996a; Van Loey et al., 1996b).

2.3.2. Recent Developments on TTIs

Recently, some studies have been undertaken on the applicability of intrinsic TTIs from thermally processed milk to assess high pressure process. The study found out that the heat markers for high pressure are different from the criteria used for the thermal treatment of milk and therefore can not be used to evaluate the efficiency of high pressure processes (Claeys et al., 2002).

New types of TTIs that require smaller analysis time are under development. This is the case of a TTI based on an amperiometric glucose oxidase biosensor. The TTI consists of enzymes trapped within an electronically generated poly o-phenylenediamine film. This film is coated on the interior wall of a capsule made of platinum or platinized stainless steel. After the thermal treatment, the TTIs were rapidly analysed (10 min) by an amperometer. Some isothermal treatments were performed at low pasteurisation temperature (70˚C - 80˚C). The enzyme followed a first order reaction and *z* values of 6˚C -7˚C were found. These TTIs showed a potential for fast assessments of pasteurisation processes where *E. coli* and *Listeria monocytogenes* are the targeted micro organisms (Reyes-De-Corcuera et al., 2005).

As stated previously, several studies have been undertaken on the production of sterilisation TTIs. Industrialists who produce various kinds of sterilised products are keen on the future development of a TTI for sterilisation temperatures. Several studies have been published on sterilisation TTIs that are made from the α-amylase from the *Bacillus subtilis* and the *Bacillus licheniformis*. This α-amylase was used at reduced water content and was immobilised on glass beads. These TTIs (*z* value of 9.4°C) that were equilibrated at a_w = 4.8 could monitor process values at 121.1°C in the range of 0-30 min under non isothermal conditions with an absolute error of 14%. The results showed that this TTI can be used for the determination of heat treatment efficiency of sterilisation processes (Guiavarc'h et al., 2004a). The same type of enzyme was used by the same group of researchers for an industrial application of sterilisation processes (Guiavarc'h et al., 2005b). The continuous rotary processing of canned ravioli was investigated by using both temperature probes and TTIs together. The *Bacillus licheniformis* α – amylase and the *Bacillus subtilis* α – amylase were used to make the TTIs which have *z* values of 13.9 and 16.4°C respectively. These 2 enzymes have different *z* values above 10°C; therefore they were used together as multi component TTIs For the same thermal process, the multi component TTI inserted inside ravioli gave a *F* value at 121.1°C of 23.6 min with a standard deviation of 1.5 min while the temperature probe measuring the sauce temperature gave a *F* value at 121.1°C of 33.6 min. This study showed that by measuring the temperature in the sauce, the impact of the thermal process is overestimated, causing food safety issues. The drawback of the use of these TTIs was the complicated preparation and the need for technical skills to produce and analyse the TTIs.

The sterilisation TTI made from the α-amylase of the *Pyrococcus furiosus* seems to be a better alternative since its preparation and use is similar to the pasteurisation TTIs. However, the culture of this micro organism is difficult and α-amylase production is difficult to control (Tucker et al., 2007 ; Tucker et al., 2005b).

2.4. Reliability of the TTIs as Indicators of the Efficiency of a Thermal Process

Investigations have been undertaken on the reliability of TTIs under non isothermal conditions (Taoukis and Labuza, 1989). Non isothermal experiments of 2 types were performed: the first experiment was performed using high temperature for a long time followed by storage at low temperature to investigate the 'history effect' and the second experiment was done by storage at controlled sinusoidally varying temperatures. The result of this work demonstrates that there is a direct correlation between the TTIs' responses and the quality of the food under variable time temperature profiles. Guiavarc'h et al. (2002b) used TTIs to monitor thermal impact inside a solid/liquid model food and showed that incorrect conclusions could be drawn using F values for solid pieces obtained just by using temperatures recorded at the centre of the solid at the tip of a thermocouple. Nevertheless, these results show the potential of TTIs in the evaluation of the thermal process efficiency. Yet, currently published literature on their use is limited. In general, there is a lack of knowledge on their accuracy and efficiency. Errors in the measurement arise from a number of factors, including variability in the manufacture of the TTIs, errors in determining the final and initial value of activity, as well as non-linearities and variations in the kinetics of the enzyme. Some understanding of the inherent accuracy of the devices is needed in order to develop TTIs as effective process probes. TTIs have been used for various industrial applications and are described in Table 2, which shows that it is not clear if the variability of the TTIs is due to the process variability or due to the TTI itself.

Table 2. Example of TTI application

	Study 1	Study 2	Study 3
Application	P-values for fruits processed in a tubular heat exchanger	F values for Raviolis inside cans processed in a spiral retort (Sterilmatic, Belgium)	P values in a stirred vessel (250l) – 200l batch of 5% Colflo 67 starch
Number of TTIs used	70	7 cans with 10 TTIs in each	50
Mean P value or F value	$P_{85°C}$~ 60 min	$F_{121.1°C}$~ 18.8 min	$P_{85°C}$~ 3.9 min
Variability	$P_{85°C}$ vary from 33 min to 109 min	$F_{121.1°C}$ vary from 10 min to 29.1 min	$P_{85°C}$ vary from 1.2 min to 11.3 min
Reference	Tucker et al., 2002	Guiavarc'h et al., 2005	Lambourne et Tucker, 2001

More recently a technique using a Peltier stage device has been developed to generate a range of temperature-time profiles which have been then used to determine the accuracy of Time Temperature Indicators (Mehauden et al. 2007). In this work α-amylase from *Bacillus amyloliquefaciens* and from *Bacillus licheniformis* was used to prepare the TTIs. A range of temperature profiles relevant to industrial processes has been used (Figure 3). This study demonstrates that the TTIs responses correlate well with the thermocouples measurements for a range of non isothermal processes. One can also see that when the holding time increases, the variabilty in *P* values estimated from both TTIs and thermocouples increase. The increase in variability is lower in the case of thermocouples. Overall the error in the evaluation of thermal processes is less than ± 20% when TTIs are used, over a wide range of temperature histories.

In practice, the accuracy of the TTIs will be constrained by (i) a lower limit of P, where there is sufficient thermal lag between the TTIs and the process, so that the TTI value is not accurate, and (ii) a higher limit of P, where the value of the enzyme activity is so low that it is not sensitive to the change in P. In between lies the operational window in which measurements can be taken with sufficient accuracy which is approximately 2 to 8 minutes at 85°C for Bacillus amyloliquefaciens and 5 to 30 minutes at 85°C for Bacillus licheniformis.

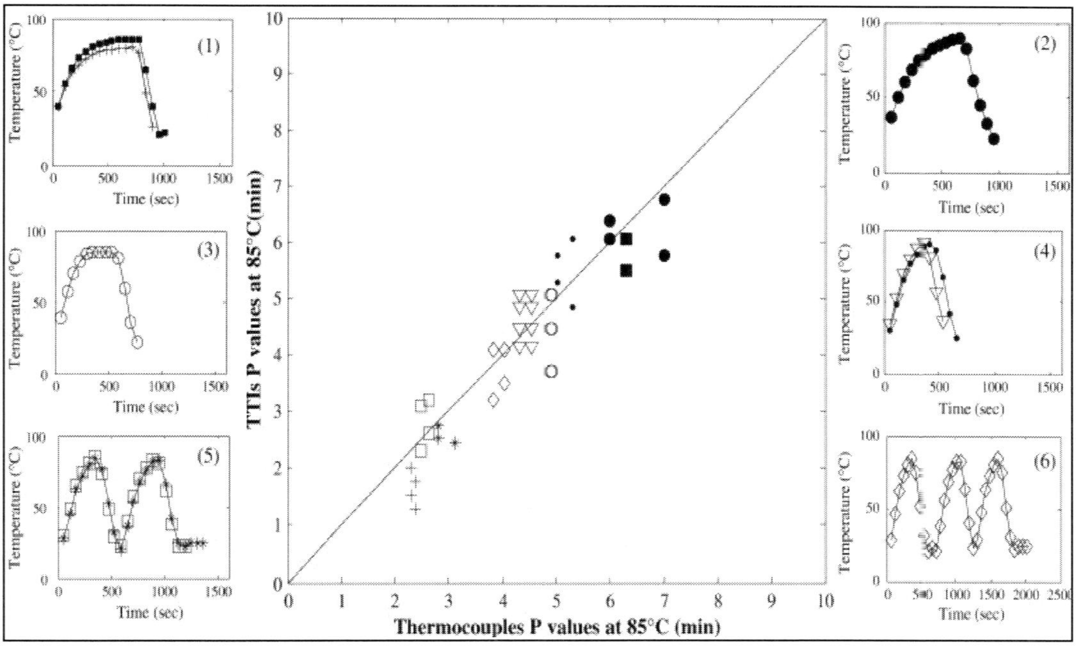

Figure 3. P values of the thermocouples versus the P values of the TTIs for industrial time (with permission from Mehauden, Cox et al. 2007).

3. An Example: The Vezuvio Case

Despite the fact that TTIs could be used to characterise thermal processes there is relatively little work published in this area. While thermocouples are commonly used for process control and validation as it was previously mentioned this measurement technique has a limitation as there is little possibility to evaluate the thermal treatment in the bulk of the fluid, especially if rotating parts are used.

In a recent work experiments were performed on a vertical jacketed mixing vessel ('Vesuvio' vessel) manufactured by Giusti Ltd (Burton on Trent, UK) shown in Figure 4 (Mehauden et al. 2008). The maximum capacity of the vessel was 250 litres and was equipped with a horizontal rotating agitator with polytetrafluoroethylene (PTFE) scrapers attached to the steel blades which make contact with the vessel wall, designed to perform an efficient and gentle mix for highly shear-sensitive products. Agitation speeds used for experiments range from 3 to 15 rpm. The vessel has the ability for indirect heating and cooling by introduction of either steam at 3 bar gauge or cooling water into the vessel jacket respectively. Direct heating can be also provided by steam injection into the vessel contents and evaporative cooling by pressure reduction of the vessel contents to below atmospheric pressure. The vessel has a heating phase, using either indirect heating or direct steam injection, until the set temperature was reached. Temperature was measured using a calibrated PT 100 thermocouple (± 0.15% at 100°C), situated in a housing at the bottom of the vessel. Estimation of Reynolds numbers (100-18000) indicated that the flow is in the laminar to transitional regime.

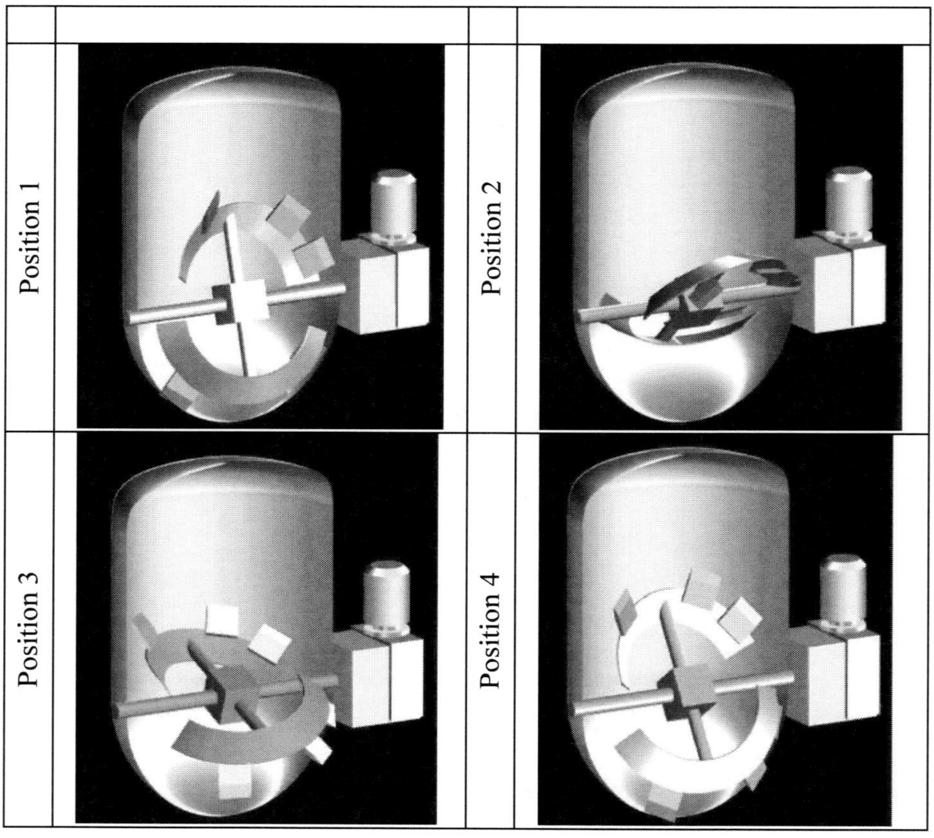

Figure 4. Schematic of the vesuvio model (D (Impeller diameter) = 754 mm, T (Vessel diameter) = 794 mm, C (Clearance) = 40 mm).

During typical food processing products could start with a water like viscosity (0.001 Pas) and end with a much higher viscosity (5 Pas). This is particularly true in the case of where transformations such as starch gelatisation occur. This presents a challenge as convective heat transfer would significantly decrease as viscosity increases. In this work a range of fluids with viscosities relevant to food processes ranging from water to a starch suspension, as well as a tomato soup product were used.

20 TTIs were used to evaluate thermal treatment and variability from the process. Preliminary experiments indicated that TTIs resulted in an abnormal high P value. This was attributed to collisions with the wall. In order to study this effect and TTIs a special reasonable simple casing was developed by placing some TTIs inside either 4 cm diameter airflow golf balls with widened holes (called GB hereafter) or inside a system of 2 tie clips (called TC hereafter) stuck together to form a 5 cm diameter ball. Effort was devoted in the design of the casings to ensure that this would affect neither the flow field nor the TTI paths. In addition a wireless thermocouple (Tracksense pro, Ellab UK Ltd., King's Lynn, Norfolk UK) with a 50 cm long probe was fixed securely onto the middle of the agitator.

When the viscosity of the fluid was low (i.e. when water was used) both thermocouples resulted in similar P values, lower to those estimated using TTIs. It is worth keeping in mind that all TTIs resulting in similar P values. When steam injection was used the variability in the TTI measurements increased suggesting that some TTIs passed through the steam jet whilst others have not.

When the starch (4%), already viscous at the start of the experiments, was used the results of the TTIs are significantly different when compared to those obtained from the wall thermocouple. The difference between the P values of the 2 thermocouples is high (10 min at 85°C). During the heating time, mixing efficiency is lowered due to the fluid being viscous. Therefore the temperature is not homogenous inside the vessel. The free TTIs have high P values, similar to that of the wall thermocouple, and that of the TTIs fitted in the balls are similar to the P values of the centre thermocouple. The GB TTIs and the TC TTIs have lower P values due to the lack of contact with the vessel wall. The heat up time is energy consuming stage and viscous fluids are difficult to process without hot and cold spots in the vessel. This can lead to local under-processing and over-processing of the fluid with consequent influences upon safety and product quality. Direct steam injection appeared to reduce temperature variability.

In Figure 5 P values using a commercially available tomato soup and a 5% starch suspension are shown. The P values for the soup and starch suspension are similar. As the TTIs added after the heating time (marked HT TTIs) have a similar P value to those added at the start of the experiment, one can conclude most of the heat treatment is achieved during the holding time. Once steam injection is used the difference between the *P* values of the thermocouples located on the wall and in the centre decreased, while the P values measured by the TTIs and the wall thermocouple increased, indicating a more uniform temperature field. As the TTIs that are in the cases (marked GB TTIs and TC TTIs) are more protected from direct steam the P values are significantly lower to those measured from the free TTIs.

When a high viscosity fluid is used (Figure 5 (c) and (d)) wall thermocouple tends to overpredict the thermal treatment experienced from the fluid and the centre thermocouple. This was expected as a high viscosity fluid would result in a low convective heat transfer. Steam injection appears to improve the heat transfer performance resulting in an increasing P value recorded by the TTIs.

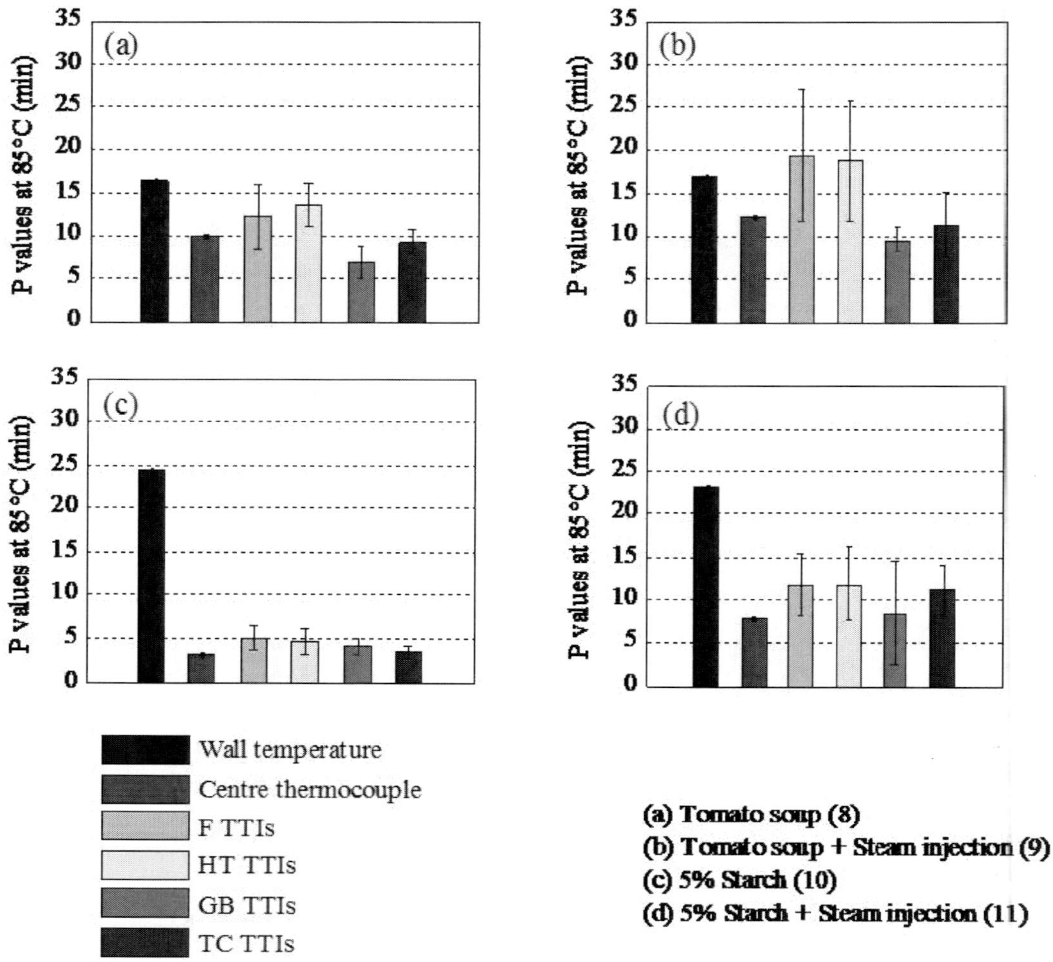

Figure 5. Experiments performed with tomato soup and with a holding temperature of 83°C and experiments performed with 5% gelatinised starch and with a holding temperature of 85°C. Error bars show one standard deviation (Mehauden 2008).

This work demonstrates how different types of TTI can be used to study the effects of a process. The results suggest that:

- TTI particles that are free to be close to the steam-heated wall show the highest P value; these are likely to be representative of the processing the fluid has received.
- TTI particles that have constrained to be away from the wall show lower values, indicating the processing received by fluid that is not heated by the wall.
- The process should be designed so that the two sets of results give *P* values that are as close as possible. The influence of overfilling shows that mixing profiles changes and that greater variation was observed; it is probable that the TTI results seen here show the overprocessed fraction only, and that underprocessing can also occur.

As in industrial processes convective heat transfer is of importance, it is of particular interest to quantify the flow field in the vessel. Of particular importance when using TTIs is to ensure that TTIs (i) have flow behaviour similar to the fluid, i.e. are isokinetic to the fluid (ii) move through all locations inside the processing equipment. If either of the above does not hold evaluation of the thermal efficiency of a process using TTIs would be problematic as they would not be represent of a typical fluid element.

A scaled down model of the Vesuvio vessel was used to investigate fluid flow using PIV and TTI parhs using PEPT (Mehauden et al. 2009). The authors demonstrated that the flow is laminar/transitional the through bulk of vessel. The only regions where significant flow instabilities were generated were located at the free surface and at the trailing edge of the impeller. Mixing throughout the bulk was therefore be expected to occur by laminar mechanisms with some mixing by eddy diffusion present at the free surface.

The velocity profiles in the vessel were generally unaffected by the viscosity and the agitation speed. However, differences are present for the overfilling case during to a vortex formation which is notable for the case of a low viscosity fluid.

The experimental set up and the coordinate system used in PEPT is shown in Figure 6

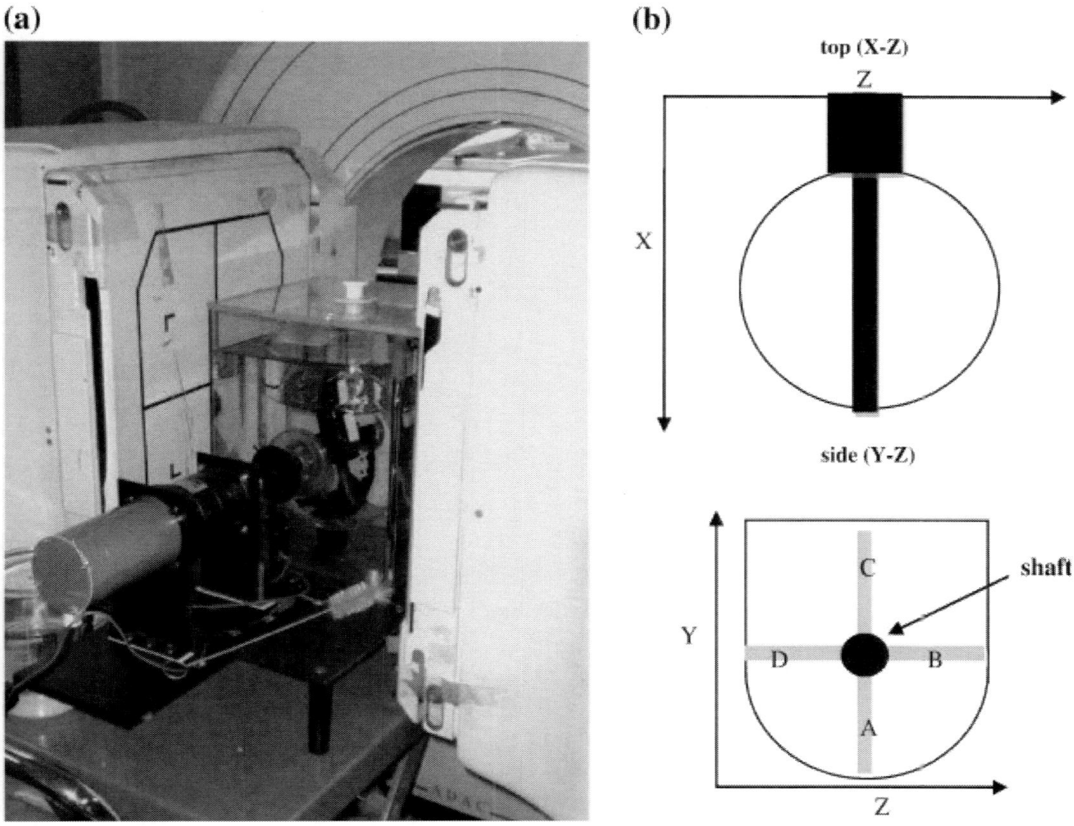

Figure 6. Experimental set up used for (a) PEPT experiments and (b) coordinate system used and planes used to show velocity profiles (with permission from Mehauden, Cox et al. 2009).

Occupancy plots obtained using PEPT indicated that TTI pass through all the regions of the tracers. Occasionally though the tracers appeared to be spent significant amount of time in certain locations, either in the surface or in close to the impeller. This indicates that it is possible for the TTI to be stuck at these locations. This is significant from an industrial perspective as these TTIs would not receive a representative treatment. This makes imperative to use a sufficient number of TTIs in order to identify such as outliers using statistical methods (Tucker et al., 2002). In Figure 7 typical velocity profiles obtained using PEPT are shown. As one can see the velocity obtained for TTIs are very similar to those of the free tracers that represent the velocity of the fluid. Differences are observed across position D where TTIs appear to move downwards faster than the fluid. This was attributed to a sedimentation effect.

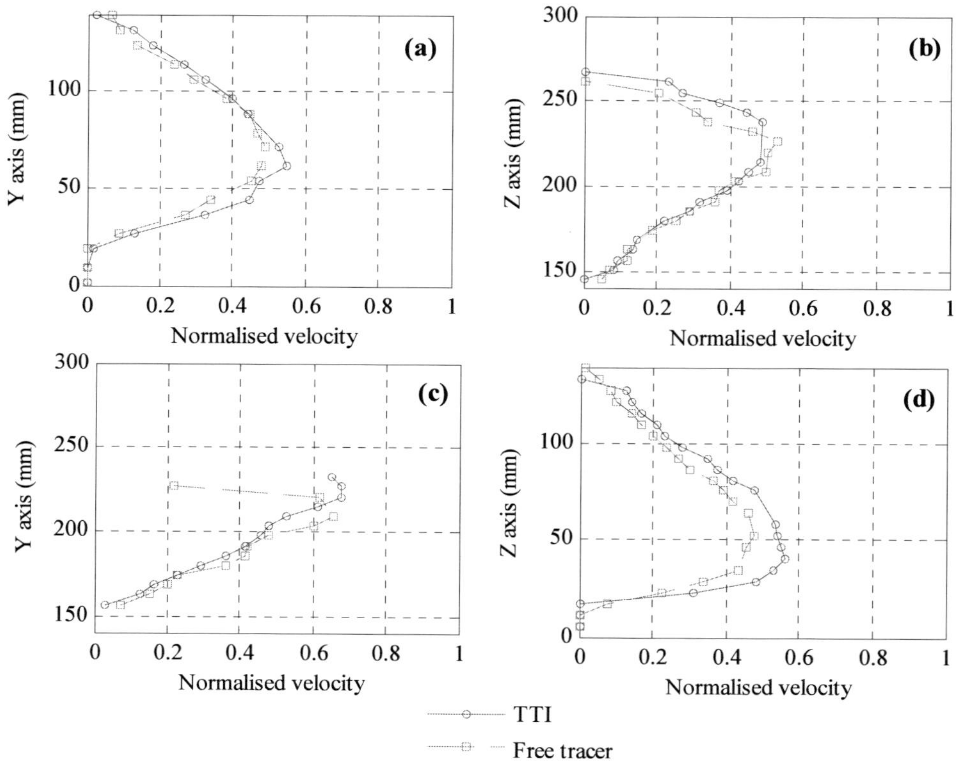

Figure 7. Comparison of the velocity fields of the free tracer and the TTI tracer for the experiment performed with 4% starch at 33 rpm. Each subfigure show the various positions of the velocity measurement (a) Position A, (b) Position B, (c) Position (C) and (d) Position D (with permission from Mehauden, Cox et al. 2009).

Overall this work demonstrated TTIs can be used to evaluate thermal processes. The results would be reliable if TTIs are isokinetic with the fluid and follow fluid streamlines. Significant differences in the path taken by the TTI and the fluid were observed when the TTI had a significant settling velocity in the fluid. Hence TTIs cannot be assumed to give reliable results in low viscosity fluids (e.g. water) or at very low agitation speeds.

Conclusion

As the food manufacturers are required to deliver safe high quality products it is essential to develop methodologies that will accurately and reliably evaluate thermal processing. Traditional techniques such as thermocouples have significant limitations as they are not able to represent a typical fluid element. Time Temperature Integrators (TTIs) have been developed and evaluated over the last decades as an alternative method. Despite being a mature technique aspects of their behaviour during processing such as flow behaviour has yet to be fully understood. For viscous fluids there is little difference between the flow of the TTIs and of the fluid, so the results are relevant. Specially designed TTIs allow the effect of local wall heating to be assessed – here the need is to ensure that the path of the TTI and the fluid is the same.

TTIs are relatively easy to use in practice, but their use on real plant is complex and requires some practice; factors such as extraction of the TTIs from the fluid can make experiments messy to do! The method is probably best suited to initial process start-up, when process schedules are being set, rather than as a day-to-day operation.

Overall TTIs present a unique tool for food manufacturers to design new processes and validate existing processes that can be applied to most food heating and mixing processes.

References

Bakalis, S., Cox, P. W. & Fryer, P. J. (2001). Modelling thermal processes: heating. Food process modelling. L. M. M. Tijskens, M. L. A. T. M. Hertog and B. M. Nicolaï. Cambridge, UK, *Woodhead publishing*: 341-363.

Bakalis, S. & Karwe, M. V. (1998). Measurement of three-dimensional velocity field in the translational region of a co-rotating twin-screw extruder. *Polymer Engineering and Science* 38: 1549-1558.

Ball, C. O. & Olson, F. C. W. (1957). *Sterilization in Food Technology*. New York, McGraw-Hill.

Barigou, M., Mankad, S. & Fryer, P. J. (1998). Heat transfer in two-phase solid-liquid food flows: A review. *Food and Bioproducts Processing* 76: 3-29.

Bird, R. B., Stewart, W. E. & Lightfoot, E. N. (2007). *Transport Phenomena*, John Wiley & Sons Inc.

Claeys, W. L., Van Loey, A. M. & Hendrickx, M. E. (2002). Intrinsic time temperature integrators for heat treatment of milk. *Trends in Food Science & Technology* 13: 293-311.

Cox, P. W., Bakalis, S., Ismail, H., Forster, R., Parker, D. J. & Fryer, P. J. (2003). Visualisation of three-dimensional flows in rotating cans using positron emission particle tracking (PEPT). *Journal of Food Engineering* 60: 229-240.

De Roeck, A., Mols, J., Duvetter, T., Van Loey, A. & Hendrickx, M. (2010). Carrot texture degradation kinetics and pectin changes during thermal versus high-pressure/high-temperature processing: A comparative study. *Food Chemistry* 120: 1104-1112.

Fennema, O. R. (1996). *Food Chemistry*, SRS Press.

Fischer, P., Ouriev, B. & Windhab, E. J. (2009). Macroscopic Pipe Flow of Micellar Solutions Investigated by Ultrasound Doppler Velocimetry. *Tenside Surfactants Detergents* 46: 140-144.

Gabriele, A., Nienow, A. W. & Simmons, M. J. H. (2009). Use of angle resolved PIV to estimate local specific energy dissipation rates for up- and down-pumping pitched blade agitators in a stirred tank. *Chemical Engineering Science* 64: 126-143.

Giannakourou, M. C., Koutsoumanis, K., Nychas, G. J. E. & Taoukis, P. S. (2005). Field evaluation of the application of time temperature integrators for monitoring fish quality in the chill chain. *International Journal of Food Microbiology* 102: 323-336.

Guerrero-Beltran, J. A., Sepulveda, D. R., Gongora-Nieto, M. M., Swanson, B. & Barbosa-Canovas, G. V. (2010). Milk thermization by pulsed electric fields (PEF) and electrically induced heat. *Journal of Food Engineering* 100: 56-60.

Guiavarc'h, Y., Sila, D., Duvetter, T., Van Loey, A. & Hendrickx, M. (2003). Influence of sugars and polyols on the thermal stability of purified tomato and cucumber pectinmethylesterases: a basis for TTI development. *Enzyme and Microbial Technology* 33: 544-555.

Guiavarc'h, Y., Van Loey, A., Zuber, F. & Hendrickx, M. (2004). Bacillus licheniformis [alpha]-amylase immobilized on glass beads and equilibrated at low moisture content: potentials as a Time-Temperature Integrator for sterilisation processes. *Innovative Food Science & Emerging Technologies* 5: 317-325.

Guiavarc'h, Y., Van Loey, A., Zuber, F. & Hendrickx, M. (2004). Development characterization and use of a high-performance enzymatic time-temperature integrator for the control of sterilization process' impacts. *Biotechnology and Bioengineering* 88: 15-25.

Guiavarc'h, Y., Van Loey, A. M. & Hendrickx, M. E. (2005). Extended study on the influence of z-value(s) of single and multicomponent time-temperature integrators on the accuracy of quantitative thermal process assessment. *Journal of Food Protection* 68: 384-395.

Guiavarc'h, Y., Zuber, F., Van Loey, A. & Hendrickx, M. (2005). Combined use of two single-component enzymatic time-temperature integrators: Application to industrial continuous rotary processing of canned ravioli. *Journal of Food Protection* 68: 375-383.

Guiavarc'h, Y. P., Deli, V., Van Loey, A. M. & Hendrickx, M. E. (2002). Development of an enzymic time temperature integrator for sterilization processes based on Bacillus licheniformis alpha-amylase at reduced water content. *Journal of Food Science* 67: 285-291.

Haentjens, T. H., Van Loey, A. M., Hendrickx, M. E. & Tobback, P. P. (1998). The Use of [alpha]-Amylase at Reduced Water Content to Develop Time Temperature Integrators for Sterilization Processes. *Lebensmittel-Wissenschaft und-Technologie* 31: 467-472.

Hendrickx, M., Maesmans, G., Decordt, S., Noronha, J., Vanloey, A. & Tobback, P. (1995). Evaluation of the integrated time-temperature effect in thermal-processing of foods. *Critical Reviews in Food Science and Nutrition* 35: 231-262.

Hendrickx, M., Weng, Z., Maesmans, G. & Tobback, P. (1992). Validation of a time-temperature-integrator for thermal-processing of foods under pasteurization conditions. *International Journal of Food Science and Technology* 27: 21-31.

Holdsworth, S. D. (1985). Optimization of thermal processing a review. *Journal of Food Engineering* 4: 89-116.

Hulsheger, H., Potel, J. & Niemann, E. G. (1981). Killing of bacteria with electric pulses of high-field strength. *Radiation and Environmental Biophysics* 20: 53-65.

Jousse, F., Jongen, T., Agterof, W., Russell, S. & Braat, P. (2002). Simplified Kinetic Scheme of Flavor Formation by the Maillard Reaction. *Journal of Food Science* 67: 2534-2542.

Lareo, C., Fryer, P. J. & Barigou, M. (1997). The fluid mechanics of two-phase solid-liquid food flows: A review. *Food and Bioproducts Processing* 75: 73-105.

Lareo, C., Nedderman, R. M. & Fryer, P. J. (1997). Particle velocity profiles for solid-liquid food flows in vertical pipes .2. Multiple particles. *Powder Technology* 93: 35-45.

Maesmans, G., Hendrickx, M., Decordt, S. & Tobback, P. (1995). Theoretical consideration of the general validity of the equivalent point method in thermal-process evaluation. *Journal of Food Engineering* 24: 225-248.

Maesmans, G., Hendrickx, M., Decordt, S., Vanloey, A., Noronha, J. & Tobback, P. (1994). Combined use of the equivalent point method and a multicomponent time-temperature integrator in thermal-process evaluation - influence of kinetic characteristics and reference temperature. *Food Control* 5: 249-256.

Maesmans, G., Hendrickx, M., Decordt, S., Vanloey, A., Noronha, J. & Tobback, P. (1994). Evaluation of process value distribution with time-temperature integrators. *Food Research International* 27: 413-423.

Marra, F. & Romano, V. (2003). A mathematical model to study the influence of wireless temperature sensor during assessment of canned food sterilization. *Journal of Food Engineering* 59: 245-252.

Mehauden, K. (2008). Evaluation of the thermal and mixing performance of an agitated vessel for processing of complex liquid foodstuffs. School of Chemical Engineering. Birmingham, UK, University of Birmingham.

Mehauden, K., Bakalis, S., Cox, P. W., Fryer, P. J. & Simmons, M. J. H. (2008). Use of Time Temperature Integrators for determining process uniformity in agitated vessels. *Innovative Food Science & Emerging Technologies* 9: 385-395.

Mehauden, K., Cox, P. W., Bakalis, S., Fryer, P. J., Fan, X., Parker, D. J. & Simmons, M. J. H. (2009). The flow of liquid foods in an agitated vessel using PEPT: Implications for the use of TTI to assess thermal treatment. *Innovative Food Science & Emerging Technologies* 10: 643-654.

Mehauden, K., Cox, P. W., Bakalis, S., Simmons, M. J. H., Tucker, G. S. & Fryer, P. J. (2007). A novel method to evaluate the applicability of time temperature integrators to different temperature profiles. *Innovative Food Science & Emerging Technologies* 8: 507-514.

Miri, T., Tsoukalas, A., Bakalis, S., Pistikopoulos, E. N., Rustem, B. & Fryer, P. J. (2008). Global optimization of process conditions in batch thermal sterilization of food. *Journal of Food Engineering* 87: 485-494.

Parker, D. J., Broadbent, C. J., Fowles, P., Hawkesworth, M. R. & McNeil, P. (1993). Positron emission particle tracking - a technique for studying flow within engineering equipment. *Nuclear Instruments & Methods in Physics Research Section a-Accelerators Spectrometers Detectors and Associated Equipment* 326: 592-607.

Rastogi, N. K., Raghavarao, K., Balasubramaniam, V. M., Niranjan, K. & Knorr, D. (2007). Opportunities and challenges in high pressure processing of foods. *Critical Reviews in Food Science and Nutrition* 47: 69-112.

Riva, M., Piergiovanni, L. & Schiraldi, A. (2001). Performances of time–temperature indicators in the study of temperature exposure of packaged fresh foods. *Packaging Technology and Science* 14: 1-9.

Sastry, S. K., Heskitt, B. F. & Blaisdell, J. L. (1989). Experimental and modeling studies on convective heat-transfer at the particle liquid interface in aseptic processing systems. *Food Technology* 43: 132-&.

Shimoni, E., Anderson, E. M. & Labuza, T. P. (2001). Reliability of time temperature indicators under temperature abuse. *Journal of Food Science* 66: 1337-1340.

Simpson, S. G. & Williams, M. C. (1974). Analysis of high temperature short time sterilization during laminar-flow. *Journal of Food Science* 39: 1047-1054.

Taoukis, P. S. & Labuza, T. P. (1989). Applicability of Time-Temperature Indicators as Shelf Life Monitors of Food Products. *Journal of Food Science* 54: 783-788.

Tucker, G. (1999). A novel validation method: Application of time-temperature integrators to food pasteurization treatments. *Food and Bioproducts Processing* 77: 223-231.

Tucker, G. S., Brown, H. M., Fryer, P. J., Cox, P. W., Poole, F. L., Lee, H. S. & Adams, M. W. W. (2007). A sterilisation Time-Temperature Integrator based on amylase from the hyperthermophilic organism Pyrococcus furiosus. *Innovative Food Science & Emerging Technologies* 8: 63-72.

Tucker, G. S., Lambourne, T., Adams, J. B. & Lach, A. (2002). Application of a biochemical time-temperature integrator to estimate pasteurisation values in continuous food processes. *Innovative Food Science & Emerging Technologies* 3: 165-174.

Van Boekel, M. A. (2008). Kinetic Modeling of Food Quality: A Critical Review. *Comprehensive Reviews in Food Science and Food Safety* 7: 144-158.

Van Loey, A., Hendrickx, M., De Cordt, S., Haentjens, T. & Tobback, P. (1996). Quantitative evaluation of thermal processes using time-temperature integrators. *Trends in Food Science & Technology* 7: 16-26.

VanLoey, A., Hendrickx, M., Smout, C., Haentjens, T. & Tobback, P. (1996). Recent advances in process assessment and optimisation. *Meat Science* 43: S81-S98.

Vanloey, A., Ludikhuyze, L., Hendrickx, M., Decordt, S. & Tobback, P. (1995). Theoretical consideration on the influence of the z-value of a single-component time-temperature integrator on thermal-process impact evaluation. *Journal of Food Protection* 58: 39-48.

Wang, W., Walton, J. H. & McCarthy, K. L. (2000). Mixing in scraped surface heat exchanger geometry using MRI. *Journal of Food Process Engineering* 23: 403-418.

Wanihsuksombat, C., Hongtrakul, V. & Suppakul, P. (2010). Development and characterization of a prototype of a lactic acid-based time-temperature indicator for monitoring food product quality. *Journal of Food Engineering* 100: 427-434.

Wells, J. H. & Singh, R. P. (1988). Application of Time-Temperature Indicators in Monitoring Changes in Quality Attributes of Perishable and Semiperishable Foods. *Journal of Food Science* 53: 148-152.

Welt, B. A., Sage, D. S. & Berger, K. L. (2003). Performance specification of time-temperature integrators designed to protect against botulism in refrigerated fresh foods. *Journal of Food Science* 68: 2-9.

Weng, Z. J., Hendrickx, M., Maesmans, G. & Tobback, P. (1991). Immobilized peroxidase - a potential bioindicator for evaluation of thermal-processes. *Journal of Food Science* 56: 567-570.

In: Progress on Quantitative Approaches in Thermal … ISBN: 978-1-62100-842-2
Editors: V. P. Valdramidis and J. F. M. Van Impe © 2012 Nova Science Publishers, Inc.

Chapter 12

SETTING OF THERMAL PROCESSES IN A CONTEXT OF FOOD SAFETY OBJECTIVES (FSOS) AND RELATED CONCEPTS

Jeanne-Marie Membré[1,2,1]
[1]INRA, UMR1014 SECALIM, Rue de la Géraudière, NANTES Cedex 3, France
[2] LUNAM Université, ONIRIS, Nantes, France

Abstract

This chapter describes how the heat-treatment setting is articulated with Food Safety Objective (FSO) and related concepts. After a short introduction to risk analysis and risk assessment, the key components of the FSO concept are presented. Several case-studies are reviewed in details to demonstrate how a sterilization, or a pasteurization value, is set knowing an FSO. This review encompasses deterministic and probabilistic techniques. Next, a methodology to combine the reduction effect of thermal process with the inhibition effect of the product formulation is introduced. Finally, some additional information on quantitative methods useful when setting thermal process parameters in a risk-based framework is presented.

Definition and Terminology Utilised in This Chapter

Government: Competent authority in a country in charge of food safety management measure and procedures

Industry: All professionals involved in food production throughout the food chain, e.g. from primary production, distribution, processing and manufacture, packaging, storage, retail, catering and food services

Appropriate level of protection (ALOP): statement of the degree of public health protection that is to be achieved by the food safety systems implemented within a country

Food Safety Objective (FSO): maximum frequency and/or concentration of a hazard in a food at the time of consumption that provides or contributes to the ALOP

[1] E-mail address: jeanne-marie.membre@oniris-nantes.fr

Performance Objective (PO): maximum frequency and/or concentration of a hazard in a food at a specified step in the food chain before the time of consumption that provides or contributes to an FSO or ALOP, as applicable

Performance Criteria (PC): effect required of one or more control measure(s) working in concert to meet a PO, e.g. inactivation and inhibition

Process criteria (PrC): control parameters (e.g. heat-treatment time and temperature) at a step or combination of steps that can be applied to achieve a PC

1. Introduction

1.1. Food Safety Management into a Risk Analysis Framework

The World Health Organization, WHO, and the Food and Agriculture Organization of the United Nations, FAO, have called upon countries to apply modern international food safety and quality standards to protect consumer health. They are in the forefront of the development of risk-based approaches for the management of public health hazards in food. Appreciating the complexity of the current food supply within and across countries, these intergovernmental organisations have adopted risk analysis as the common framework for building food safety control programmes.

Risk analysis consists of three interconnected activities: risk assessment, risk management, and risk communication.

- ✓ Risk assessment is a scientific method to determine if exposure to hazards is expected to cause adverse effects, and the potential magnitude of such effects. It is used increasingly to better understand the complex interactions between hazards, products, ingredients and organisms. Risk assessment consists of four steps: hazard identification, exposure assessment, hazard characterization and risk characterization. The knowledge in each step is combined to represent a cause-and-effect chain from the prevalence and concentration of the pathogen to the probability and magnitude of health effects [22].
- ✓ Risk management is the process whereby information related to a risk, including the results of a risk assessment if available, are used to make decisions on how the risk will be controlled and then how that decision is implemented.
- ✓ Finally, risk communication is the process by which information related to the assessment and management of a risk is exchanged among the individuals affected by the hazard [6].

Through Codex Alimentarius, FAO and WHO are developing guidelines and reports providing detailed advice on the various aspects of risk analysis [12].

1.2. Articulation of FSO - and Related Concepts- with Risk Assessment

In the context of food safety, an Appropriate Level of Protection (ALOP) is a statement of the degree of public health protection that has to be achieved by the food safety systems implemented within a country. For example, an ALOP may be stated for a particular country as incidence of salmonellosis attributable to poultry to be less than 10 cases per 100 000 population.

Whilst expression of an ALOP in terms relevant to public health, e.g. the reported number of cases per 100 000 population serves to inform the public, the ALOP is not a useful measure in the actual implementation of food controls throughout the food chain. Instead, a measurable target for producers, manufacturers and control authorities is required; this is the basis of the Food Safety Objective (FSO) concept.

An FSO corresponds to the maximum frequency and/or concentration of a hazard in a food at the time of consumption that provides or contributes to the ALOP. Expression of an FSO as the frequency and/or concentration of a hazard in the food, is quantitatively linked to the ALOP by integrating the variability distribution of the exposure assessment with a dose-response curve i.e. the risk characterization [12].

Only Government (i.e. competent authority in a country) set FSOs, while mainly Industry (i.e. all professionals involved in food production) sets "in-chain" targets as part of their Food Safety Management systems (Figure 1).

In addition, for some food hazards, the FSO is likely to be very low. For instance, for a processor that makes ingredients or foods that require cooking prior to consumption, this level may be very difficult to use as a guideline in the factory [21]. Therefore, it is often required to set a level that must be met at earlier steps in the food chain. This level is called a performance objective (PO). A PO corresponds to a maximum frequency and/or concentration of a hazard in a food at a specified step in the food chain before the time of consumption that provides or contributes to an FSO or ALOP, as applicable (Figure 2).

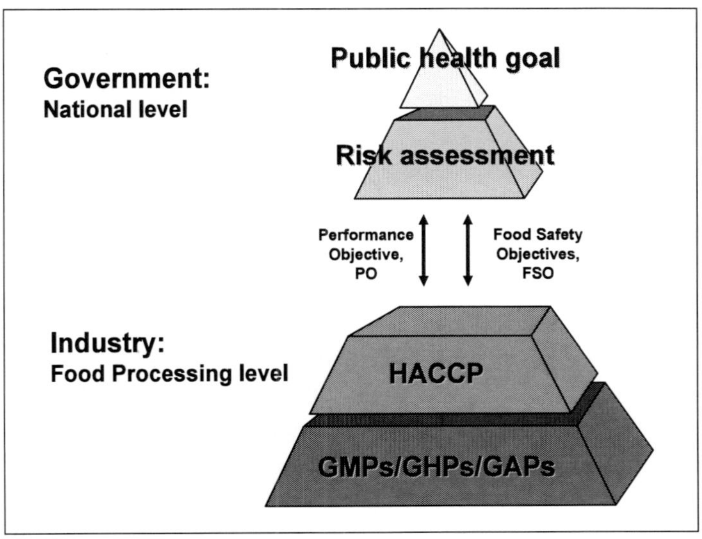

Figure 1. Articulation of food safety management and risk-based approach, set at Government and Industry levels. Figure adapted from [21].

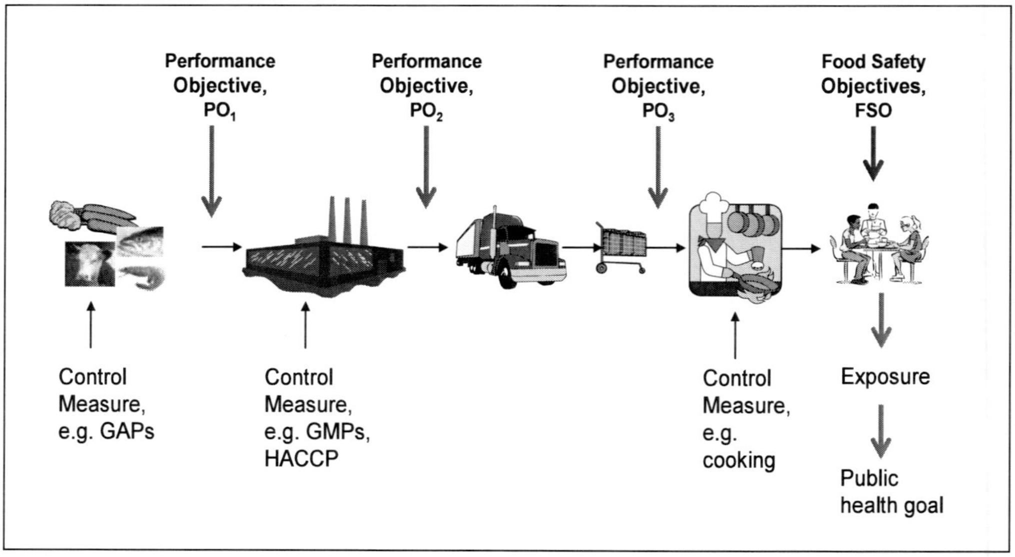

Figure 2. Position of Food Safety Objective (FSO) and derived Performance Objectives (PO_1, PO_2, PO_3) in the food supply-chain. Illustration based on [21].

1.3. Current Status of FSO Concept in Food Safety Management

Food Safety Objectives and Performance Objectives have been presented and discussed rather extensively in the scientific community [18, 45, 46, 54, 55, 57]. However, it is important to bear in mind that FSO are not yet used by regulatory authorities, and consequently not yet applied by food manufacturers. As emphasized by the European Food Safety Agency, "the practical application of these new metrics is still at a very early stage of development" [9]. Nevertheless, the risk-based approach looks promising to make food safety transparent and quantifiable [57].

In a near future, Industry will need to be able to provide evidence that their foods at the moment they are eaten, comply with an FSO or a PO. This will require Industry to become familiar with the new concepts, how they are established and, where appropriate, used in practice [16].

In the field of thermal process, there are only few applications of the FSO concepts (some examples will be developed in this Chapter) to the heat-treatment settings. In fact, thermal setting is generally based upon long-time accepted rule such as the botulinum cook , with a F_0 3 minimum, calculated with a reference temperature of 121.1°C and a Z value of 10°C for low-acid ambient stable product [1, 47]. A common practice in manufacturing is to set the sterilization value on a somewhat arbitrary standard value higher than F_0 3, to guarantee the commercial sterility of ambient stable products. This practice has some disadvantages because it almost always recommends a very high sterilization value which leads to an over-cooked product with less nutritional and organoleptic quality for the consumer. It is then tempting to explore whether the risk-based approach, with the FSO and PO concept, could be extended to spoilage microorganism [30].

2. Setting a Log Reduction for a Targeted FSO

2.1. Principle as Developed by the ICMSF and Application through Examples

To get a step further in the risk-based concept, ICMSF [20] has summarised the various changes of microbiological load through the food chain with a simple equation (Eq. 1).

$$H_0 - \Sigma R + \Sigma I \leq PO \text{ or } FSO \tag{1}$$

The different elements of Eq.1 are defined as followed:

- ✓ H_0: initial level of hazard, might be prevalence and/or concentration, expressed in log count
- ✓ ΣR: quantification of microbial inactivation in one or several of the food processing steps; expressed in log count
- ✓ ΣI: the sum of any hazard increase (growth or recontamination), expressed in log count
- ✓ PO: Performance Objective, expressed in log count
- ✓ FSO: Food Safety Objective, expressed in log count

The targeted log reduction is then deduced directly from Eq. 1 as follows:

$$\Sigma R \geq +H_0 + \Sigma I - PO \text{ (or FSO)} \tag{2}$$

In words, Eq. 2 means that the targeted log reduction will be set as function of the initial level of hazard in the food chain, the potential increase of hazard by growth and/or recontamination and the desired PO or FSO.

To illustrate this risk-based concept, two examples from the literature are provided in Tables 1 and 2. This is not an exhaustive list and other examples from the literature may be useful to grab the ICMSF scheme, among them, it is worth mentioning *Staphylococcus aureus* in cream-filled baked goods [45], *Listeria monocytogenes* in cold-smoked salmon [46] and *L. monocytogenes* in a ready-to-eat food [51].

When thinking of heat-treatment process as sole log reduction measure, the risk-based approach means that a severe heat treatment is applied when the initial hazard level is high and/or the growth likely, while a less severe heat treatment is applied where growth can be prevented or low initial levels can be assured. The heat treatment is thus balanced against the level of control in the other parts of the process, or even the level of control in preceding or subsequent steps in the food processing chain. Consequently, this risk-based approach revisits the current standards, either prescribed by regulation or ingrained in industry practice, for examples a 6-log reduction of *Listeria monocytogenes* in ready-to-eat food [27], a 7-log reduction of *Salmonella* in poultry [49] or a 12-log reduction of spores of *Clostridium botulinum* in canned food [5].

Table 1. Inputs considered in the log reduction setting, example of *Escherichia coli* O157:H7 in frozen beef patties [20]

✓	The hazard was identified as *E. coli* O157:H7, the reservoir as the cattle. A very small proportion of lots of ground beef has a relatively high prevalence (5% or higher) and higher concentration of cells (1-10 g^{-1}); consequently, the initial level was estimated to be 100 cfu/g	$H_0 = 2$
✓	In term of hazard characterization, disease associated with *E. coli* O157:H7 was assessed to be moderate to severe; consequently the maximum concentration at the time of consumption was 1 cell per 250g-portion	FSO=-2.4
✓	In term of exposure assessment, it was assumed neither carcass surface contamination nor decontamination before the heat-treatment process, and, no increase under controlled chilling/fabrication operations once the patties cooked; consequently, no growth or recontamination was assumed	$\Sigma I=0$
✓	$\Sigma R \geq +H_0 + \Sigma I - PO \text{ (or FSO)}$	$\Sigma R \geq 2+0+2.4=4.4$

Table 2. Inputs considered in the log reduction setting, example of *Listeria monocytogenes* in shrimp [53]

✓	The hazard was identified as *Listeria monocytogenes*, concentration in raw shrimp before processing was known to be 100 CFU/g	$H_0 = 2$
✓	In term of hazard characterization, disease associated with *L. monocytogenes* was assessed to be moderate to severe, the serving size evaluated to 100 g (5 to 10 shrimp); consequently, the maximum concentration at the time of consumption was 1 cell per 100g-portion	FSO=-2
✓	In term of exposure assessment, it was assumed that there was no opportunity for levels to increase during cooling, packaging, distribution, and retail sale, for example if the consumer buys frozen shrimp, thaws them properly, and eats them; consequently, no growth or recontamination was assumed	$\Sigma I=0$
✓	To compensate for any errors in estimating the levels based on seasonal variations or differences in heat resistance of the bacteria, a safety factor of 1 log unit was added.	Safety factor 1
✓	$\Sigma R \geq +H_0 + \Sigma I - PO \text{ (or FSO)} + \text{Safety factor}$	$\Sigma R \geq 2+0+2+1=5$

2.2. Mathematical Difficulties when Handling the ICMSF Equation

Although straightforward, FSO, PO and related concepts are not yet applied by Government and Industry. There are several reasons for this delay in application, the main one being the extensive time required to harmonize food safety management policies between countries (e.g. to agree on ALOP) or to perform risk analysis for extensive list of hazards in the food chain.

There are also some mathematical difficulties when handling risk-based concepts in general and ICMSF equation in particular, in heat-treatment process settings:

- ✓ First of all, and probably the most difficult to deal with, the targeted log reduction is set as a minimum instead of a practical and easy-to-implement value;
- ✓ Secondly, in many examples including those provided in Tables 1 and 2, FSO (or PO) are chosen as "1 cfu/portion" to deal with calculation based on logarithm transformation, while the target should be "absence in portion", i.e. zero cfu/portion. This is particularly important if the heat treatment is the sole control measure to inhibit/inactivate the microorganisms because one has to be confident that the consumer portions will be free of bacteria or spores before manufacturing and distributing them;
- ✓ Finally, to tackle properly the ICMSF equation, a lot of pieces of information, or data, are required. The ingredient contamination has to be defined, and, a log increase (or absence of growth) before the heat treatment has to be determined. If limited information is available, this might be a source of uncertainty.

Because of the three reasons listed above, the log reduction setting tends to include safety margins and to deal with worst-case scenarios (e.g. utilize the upper bound of a 95th confidence interval [50]) and then looses transparency and even advantages over the use of current standards (such as 6-log reduction of *Listeria monocytogenes* in ready-to-eat food [27]).

One way to circle these mathematical difficulties is to deploy probabilistic techniques.

2.3. Application of Probabilistic Techniques in Heat-Treatment Setting

2.3.1. Introduction to Deterministic and Probabilistic Risk Assessment

In risk assessment, uncertainty and variability are inherent [48]. Uncertainty arises from our lack of knowledge and may be related to the model used to characterise the risk, the parameters used to provide values for the model, or both. In some cases, we can reduce uncertainty by obtaining better information (more data, better Expert elicitation), but this may not always be possible. Uncertainty implies that we might make a non optimal choice because we may expect one outcome but something quite different might actually occur.

Variability differs significantly from uncertainty. Variability refers to real and identifiable differences between individuals within a population addressed by risk assessment. For example, variability might refer to differences between individual people or individual facilities (at factory, retailer or consumer), or individual crops/species/strains/batches of manufactured products. True variability does not disappear with improved measurements. The existence of variability in a population implies that a single action or strategy that is optimal for each of the individuals, may not emerge; consequently any decision made will go too far for some and not far enough for others.

When performing risk assessment two distinct approaches exist to incorporate uncertainty and variability:

- ✓ Using the "point estimate" approach (deterministic approach), risk assessors consider the available information, select a single number for each of the model factors/inputs and then generate a point estimate (single number) of risk. By choosing single numbers for factors/inputs, risk assessors neglect or make simplifying assumptions about the natural variability and uncertainty in the risk estimate. Instead these are accounted for by applying conservative uncertainty (safety) factors.
- ✓ In contrast, using a probabilistic approach, risk assessors employ probability distributions to characterise variability, in some, or all of the model inputs and to capture uncertainty in the data sets. They propagate these input distributions through the model to produce probability distributions that convey variability and uncertainty in the risk estimate.

Both probabilistic and deterministic approaches are equally valuable in the correct context. The decision-making process may require iterations of the risk assessment and a mixture of both approaches may be used in the same assessment. For example, the process may start with a screening assessment, which uses generic numbers ("defaults") for many inputs in the deterministic approach and ultimately progress to a highly refined probabilistic analysis.

Probabilistic risk assessment has been already applied in food safety domain for various foodborne hazards, for instance *Clostridum perfringens* [13], *Escherichia coli* [8], *Campylobacter* [25, 39] or *Listeria monocytogenes* [56]. In contrast, modelling of thermal processes and their effect on lethality of microbial populations has been traditionally carried out using deterministic approaches [2]. Assumptions about the processing conditions and microbial populations are made which enable the application of exact mathematical models, with purely quantitative objectives. Less frequently, studies using stochastic approaches or combinations of deterministic and stochastic approaches have been reported [44]. Definitely, heat transfer models, which predict the spatial distribution of temperature as a function of time and processing conditions, provide accurate inputs for a probabilistic model.

An example of an application of probabilistic technique to determine the effect of thermal process on lethality of microbial population has been reported in the literature [31]. Assessment of *Bacillus cereus* spore load after in-pack sterilization has been performed; the variability associated with i) raw material contamination, ii) mesophilic and psychropilic spore heat-resistance and iii) process characteristics, have been incorporated in the analysis. The spatial distribution of heat was taken into account: the time equivalent spent at 90°C was estimated to be in a range of 25 to 50 min, due to heat transfer characteristics. The difference between deterministic and probabilistic approaches in this case could be illustrated by considering that, in the deterministic approach the coldest point of the pack (25 min) would be used as point estimate (single number) whilst with a probabilistic approach, the 25-min time is still included

Inputs	Deterministic approach	Probabilistic approach
FSO	Hypothetical FSO chosen as "less than 1 cfu per portion" with portion of 25g, FSO=-1.4	
ΣI	No growth possible after manufacturing until the consumer eats. During manufacture recontamination is effectively excluded, ΣI =0.	
H_0	From studies that reported on the concentration of *Salmonella* spp. in raw poultry, a conservative worst case level of contamination was set at 1500 CFU/g of chicken meat (H_0=3.18 log cfu/g).	A Pert distribution [52] was built with H_0 (in log cfu/g) set up as minimum = 0, most likely = 1, maximum = 3.18. 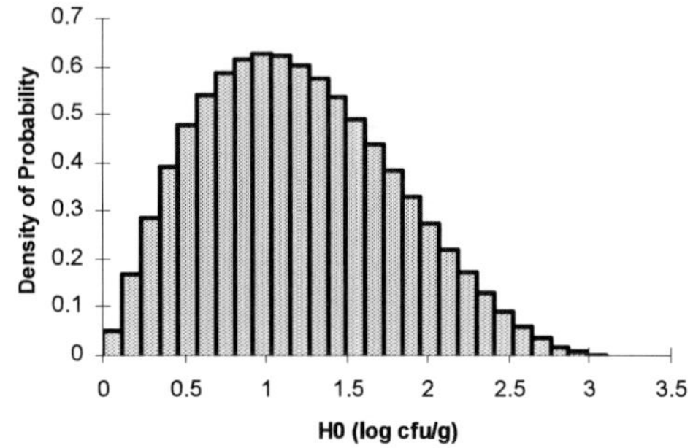
	Prevalence of *Salmonella* in the raw poultry meat of 100%.	Prevalence of *Salmonella* in the raw poultry meat of either 100% or 5.7%
D-value at 70°C incorporated in the log normal heat-treatment model [3]	Either D_{70}=0.093 min using the upper bound of the 95th confidence interval [50] or D_{70}=0.057 min using the mean value [53]	log *D* values were estimated by a Normal distribution with mean= -1.247 and standard deviation = 0.1319 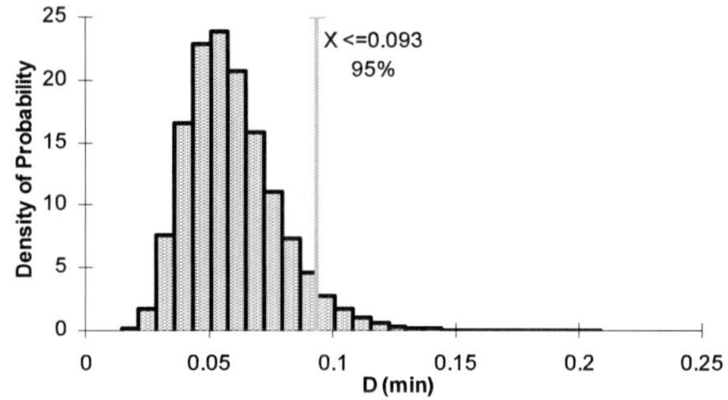

Figure 3. List of inputs incorporated in the model applied to *Salmonella* thermal inactivation: comparison of deterministic and probabilistic settings. Figure adapted from [32].

in the analysis, but it is associated with the percentage of the pack predicted to receive this time equivalent (5%). By deploying the probabilistic analysis, it was established that there was 11% of packs containing psychrotrophic *Bacillus cereus* after the cooking process.

2.3.2. Example of a Probabilistic Technique Applied to Pasteurisation Setting

An example from the literature [32] is presented in this section to highlight the differences between deterministic and probabilistic techniques. The heat-treatment time required to pasteurise the poultry meat at 70°C was calculated using the ICMSF equation (Eq. 1). In Figure 3, the values of the model inputs are listed for both deterministic and probabilistic approaches.

Note that the heat-treatment time is named Process Criteria (PrC) in the risk-based terminology (see also section 3.1):

- ✓ With deterministic approaches, the heat-treatment time was estimated to be 0.32 or 0.43 min, using a 1 log safety margin and the mean of *D*-value [53], or, no safety margin and the upper bound of the 95% confidence interval of *D*-value [50], respectively.
- ✓ With the probabilistic approaches, the output of the analysis was no longer a single value but a density of probability function (Figure 4A), whose the 95th percentile value was 0.27 min.

To make the comparison between the methods possible, these three values were obtained with a scenario of 100% prevalence of *Salmonella* in raw poultry meat, and an FSO of less than 1 cfu/portion (i.e. in log, -1.4). Whatever the approach, the pasteurization value was found to be definitively lower than the current standard for pasteurisation of poultry meat (70°C for 2 min).

To get a step further, the probability to meet an FSO chosen as less than 1 cfu/portion or 0 cfu/portion was calculated for various heat-treatment times. Results are presented in Figure 4B. They were obtained by carrying out the following methodology:

- ✓ The probability, p, for one bacterium to survive the heat treatment is deduced from the heat-treatment time and the D-value [38] as indicated in Eq. 3.

$$p = 10^{(-\text{time} / D)} \tag{3}$$

- ✓ In absence of growth or recontamination ($\Sigma I=0$), the number of bacteria in a 25g-portion is $N_0=25g \times 10^{H_0}$
- ✓ Each bacterial cell in the portion has a probability p to survive the heat treatment, then, the number of bacteria in the 25g-heat treated portions follows a Binomial distribution, with parameters N_0 and p [37, 38], as indicated in Eq. 4.

$$N \sim \text{Binomial}(N_0, p) \tag{4}$$

✓ Eventually, to facilitate the estimation process, and under the hypothesis of N0 large and p small, a Binomial distribution can be approximated by a Poisson distribution [52] as indicated in Eq. 5.

$$N \sim \text{Poisson}\ (25 \times 10^{H_0} \times p) \tag{5}$$

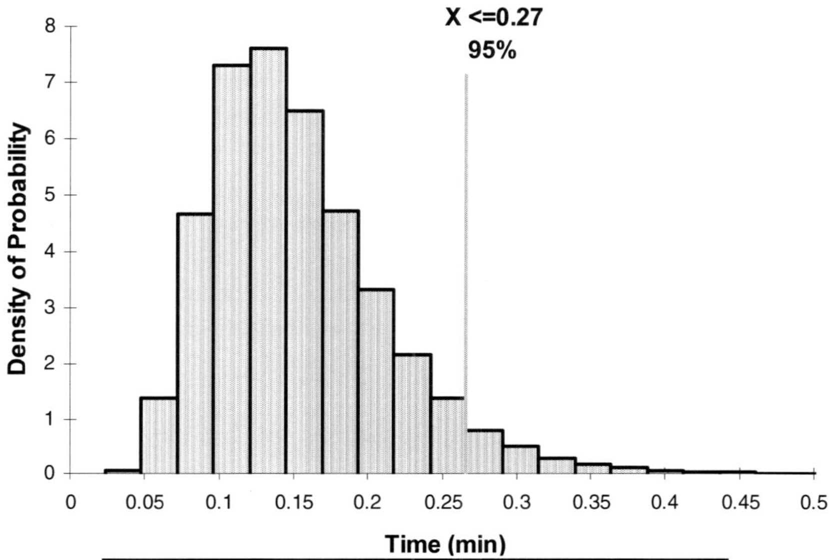

A

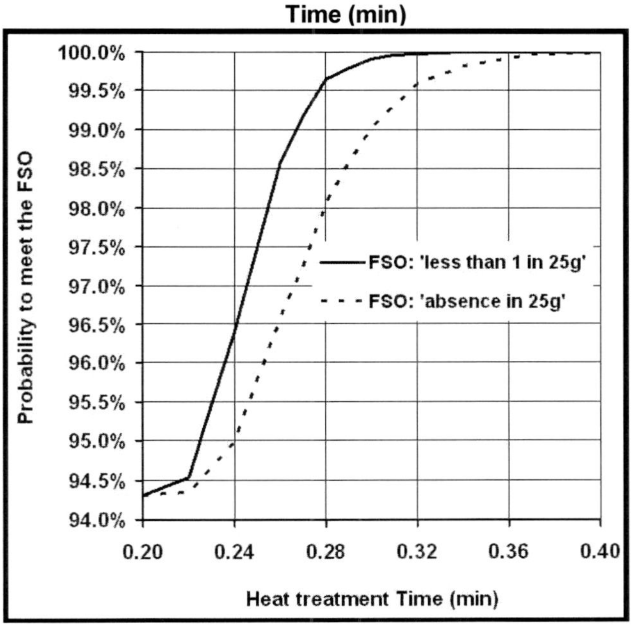

B

Figure 4. Outputs from the probabilistic model applied to *Salmonella* thermal inactivation: **A**: probability distribution function associated with the heat-treatment time at 70°C, 100% prevalence of *Salmonella* in raw poultry meat assumed, and FSO as "less than 1 cell/portion". **B**: probability to meet an FSO interpreted as "less than 1 cell" or "0 cell" per portion, with inputs D=0.057 min, H_0=3.18, prevalence of 5.7%. Figure adapted from [32].

For reporting the outputs, several formats are possible, the most convenient is likely to be the cumulative distribution function since the probability to get less than a specific value is directly calculated. By choosing zero or one as values of interest, the probability to meet an FSO is directly calculated. For instance, with a heat-treatment time of 0.32 min as obtained using the method recommended by Walls [53], the probability of meeting the FSO set as 'absence in 25g' is 99.5% (Figure 4B). In this Figure, results are shown with D and H_0 as single-point estimate (D=0.057 min, H_0=3.18) and with a prevalence of 5.7%.

Note that up to now, the percentile (95th, 99th, 99.9th, others) with which an FSO has to be achieved has not been stated by competent authorities or by scientific communities; obviously, this should be clarified before advocating the use of FSO and related concepts in food safety management.

2.3.3. Some Limitations in the Use of Probabilistic Techniques

Generally speaking, although the probabilistic approach offers many potential advantages over the point-estimate approach, Government and Industry have been slow to encourage or even consider its use. Several factors may explain this limited use:

- ✓ The lack of guidance. Currently there are no technical support documents on how to deploy probabilistic techniques and present the results. Guidance for decision-makers who receive the outputs is also lacking.
- ✓ Let's take a simplified example. An FSO of *Listeria monocytogenes* is set to 100 cfu/g in an in-pack pasteurized, ready-to-eat product, with a short shelf-life (no post- heat treatment growth or recontamination). Assuming a constant heat-treatment temperature and a log-linear inactivation model [3], the heat-treatment time should be easily calculated from the ICMSF equation (Figure 5). If a probabilistic analysis is carried out, the left hand side of the equation will be represented as a probabilistic distribution, while the right hand side will remains as a fixed value (an FSO is a target, not a calculated value). Then, one of the difficulty, in absence of guidance for decision-makers, is to define the probability x% of exceeding the FSO (Figure 5). It might be useful at this point to emphasise that choosing 'x%=0' is not relevant as the true maximum of the probability density function might be the infinity and also because the probability distribution functions are generally calculated by approximation (for instance by using the Monte Carlo simulation technique, see section 4.2) leading to obtain a maximum depending on the number of iterations.
- ✓ Inexperience with using probabilistic results. In addition to the lack of guidance, the lack of experience that risk managers have with probabilistic outputs means that understanding these requires more effort than that needed for point-estimate results. Decision making is also more complicated because risk managers cannot simply compare the results against "established" distributions.
- ✓ Resource demanding. Probabilistic techniques require definitively time/computing power and also human resources with modelling expertise, what is generally not available in R&D departments of small and medium enterprises.
- ✓ Data demanding. The quantity of data required to perform a probabilistic analysis might be seen also as a limitation for its implementation, even if the use

of Bayesian inference with Experts' information incorporated into the prior distribution (see section 4.3) reduces this issue.

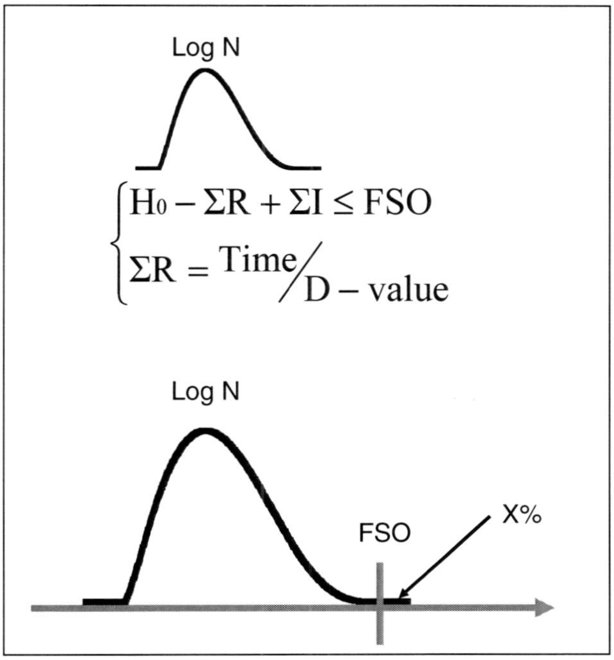

Figure 5. Summary of modelling framework under which a probabilistic analysis of heat-treatment setting operates.

3. Combining Several Process Criteria for Achieving a Targeted PO or FSO

3.1. Introduction to Process Criteria

Process criteria (PrC) are the control parameters at a step or combination of steps that can be applied to achieve a desired reduction (ΣR) or a desired limited growth and contamination (ΣI). For example, the PrC to achieve at least a 6 decimal reduction of *L. monocytogenes* in milk are 71.7°C for 15 sec [19]. Another example is a specified pH of 4.6 or less in canned shelf stable acidified foods which has proven effective for controlling the growth of *Clostridium botulinum* and preventing botulism. Such PrC are the critical limits in a HACCP plan when the control occurs at a Critical Control Point.

Note that in this chapter, the term 'Process Criteria' is used in a broader sense since it includes control measure associated with the process (e.g. heat treatment time) and those associated with the product formulation (e.g. pH). In literature, the latter are also named 'Product criteria'.

Determining a PrC for a given target such as the log reduction ΣR might require only one control measure, i.e. the heat-treatment characteristic. For example, the United Kingdom Advisory Committee on the Microbial Safety of Food (ACMSF) recommend a PrC of 70°C for 2min to achieve a 6D reduction of *E. coli* 0157H7, *Salmonella* spp. and *L. monocytogenes* in meat products.

However, there are some advantages of combining more than one PrC to achieve the desired ΣR, PO or FSO: it allows a better use of combination preservation conditions that are less damaging for product quality while still delivering the required level of safety. The importance of such interactions has been recognized for controlling mesophilic spore-formers in ambient-stable products. Braithwaite and Perigo [4] produced a set of contour plots where limiting F_0 values were plotted against a_w and pH, clearly showing combinations of conditions that may only require relatively mild heat treatments (e.g. F_0 0.3) to deliver ambient stability. Mafart and Leguerinel group have worked intensively on modelling approaches to include pH and a_w beside the temperature on thermal pasteurization and sterilization values [24, 28].

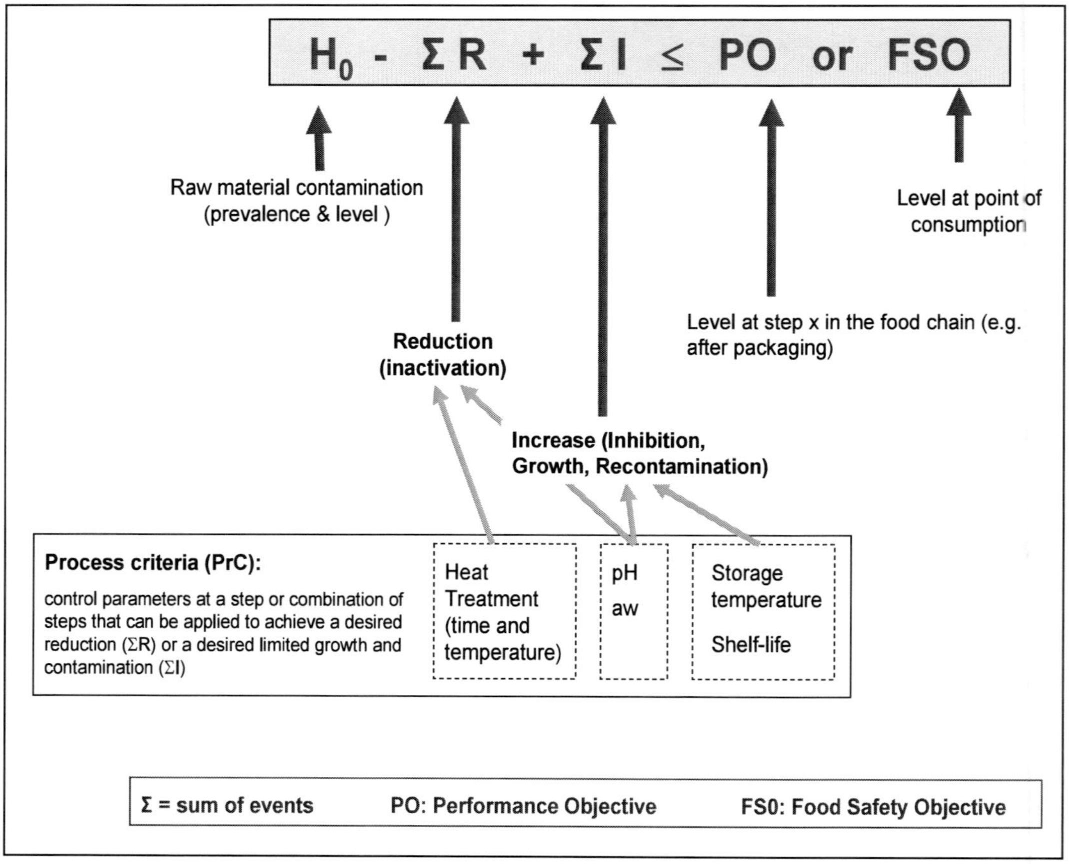

Figure 6. Articulation between various risk-based concepts introduced by the ICMSF [20] and adopted by Government. Illustration of the key role played by the PrC on the control of microbial hazard propagation and consequently on the FSO compliance.

Going a step further, it is tempting to articulate pH and a_w effects on inactivation (ΣR) with inhibition or limitation of growth (ΣI) to end up with a cumulative control measure effect (Figure 6). That is the basis for the work currently performed by the National Center for Food Safety and Technology, in US, on 'Food Safety Objective (FSO) Approach to Controlling

Clostridium botulinum Toxin in Commercially Sterile Foods' (more information at http://www.iit.edu/ncfst/world_class_food_science/food_processing_and_packaging/current_projects.shtml, accessed on 27/05/2011).

Likewise, other controlled measures such as storage temperature (particularly with chilled products) can participate to achieve a targeted PO or FSO, by extending the growth lag time and reducing the growth rate. Two examples of application of combined PrC are provided in the following section.

3.2. Application to Refrigerated, Processed Food of Extended Durability

The use of risk-based approaches helps in conceptualize the role of product formulation, storage condition (temperature, shelf life) beside the heat-treatment process in maintaining product quality and safety. Refrigerated, processed food of extended durability (REPFEDs) relies on a combination of mild heat treatment and refrigerated storage, sometimes in combination with other hurdles such reduced pH and a_w to ensure a high quality in terms of taste and aroma while guaranteeing consumer safety. Shelf-life of REPFEDs is generally in a range of two/three up to five/six weeks as function of the process and formulation conditions and their market is increasing, following the consumer demand for convenient and high organoleptic quality products.

3.2.1. Process Criteria to Achieve a 6 Log Reduction and/or Inhibition of C. Botulinum Spores

In REPFEDs, non-proteolytic *Clostridium botulinum*, is considered as a serious hazard because of its ability to grow at temperatures as low as 3ºC. Heat treatments affect the outgrowth of spores of non-proteolytic *C. botulinum* in two ways: inactivation and injury. It is generally agreed that a temperature of 90°C for 10 min will deliver a 6D inactivation of non-proteolytic *C. botulinum* and this is a commonly used performance standard for the heat processing of chilled foods [17]. In addition to this killing effect and depending on the heat treatment applied, some spores remaining after the heat treatment may be injured. Due to this injury, the time for the spores to recover is prolonged and the length of the lag time is increased [41]. This injury heat-treatment effect depends also upon the recovery conditions such as temperature (chilled conditions), pH and a_w. If the shelf-life of the REPFED is short enough, it is biologically possible that the injured spores of *C. botulinum* have insufficient time to recover, germinate, multiply and potentially produce toxins.

In the example presented below [35], the authors have incorporated more than one PrC to achieve the target. Indeed, instead of designing the process to achieve 6 log reduction by the sole heat-treatment control measure (i.e. applying the recommended standard of 90°C for 10 min), they have explored the possibility of combining the pH, storage temperature and shelf-life with the heat-treatment, to obtain a 6 log reduction and/or inhibition. The main advantage of such approach was obviously to gain in product quality by applying milder heat treatment without compromising public health safety. The calculations were based upon the interpretation of the heat-treated spore lag time as a probability of delaying the growth, and the incorporation of this probability into an overall thermal effect [26].

The approach is detailed here, using the ICMSF terminology [20]:

- ✓ ΣR corresponds to the log reduction (inactivation, lethality) while ΣI to limited growth or even inhibition; algebraically, ΣI is positive if growth occurs and negative if inhibition is observed.
- ✓ Both terms have been expressed as the decimal logarithm of the reciprocal of a probability (Eqs 6 and 7).

$$\text{Reduction} = \Sigma R = \log(1/Pr) \tag{6}$$

In Eq. 6, Pr is the probability that one spore survives the heat treatment [38]. The heat-treatment effect was assessed using a log-linear inactivation model [3] with a D_{90} of 1.67 min and a Z value of 7°C.

$$\text{Inhibition} = -\Sigma I = \log(1/Pi) \tag{7}$$

In Eq. 7, Pi is the probability that lag time (recovery time) is shorter than the shelf-life, SL (Eq. 8).

$$Pi = \text{Probability}\{\text{Lag} \leq SL\} \tag{8}$$

For example, if Pi = 0.1 (moderate inhibition), the Inhibition = +1 (ΣI=-1).

- ✓ To articulate the thermal effect on inactivation (ΣR) with inhibition of growth (-ΣI) and to end up with a cumulative control measure effect as schematically pictured in Figure 6, ΣR and ΣI have to be associated (Eq. 9). Note that these two entities correspond to the Performance Criteria (PC): the effect required of one or more control measure(s) working in concert to meet a PO [16].

$$PC = \text{Reduction} + \text{Inhibition} = \Sigma R - \Sigma I \tag{9}$$

- ✓ In the example of *C. botulinum* in REPFEDs, the overall thermal effect (inactivation and injury) is summarized as targeting PC=6 log. To achieve, this target instead of having a heat-treatment which deliver 6 log inactivation, ΣR=6, by applying a sole intervention of thermal process (delivering 90°C for 10 min), a milder heat-treatment might be applied as long as they follow the rules illustrated in Eq. 10.

$$\begin{cases} \Sigma R = x \\ -\Sigma I = 6 - x \end{cases} \tag{10}$$

Data on the impact of heat-treatment, pH, a_w, storage temperature on *C. botulinum* lag time were gathered and utilised to develop a predictive model [35]. The model equations are reported in Table 3. The lag model was run with a realistic range of refrigerator temperatures based on a survey [14], incorporating as probabilistic input.

Once the lag time predicted and the heat-treatment effect known, it was possible to define a set of PrC combinations, i.e. a set of process and formulation characteristics and shelf-life, which deliver a targeted PC and eventually a defined PO or FSO.

The shelf life, the heat-treatment temperature (applied for 10 min) and the pH were the three PrC chosen to achieve a 6 log PC (Figure 7). Obviously, there was more than one combination of these three PrC which achieve the same PC as the term x in Eq. 10 may vary from 0 to 6.

Table 3. Set of lag model equations utilised to predict the lag time (recovery, germination and outgrowth) of heat-injured spores of non-proteolytic *C. botulinum* [35]

$$\sqrt{Lag} = a_1 \times f_1(T) \times f_2(pH) \times f_3(a_w) \times f_4(HT) + \varepsilon$$

$$f_1(T) = \left(\frac{Topt - Tmin}{T - Tmin}\right)^{p1} \quad \text{for } T \geq Tmin$$

$$f_2(pH) = \left(\frac{pHopt - pHmin}{pH - pHmin}\right)^{p2} \quad \text{for } pH \geq pHmin$$

$$f_3(a_w) = \left(\frac{a_w opt - a_w min}{a_w - a_w min}\right)^{p3} \quad \text{for } aw \geq awmin$$

$$f_4(HT) = (HTT - HTT_{min})^{p41} \times HT\_t^{p42} \quad \text{for } HTT \geq HTTmin$$

p1, p2, p3, p41 and p42 are shape parameters

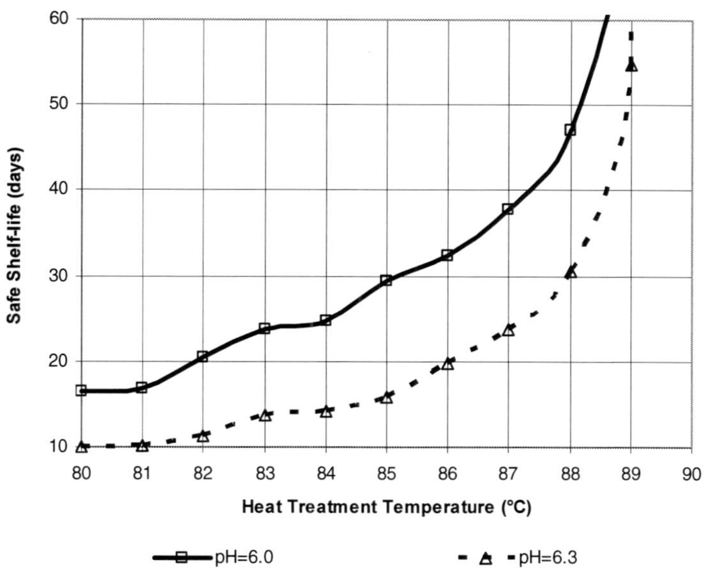

Figure 7. Combination of three Process criteria (shelf life, heat-treatment temperature -applied for 10min- and pH) applied to achieve the Performance Criteria '6 log reduction/inhibition' of non-proteolytic *C. botulinum* in REPFEDs. Figure adapted from [35].

For example, to achieve a PC of 6 log in a product at pH 6.3 having a maximum shelf life of 30 days, a heat-treatment of 88ºC for 10 min is required. On the other hand, if the pH can be dropped to 6.0, a heat treatment of 85°C for 10 min is sufficient to guarantee a 6 log reduction/inhibition in a product having the same shelf-life (30 days). Note that in this example, the storage temperature is not a PrC since it has been chosen as varying in a realistic range, it is then not considered as a control measure.

3.2.2. Process Criteria to Obtain Less than 5 Log Cfu/G of B. Cereus at the Point of Consumption

Setting PrC for REPFEDs may be done also toward the *Bacillus cereus* hazard. Indeed, traditionally, these products are given an in-pack thermal process equivalent to at least 90°C for 10 min to control non-proteolytic *C. botulinum* (see section 3.2.1). In such a case, at pasteurisation temperature above 90°C for 10 min, spores of psychrotrophic *B.cereus* might survive and germinate, and their subsequent vegetative cells grow up to a hazardous level. In this context, beside, heat-treatment intensity, the formulation (pH, organic acid) and the storage conditions (shelf-life and chilled temperature) are fundamental factors to prevent *B. cereus* outgrowth [10, 11, 23, 29].

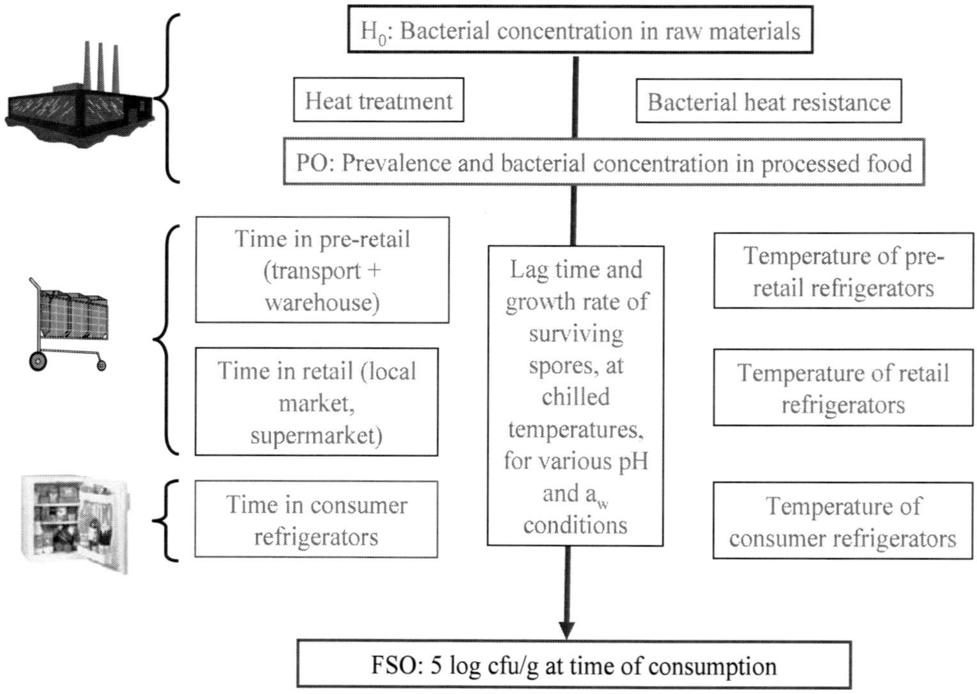

Figure 8. Exposure assessment framework developed by Membré et al. [34] to calculate the probability of achieving the FSO 'less than 5 log cfu/g of *Bacillus cereus* in REPFEDs'. Figure adapted from [34].

Recently, a probabilistic exposure assessment model was developed in industry to calculate the probability of having 10^5 cfu/g or more of *Bacillus cereus* at the point of consumption, knowing the process, the formulation and the distribution supply-chain of a REPFED product [33, 34]. Again, this example will be revisited in the light of the FSO concept. The factors considered in this assessment included raw material contamination, heat treatment, bacterial spore heat resistance, injured spore lag time and vegetative cell growth, chilled supply-chain market and consumers' habits (Figure 8).

The food supply-chain analysis is split into two successive steps: before and after the heat-treatment process:

- ✓ Before the in-pack heat-treatment process, Eq. (1) is applied as indicated in E. 11:

$$H_0 \quad - \quad \Sigma R \quad + \leq PO \tag{11}$$

  H_0 is the bacterial concentration in raw materials, ΣR the log reduction due to the heat treatment and PO the prevalence and bacterial concentration in the food just after the heat-treatment process (Figure 8).
- ✓ After the heat-treatment process and particularly during the supply-chain distribution, the potential growth observed subsequently to the recovery and germination of spores (ΣI) is included in the analysis as indicated in Eq. 12.

$$PO \quad + \quad \Sigma I \quad + \leq FSO \tag{12}$$

- ✓ There is no FSO for *B. cereus*, however an hypothetical value of 5 log cfu/g was chosen as this concentration is generally considered as the hazardous threshold for *B. cereus*.

The exposure assessment was set up with a probabilistic model to include variability and/or uncertainty associated with the following inputs: the bacterial concentration in raw materials (used to characterise H_0, Eq. 11), the bacterial heat-resistance (used to characterise ΣR, Eq. 11), the time and chilled temperature in pre-retail, retail and consumer refrigerators (used to characterise ΣI, Eq. 12).

As in the previous example, since the refrigerator temperatures were incorporated in the exposure model as distributions of probabilities reflecting the variability observed among the refrigerators, the storage temperatures was not considered as control measures and consequently not as a PrC. Consequently, the PrC were the heat-treatment process (time and temperature), the intrinsic product characteristic such as pH and/or a_w and the maximum shelf-life.

With H_0, ΣR and ΣI set as probabilistic variables and FSO a single-value threshold, the interpretation of the exposure assessment model output suffers of the same lack of guidance as previously described (see section 2.3.3 and Figure 5). The concentration of *B. cereus* at the time of consumption, log N, is a probabilistic output which has to be compared to a single-value FSO. The probability of exceeding the FSO, Pr (log N > FSO) corresponds to the value x in Figure 5. The compliance with the FSO is directly deduced from this probability: 1-x.

In Figure 9, the exposure model output is presented for two heat-treatment temperatures and three heat-treatment times, at pH 6 and a_w 0.996. The 95% confidence interval bars in this graph capture the uncertainty in the model output (see section 4.5). With a shelf-life of 40 days, an FSO of 5 log is achieved (99% compliance) by applying a PrC as mild as 90°C for 10 min, or alternatively 95°C for 5 min. If a more stringent compliance is required, for example 99.9%, the PrC has to be set to 95°C for 10 min (or 95°C for 15 min) to achieve the targeted FSO.

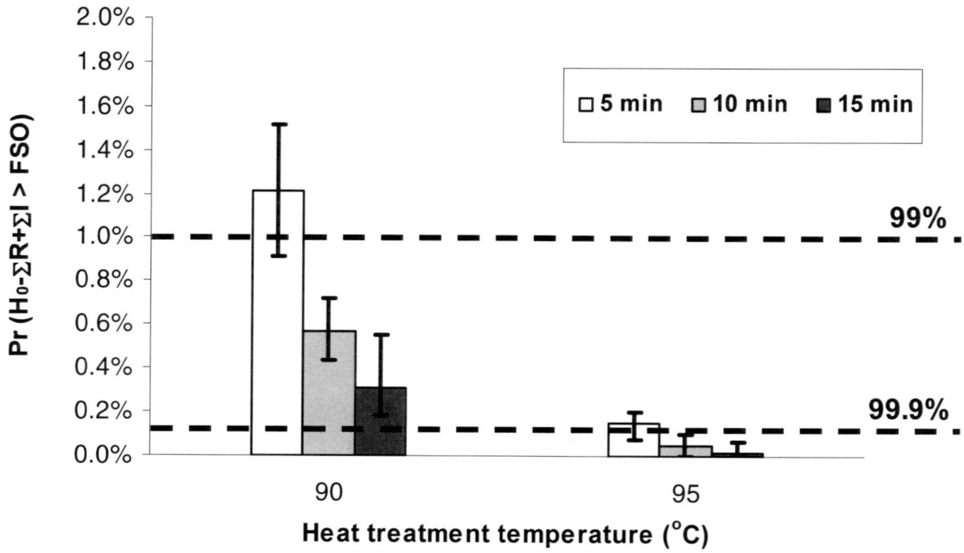

Figure 9. Probability of exceeding an FSO set to 5 log cfu/g of *Bacillus cereus* in REPFEDs at the point of consumption, for various heat-treatment settings. The probability of achieving the FSO target is also provided for two particular percentiles. Figure adapted from [34].

4. Further Information: Helpful Mathematical Toolbox

In this section, a short introduction to useful techniques utilized in probabilistic risk assessment is provided. The list is not exhaustive and Readers are advised to seek additional information in specialized literature.

4.1. Fitting Data by Probabilistic Distribution Functions

When a probabilistic risk assessment is performed, each input is represented by a distribution of values. There is no unique way to define these distributions. This will depend on the type and quantity of information available and the knowledge we have about what could control the distribution of each variable.

Experimental data can be original data or external data (publication, books, official web-based datasets,... etc.).

Data gathering can be complemented by Expert judgement. Expert judgement refers to inferential opinion of a specialist or group of specialists within an area of their expertise. When there is uncertainty associated with an input variable, such as a data gap or poor quality data, Expert opinion may be appropriate for obtaining distributions. Methodologies for eliciting Expert information are not detailed in this chapter. Expert information can be combined with data using Bayesian techniques (see section 4.3).

When selecting a probability distribution function, the following points should be considered:

- ✓ The variable is discrete (e.g. counts of head in coin tossing) or continuous (generally represent measurements, e.g. weight, concentration),

- ✓ The variable has some intrinsic or imposed bounds (e.g. a European adult weight is likely to be less than 150kg and more than 30 kilos). Many variables cannot have values less than zero
- ✓ The distribution is skewed (positively or negatively) or symmetric.

Once the probabilistic distributions are chosen, data can be fitted using several pieces of Software. The Excel add-in package @Risk (information available at http://www.palisade.com/risk/) is one of them. The package fitdistrplus of R software is another one (information available at http://www.r-project.org/).

4.2. Monte Carlo Procedure

Monte Carlo methods have been used for centuries, but only in the past several decades has the technique gained the status of a full-fledged numerical method capable of addressing the most complex applications.

Statistical simulation methods may be contrasted to conventional numerical discretisation methods, which typically are applied to ordinary or partial differential equations that describe some underlying physical or mathematical system. In many applications of Monte Carlo, the physical process is simulated directly, and there is no need to even write down the differential equations that describe the behaviour of the system. The only requirement is that the physical (or mathematical) system be described by probability density functions.

The Monte Carlo simulation proceeds by sampling, generally with the Latin hypercube sampling method. Latin hypercube sampling divides the range of the probability density into a number of intervals, equal to the number of iterations to be performed. A simple random sample is chosen from within each interval. Each interval is only selected once during a simulation. As a result, Latin hypercube sampling ensures that values from the entire range of the distribution will be sampled proportionally to the probability density function.

There are several software packages to run Monte Carlo simulations with Latin hypercube sampling method. One of the most commonly used is @Risk (information available at http://www.palisade.com/risk/).

4.3. Bayesian Inference and Markov Chain Monte Carlo Procedure

> "The essence of the Bayesian approach is to provide a mathematical rule explaining how you should change your existing beliefs in the light of new evidence. In other words, it allows scientists to combine new data with their existing knowledge or expertise. The canonical example is to imagine that a precocious newborn observes his first sunset, and wonders whether the sun will rise again or not. He assigns equal prior probabilities to both possible outcomes, and represents this by placing one white and one black marble into a bag. The following day, when the sun rises, the child places another white marble in the bag. The probability that a marble plucked randomly from the bag will be white (ie, the child's degree of belief in future sunrises) has thus gone from a half to two-thirds. After sunrise the next day, the child adds another white marble, and the probability (and thus the degree of belief) goes from two-thirds to three-quarters. And so on. Gradually, the initial belief that the sun is just as likely as not to rise each morning is modified to become a near-certainty that the sun will always rise." (In praise of Bayes, The Economist, September 2000).

In other words, Bayesian analysis allows the current state of knowledge, the prior information, expressed as a probability distribution, to be formally combined with new data, the likelihood function, to reach an updated information state, the posterior distribution (Eq. 13). The posterior distribution reflects the reduction in uncertainty arising from the new information.

$$\text{Posterior} \equiv \text{Prior x Likelihood} \tag{13}$$

Computing Eq. 13 is often analytically impossible. A variety of methods have been developed to carry out a modelling approach based on Bayesian inference. One of the most popular one is the Markov Chain Monte Carlo technique in which a Markov Chain is used to sample for the posterior distribution [7, 40].

Bayesian Markov Chain Monte Carlo techniques can be run by various software, for example with the Winbugs package (more information at http://www.mrc-bsu.cam.ac.uk /bugs/welcome.shtml).

4.4. Sensitivity Analysis

Sensitivity Analysis is widely used in quantitative risk assessment to i) identify the most influential variables in a model, ii) provide a better understanding and interpretation of the analysis, and iii) identify data gaps and then prioritise future research. Sensitivity analysis might be seen as well as a prerequisite for model building to test the robustness and the relevance of the model [43].

Deterministic methods might be of interest, at least at a first stage of model development [58]. Frey and Patil [15] have reported ten different sensitivity analysis methods, grouped in three approaches: mathematical, statistical and graphical methods. They did not recommend the use of one or another, but instead to combine several approaches to increase confidence in the ranking of key inputs.

Methods based on statistics (e.g. rank order correlation) are generally implemented in software packages which run Monte Carlo simulations.

Below is a description of a little more sophisticated method implemented in the Excel add-in @risk (option 'advanced sensitivity analysis') which is very useful when interpreting probabilistic results:

- ✓ A set of seven percentiles, P = {1st-ile, 5th-ile, 25th-ile, 50th-ile, 75th-ile, 95th-ile, 99th-ile} is defined,
- ✓ For each probabilistic input considered in the sensitivity analysis, values corresponding to the set of percentiles are calculated: f(P),
- ✓ A simulation of the model is run by Monte Carlo technique, however, one of the input is not sampled from its distribution but set successively to its specific percentile values f(P),
- ✓ The process is repeated until each input has been analysed through its specific percentile values f(P).
- ✓ Results of simulation are collected and summarised or illustrated in Figures such as a Tornado graph and Percentile graph.

Let's take an example. For a given microorganism and product, the food supply-chain was analysed using a risk-based approach. A model was developed to calculate the microbiological concentration at the point of consumption, log *N* and then to compare it to an FSO as explained in Figure 5. The probabilistic model inputs were: the initial microbial (H_0), the logarithm of *D*-values (log *D*), the heat-treatment time (*HT* time), the storage temperature in the supply-chain distribution, the minimum temperature where the growth is observed (T_{min}) and the optimal growth rate. Log linear models for inactivation (ΣR) and growth (ΣI) were applied. No post-process recontamination was assumed. If an FSO of 3 log has been set as the target, the model output indicates that there is 99% chance of obtaining a microbiological concentration at the point of consumption, log *N*, below the FSO (Figure 10), with this model setting.

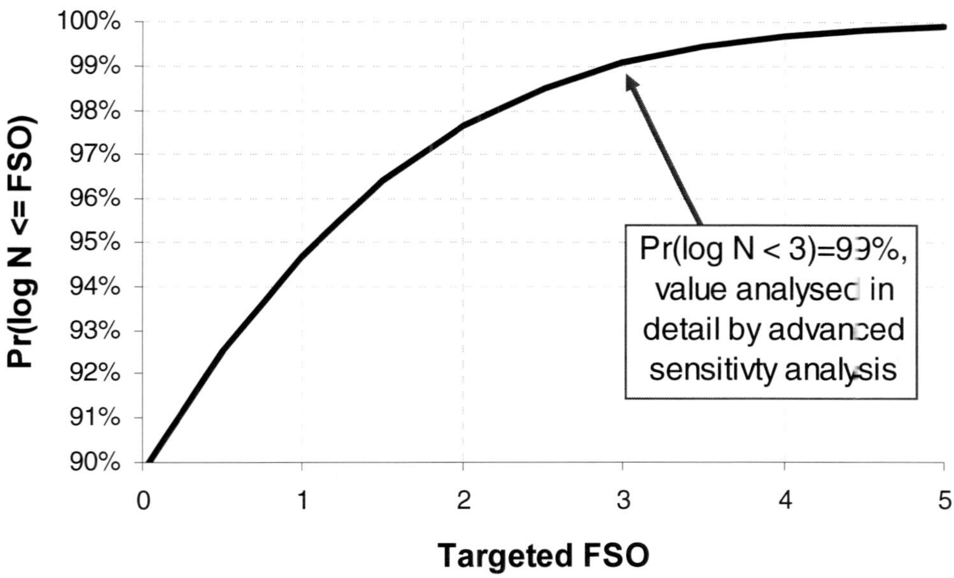

Figure 10. Probabilistic assessment model output; hypothetical example to illustrate the advanced sensitivity analysis technique.

A sensitivity analysis helps in interpreting this result and particularly helps in assessing which input has the most influence on the FSO compliance. Results are provided in Figure 11 with a Tornado graph (A) and then a percentile graph (B). The Tornado graph indicates that log *D* has the most influence on the 99th percentile of the cumulative probability function of log *N*, in other words, when log *D* was changing from its 1st to its 99th percentile, the 99th percentile of log *N* was changing the most. To respect a compliance with an FSO of 99%, log *D* is definitively an important input to consider. However, the input which might be a concern to respect the compliance is the storage temperature because if the overall range of variation is narrower than with log *D*, the variation goes up to 5 log. The percentile graph (Figure 11 B) and the Table 4 helps in understanding the impact of the storage temperature. When the storage temperature is set to its median (50th percentile, 6°C), log *N* reaches 1.53 (99th percentile) with a mean of -4.76; whereas with a storage temperature of 8.33 (99th percentile), log *N* reaches 5.43 (99th percentile) with a mean of -0.89. This means that the FSO target (with a compliance rate of 99%) might be missed by more than 2 log because of the high storage temperature values.

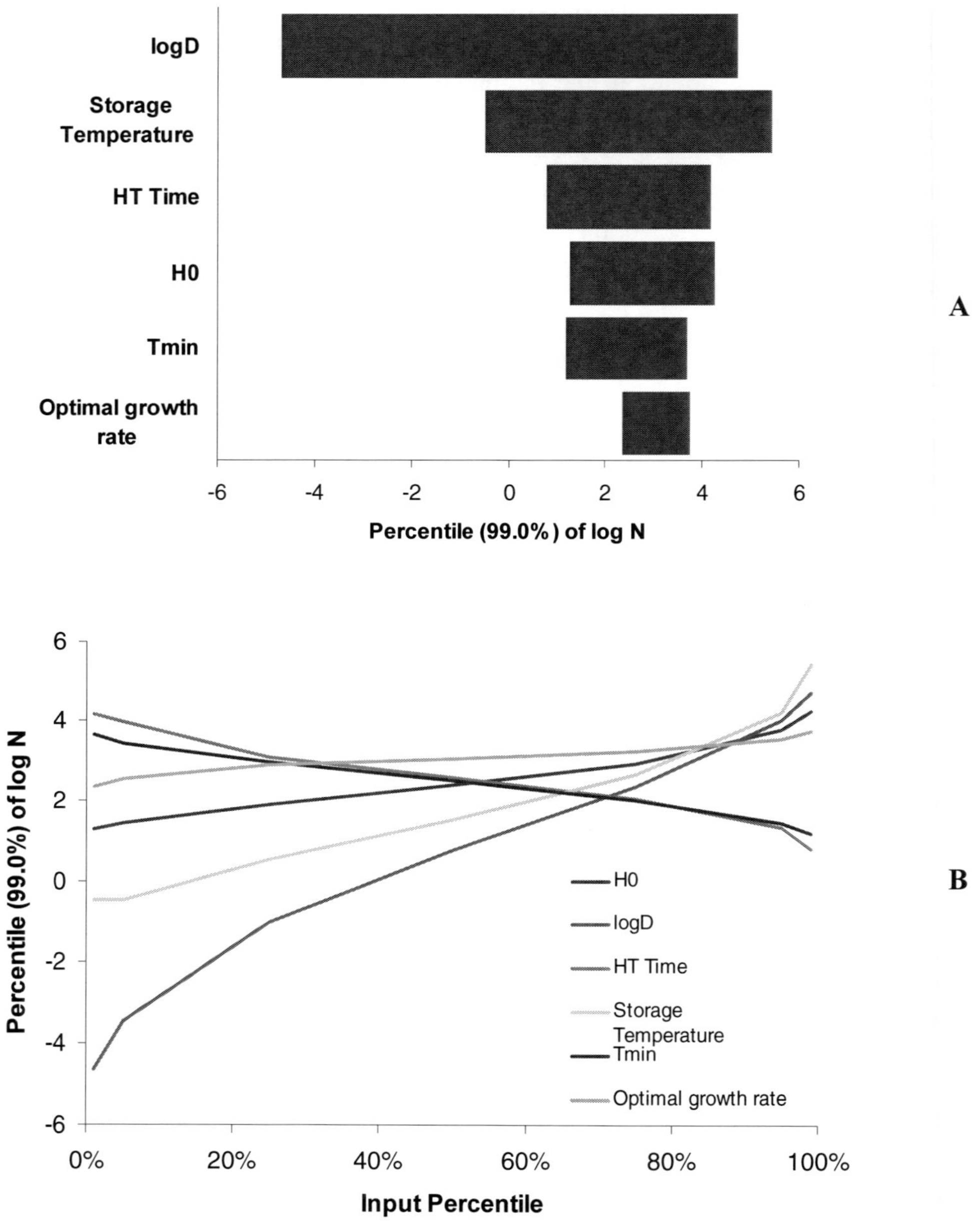

Figure 11. Sensitivity analysis outputs: impact of six inputs on the bacterial concentration at the point of consumption, log N. Hypothetical example. A. Tornado graph, B. Percentiles graph.

Table 4. Illustration of sensitivity analysis technique, based on an hypothetical example: variation of bacterial concentration at the time of consumption, log N, for various percentiles of the input 'Storage Temperature'

INPUT: Storage Temperature		OUTPUT: log N			
Percentile P	Input Value f(P)	Mean	5th percentile	95th percentile	99th percentile
Perc%: 1%	3.67	-6.43	-12.96	-1.66	-0.48
Perc%: 5%	4.36	-6.42	-12.96	-1.65	-0.47
Perc%: 25%	5.33	-5.82	-12.59	-0.87	0.51
Perc%: 50%	6.00	-4.76	-11.57	0.15	1.53
Perc%: 75%	6.67	-3.64	-10.48	1.27	2.66
Perc%: 95%	7.64	-2.02	-8.86	2.89	4.21
Perc%: 99%	8.33	-0.89	-7.78	4.02	5.43

4.5. Additional Probabilistic Technique: Two Dimension Sensitivity Analysis

The difference by nature and by interpretation between uncertainty and variability might justify separating them in risk assessment analysis. As already mentioned:

- ✓ Uncertainty arises from our lack of knowledge and may be related to the model used to characterise the risk, the parameters used to provide values for the model, or both. In some cases, we can reduce uncertainty by obtaining better information (more data, better Expert elicitation), but this may not always be possible.
- ✓ Variability refers to real and identifiable differences between individuals within a population addressed by risk assessment. True variability does not disappear with improved measurements. For example, heat resistance of *B. cereus* (*D*-values at 90°C) may vary among strains. For psychrotrophic strains, the logarithm of *D* 90°C-values has been ranged from -0.04 (5th percentile) to 1.28 min (95th percentile) and for mesophilic strains from 0.20 to 2.02 min [33].

Uncertainty forces decision makers to judge how probable it is that risks will be overestimated or underestimated for every member of the exposed population, whereas variability forces them to cope with the certainty that different individuals will be subjected to risks both above and below any reference point that is chosen.

To separate uncertainty and variability, a two-dimension sensitivity analysis is recommended, several examples of its application are available in the literature [36, 42]. There are no many two-dimension sensitivity analyses applied to the heat-treatment setting. Recently, the impact of heat-treatment, storage temperature, pH and shelf-life on the probability of exceeding 5 log cfu/g of *Bacillus cereus* in REPFEDs was analysed using this technique [33].

Two-dimension sensitivity analysis is not straightforward and can be time consuming; it should be fit-for-purpose and deployed only when risk assessors/risk managers have identified its true benefit.

Conclusion

The framework of Risk Analysis has been proposed as the common way to evaluate risks and determine appropriate risk management interventions relating to societal public health goals. The concept of Food Safety Objective (FSO), and related concepts, strengthens the link between operational food safety management and the public health goals.

Some changes in thermal process setting may occur as a consequence of this food safety management shift toward a risk-based approach. Widely established performance standard (e.g. 12D botulinum cook) may be revisited, since milder heat-treatment process may be set if the other part of the food supply-chain is controlled, for example if the post-process microbial growth is prevented.

Setting of thermal processes in a context of FSO opens definitively new opportunities in term of product innovation. Mathematically, techniques and tools (heat transfer process modelling, predictive microbiology, probabilistic analyses) are now sufficiently mature to make the change possible. However, re-analysing a food supply-chain, gathering the data and information, fitting the model and interpret the outputs take time and expertise, and it is likely that this resource effort will be made only where the expected benefit is high (increase product quality, reduction of cost) or with a newly identified hazard. This requires also improved sensor accuracy and/or improved system control at factory level to be confident in working closer to the safety and stability edge.

References

[1] Ball, C. O. (1923). Thermal process time for canned foods. p. 37. In *Bulletin of National Research Council*, vol. 7-1. Washington, D.C. USA.

[2] Balsa-Canto, E., Banga, J. R. & Alonso, A. A. (2002). A novel, efficient and reliable method for thermal process design and optimization. Part II: applications. *Journal of Food Engineering*, 52, 235-247.

[3] Bigelow, W. D. (1921). The logarithmic nature of thermal death time curves. *Journal of Infectious Diseases*, 29, 528-236.

[4] Braithwaite, P. J. & Perigo, J. A. (1971). The influence of of pH, water activity and recovery temperature on the resistance and outgrowth of *Bacillus* spores. p. 289-302. In A.M. Barker, Gould, G.W., & Wolf, J. (Ed.), *Spore research* London, Academic Press.

[5] Brown, B. E. & Larousse, J. (1997). *Thermal processes - development, validation, adjustment and control.* New-York, Wiley-VCH.

[6] Buchanan, R. L. (2004). Principles of risk analysis as applied to microbial food safety concerns. *Mitteilungen aus Lebensmitteluntersuchung und Hygiene*, 95, 6-12.

[7] Dakins, M. E., Toll, J. E., Small, M. J. & Brand, K. P. (1996). Risk-based environmental remediation: Bayesian Monte Carlo analysis and the expected value of sample information. *Risk Analysis*, 16, 67-79.

[8] Delignette-Muller, M. L. & Cornu, M. (2008). Quantitative risk assessment for *Escherichia coli* O157:H7 in frozen ground beef patties consumed by young children in French households. *International Journal of Food Microbiology*, 128, 158-164.
[9] European Food Safety Agency. (2007). Request for updating the former SCVPH opinion on *Listeria monocytogenes* risk related to ready-to-eat foods and scientific advice on different levels of *Listeria monocytogenes* in ready-to-eat foods and the related risk for human illness. *The EFSA Journal*. [On-line serial]. 599.
[10] Faille, C., Lebret, V., Gavini, F. & Maingonnat, J. F. (1997). Injury and lethality of heat treatment of *Bacillus cereus* spores suspended in buffer and in poultry meat. *Journal of Food Protection*, 60, 544-547.
[11] Faille, C., Membré, J. M., Kubaczka, M. & Gavini, F. (2002). Altered ability of Bacillus cereus spores to grow under unfavorable conditions (presence of nisin, low temperature, acidic pH, presence of NaCl) following heat treatment during sporulation. *Journal of Food Protection*, 65, 1930-1936.
[12] FAO/WHO. Date 2002. Principles and guidelines for incorporating microbiological risk assessment in the development of food safety standards, guidelines and related texts Report of a Joint FAO/WHO Consultation, Kiel, Germany, 18 - 22 March 2002. Accessed 30:07:2010.
[13] Fazil, A. M., Ross, T., Paoli, G., Vanderlinde, P., Desmarchelier, P. & Lammerding, A. M. (2002). A probabilistic analysis of *Clostridium perfringens* growth during food service operations. *International Journal of Food Microbiology*, 73, 315-329.
[14] Food and Drug Administration. (1999). 1999 U.S. *Cold Temperature Evaluation Design and Study Summary*. Available at: http://www.foodrisk.org/exclusives/audits/. Accessed 30/07/2010.
[15] Frey, H. C. & Patil, S. R. (2002). Identification and review of senstivity analysis methods. *Risk Analysis*, 22, 553-578.
[16] Gorris, L. G. M., Bassett, J., Membré, J. M., Motarjemi, Y. & Adams, M. (2006). The role of food safety objectives in dealing with emerging pathogens In *Emerging Foodborne Pathogens* Woodhead Publishing Limited, Cambridge, UK.
[17] Gould, G. W. (1999). Sous vide foods: conclusion of an ECFF Botulinum Working Party. *Food Control*, 10, 47-51.
[18] Havelaar, A. H., Nauta, M. J. & Jansen, J. T. (2004). Fine-tuning Food Safety Objectives and risk assessment. *International Journal of Food Microbiology*, 93, 11-29.
[19] ICMSF. (1996). *Microorganisms in Foods 5: Characteristics of Microbial Pathogens*. London, Chapman & Hall.
[20] ICMSF. (2002). *Microorganisms in Foods 7: Microbiological Testing in Food Safety Management*. New York, Kluwer Academic/Plenum Publishers.
[21] ICMSF. (2005). A simplified guide to understanding and using Food Safety Objectives and Performance Objectives. Available at: http://www.icmsf.iit.edu/pdf/Simplified%20FSO9nov05.pdf. Accessed 22/09/2010.
[22] Lammerding, A. M. & Paoli, G. M. (1997). Quantitative risk assessment an emerging tool for emerging foodborne pathogens. *Emerging Infectious Diseases*, 3, 483-7.
[23] Laurent, Y., Arino, S. & Rosso, L. (1999). A quantitative approach for studying the effect of heat treatment conditions on resistance and recovery of Bacillus cereus spores. *International Journal of Food Microbiology*, 48, 149-157.

[24] Leguerinel, I., Spegagne, I., Couvert, O., Coroller, L. & Mafart, P. (2007). Quantifying the effects of heating temperature, and combined effects of heating medium pH and recovery medium pH on the heat resistance of Salmonella typhimurium. *International Journal of Food Microbiology*, 116, 88-95.

[25] Lindqvist, R. & Lindblad, M. (2008). Quantitative risk assessment of thermophilic *Campylobacter* spp. and cross-contamination during handling of raw broiler chickens evaluating strategies at the producer level to reduce human campylobacteriosis in Sweden. *International Journal of Food Microbiology*, 121, 41-52.

[26] Lund, B. M. (1993). Quantification of factors affecting the probability of development of pathogenic bacteria, in particular *Clostridium botulinum* , in foods. *Journal of Industrial Microbiology*, 12, 144-155.

[27] Lund, B. M., Knox, M. R. & Cole, M. B. (1989). Destruction of *Listeria monocytogenes* during microwave cooking. *Lancet*, 1, 218-218.

[28] Mafart, P., Leguérinel, I., Couvert, O. & Coroller, L. (2010). Quantification of spore resistance for assessment and optimization of heating processes: A never-ending story. *Food Microbiology*, 27, 568-572.

[29] Martinez, S., Borrajo, R., Franco, I. & Carballo, J. (2007). Effect of environmental parameters on growth kinetics of *Bacillus cereus* (ATCC 7004) after mild heat treatment. *International Journal of Food Microbiology*, 117, 223-227.

[30] Membré, J.-M. & van zuijlen, A. (2011). A probabilistic approach to determine thermal process setting parameters: application for commercial sterility of products. *International Journal of Food Microbiology*, 144, 413-420.

[31] Membré, J. M., Amézquita, A., Bassett, J., Giavedoni, P., Blackburn, C. d. W. & Gorris, L. G. M. (2006). A probabilistic modeling approach in thermal inactivation: Estimation of postprocess *Bacillus cereus* spore prevalence and concentration. *Journal of Food Protection*, 69, 118-129.

[32] Membré, J. M., Bassett, J. & Gorris, L. G. M. (2007). Applying the Food Safety Objective and related standards to thermal inactivation of *Salmonella* in poultry meat. *Journal of Food Protection*, 70, 2036-2044.

[33] Membré, J. M., Kan-King-Yu, D. & Blackburn, C. d. W. (2008). Use of sensitivity analysis to aid interpretation of a probabilistic Bacillus cereus spore lag time model applied to heat-treated chilled foods (REPFEDs). *International Journal of Food Microbiology*, 128, 28-33.

[34] Membré, J. M. & Lambert, R. J. W. (2008). Application of predictive modelling techniques in industry: from food design up to risk assessment. *International Journal of Food Microbiology*, 128, 10-15.

[35] Membré, J. M., Wemmenhove, E. & McClure, P. J. (2009). Exposure assessment model to combine thermal inactivation (log reduction) and thermal injury (heat-treated spore lag time) effects on non-proteolytic Clostridium botulinum p. 2295-2303. *In* S. Martorell, Soares, C.G., & Barnett, J. (Ed.), vol. 3. CRC Press, Taylor and Francis Grou, London.

[36] Mokhtari, A. & Frey, H. C. (2005). Sensitivity analysis of a two-dimensional probabilistic risk assessment model using analysis of variance. *Risk Analysis*, 25, 1511-1529.

[37] Montville, R. & Schaffner, D. (2005). Monte Carlo simulation of pathogen behavior during the sprout production process. *Applied and Environmental Microbiology*, 71, 746-753.

[38] Nauta, M. J. (2001). A modular process risk model structure for quantitative microbiological risk assessment and its application in an exposure assessment of *Bacillus cereus* in REPFED. Available at: http://www.rivm.nl/bibliotheek/rapporten/149106007.html. Accessed 30/07/2010.

[39] Nauta, M. J., Jacobs-Reitsma, W. F., Evers, E. G., van Pelt, W. & Havelaar, A. (2005). Risk assessment of *Campylobacter* in the Netherlands via broiler meat and other routes. Available at: http://www.rivm.nl/bibliotheek/rapporten/250911006.pdf. Accessed 15/01/2010.

[40] Patwardhan, A. & Small, M. J. (1992). Bayesian methods for model uncertainty analysis with application to future sea level rise. *Risk Analysis*, 12, 513-523.

[41] Peck, M. W. (1997). Clostridium botulinum and the safety of refrigerated processed foods of extended durability. *Trends in Food Science & Technology*, 8, 186-192.

[42] Pouillot, R., Beaudeau, P., Denis, J. B. & Derouin, F. (2004). A quantitative risk assessment of waterborne cryptosporidiosis in France using second-order Monte Carlo simulation. *Risk Analysis*, 24, 1-17.

[43] Saltelli, A. (2002). Sensitivity analysis for importance assessment. *Risk Analysis*, 22, 579-590.

[44] Smout, C., Van Loey, A. M. L. & Hendrickx, M. E. G. (2000). Non-uniformity of lethality in retort processes based on heat distribution and heat penetration data. *Journal of Food Engineering*, 45, 103-110.

[45] Stewart, C. M., Cole, M. B. & Schaffner, D. W. (2003). Managing the risk of staphylococcal food poisoning from cream-filled baked goods to meet a food safety objective. *Journal of Food Protection*, 66, 1310-1325.

[46] Stringer, M. (2005). Food safety objectives - role in microbiological food safety management. *Food Control*, 16, 775-794.

[47] Stumbo, C. R. (1965). *Thermobacteriology in food processing*. New-York, Academic Press Inc.

[48] Thompson, K. M. (2002). Variability and uncertainty meet risk management and risk communication. *Risk Analysis*, 22, 647-654.

[49] U. S.Department of Agriculture - Food Safety Inspection, S. (1999). Appendix A - Compliance guidelines for meeting lethality performance standards for certain meat and poultry products Available at: http://www.fsis.usda.gov/oa/fr/95033F-a.htm. Accessed 30/09/2010.

[50] Van Gerwen, S. J. C., te Giffel, M. C., t Riet, K., Beumer, R. R. & Zwietering, M. H. (2000). Stepwise quantitative risk assessment as a tool for characterization of microbiological food safety. *Journal of Applied Microbiology*, 88, 938-951.

[51] van Schothorst, M. (2005). A proposed framework for the use of FSOs. *Food Control*, 16, 811-816.

[52] Vose, D. (2000). *Risk Analysis, a Quantitative Guide*. Chichester, United Kingdom, John Wiley & Sons, Ltd.

[53] Walls, I. (2005). Achieving continuous improvement in reductions in foodborne listeriosis - A risk-based approach. *Journal of Food Protection*, 68, 1932-1994.

[54] Walls, I. & Buchanan, R. L. (2005). Use of food safety objectives as a tool for reducing foodborne listeriosis. *Food Control*, 16, 795-799.

[55] Whiting, R. C., Rainosek, A., Buchanan, R. L., Miliotis, M., LaBarre, D., Long, W., Ruple, A. & Schaub, S. (2006). Determining the microbiological criteria for lot rejection from the performance objective or food safety objective. *International Journal of Food Microbiology*, 110, 263-267.

[56] World Health Organization & Food Agricultural Organisation of the United Nations. (2004). Risk assessment of *Listeria monocytogenes* in ready-to-eat foods. Available at: http://www.who.int/foodsafety/publications/micro/mra_listeria/en/index.html. Accessed 22/09/2010.

[57] Zwietering, M. H. (2005). Practical considerations on food safety objectives. *Food Control*, 16, 817-823.

[58] Zwietering, M. H. & Van Gerwen, S. J. C. (2000). Sensitivity analysis in quantitative microbial risk assessment. *International Journal of Food Microbiology*, 58, 213-221.

INDEX

C

D

E

G

H

I

N

O

P

Q

R

S

T

U

V

W

Y